Machine Intelligence, Big Data Analytics, and IoT in Image Processing

Scrivener Publishing
100 Cummings Center, Suite 541J
Beverly, MA 01915-6106

Advances in Intelligent and Scientific Computing

Series Editors: Dr. Sujata Dash, Dr. Subhendu Kumar Pani and Dr. Milan Tuba

The series provides in-depth coverage of innovations in artificial life, computational intelligence, evolutionary computing, machine learning and applications. It is the intention for the volumes in the series to be practically relevant, so that the results will be useful for managers in leadership roles. Therefore, both theoretical and managerial implications of the research will be considered.

Submission of book proposals to
Dr. Subhendu Kumar Pani at skpani.india@gmail.com
or Pani.subhendu@gmail.com

Publishers at Scrivener
Martin Scrivener (martin@scrivenerpublishing.com)
Phillip Carmical (pcarmical@scrivenerpublishing.com)

Machine Intelligence, Big Data Analytics, and IoT in Image Processing

Practical Applications

Edited by

Ashok Kumar

Chitkara University Institute of Engineering & Technology, Chitkara University, Punjab, India

Megha Bhushan

School of Computing, DIT University, Dehradun, Uttarakhand, India

José A. Galindo

Department of Computer Languages and Systems, University of Seville, Spain

Lalit Garg

Computer Information Systems, University of Malta, Malta

and

Yu-Chen Hu

Dept. of Computer Science and Information Management, Providence University, Tai Chung, Taiwan

Scrivener
Publishing

WILEY

This edition first published 2023 by John Wiley & Sons, Inc., 111 River Street, Hoboken, NJ 07030, USA
and Scrivener Publishing LLC, 100 Cummings Center, Suite 541J, Beverly, MA 01915, USA
© 2023 Scrivener Publishing LLC
For more information about Scrivener publications please visit www.scrivenerpublishing.com.

Wiley Global Headquarters

111 River Street, Hoboken, NJ 07030, USA

For details of our global editorial offices, customer services, and more information about Wiley products visit us at www.wiley.com.

Limit of Liability/Disclaimer of Warranty

Library of Congress Cataloging-in-Publication Data

ISBN 978-1-119-86504-9

Cover image: Pixabay.Com
Cover design by Russell Richardson

Set in size of 11pt and Minion Pro by Manila Typesetting Company, Makati, Philippines

Printed in the USA

10 9 8 7 6 5 4 3 2 1

Contents

Preface   xv

Part I: Demystifying Smart Healthcare   1

1 Deep Learning Techniques Using Transfer Learning for Classification of Alzheimer's Disease   3
Monika Sethi, Sachin Ahuja and Puneet Bawa
1.1   Introduction   4
1.2   Transfer Learning Techniques   6
1.3   AD Classification Using Conventional Training Methods   9
1.4   AD Classification Using Transfer Learning   12
1.5   Conclusion   16
     References   16

2 Medical Image Analysis of Lung Cancer CT Scans Using Deep Learning with Swarm Optimization Techniques   23
Debnath Bhattacharyya, E. Stephen Neal Joshua and N. Thirupathi Rao
2.1   Introduction   24
2.2   The Major Contributions of the Proposed Model   26
2.3   Related Works   28
2.4   Problem Statement   32
2.5   Proposed Model   33
     2.5.1   Swarm Optimization in Lung Cancer Medical Image Analysis   33
     2.5.2   Deep Learning with PSO   34
     2.5.3   Proposed CNN Architectures   35
2.6   Dataset Description   37
2.7   Results and Discussions   39
     2.7.1   Parameters for Performance Evaluation   39
2.8   Conclusion   47
     References   48

3 Liver Cancer Classification With Using Gray-Level Co-Occurrence Matrix Using Deep Learning Techniques   **51**
Debnath Bhattacharyya, E. Stephen Neal Joshua and N. Thirupathi Rao
3.1 Introduction   52
   3.1.1 Liver Roles in Human Body   53
   3.1.2 Liver Diseases   53
   3.1.3 Types of Liver Tumors   55
      3.1.3.1 Benign Tumors   55
      3.1.3.2 Malignant Tumors   57
   3.1.4 Characteristics of a Medical Imaging Procedure   58
   3.1.5 Problems Related to Liver Cancer Classification   60
   3.1.6 Purpose of the Systematic Study   61
3.2 Related Works   62
3.3 Proposed Methodology   66
   3.3.1 Gaussian Mixture Model   68
   3.3.2 Dataset Description   69
   3.3.3 Performance Metrics   70
      3.3.3.1 Accuracy Measures   70
      3.3.3.2 Key Findings   74
      3.3.3.3 Key Issues Addressed   75
3.4 Conclusion   77
   References   77

4 Transforming the Technologies for Resilient and Digital Future During COVID-19 Pandemic   **81**
Garima Kohli and Kumar Gourav
4.1 Introduction   82
4.2 Digital Technologies Used   84
   4.2.1 Artificial Intelligence   85
   4.2.2 Internet of Things   85
   4.2.3 Telehealth/Telemedicine   87
   4.2.4 Cloud Computing   87
   4.2.5 Blockchain   88
   4.2.6 5G   89
4.3 Challenges in Transforming Digital Technology   90
   4.3.1 Increasing Digitalization   91
   4.3.2 Work From Home Culture   91
   4.3.3 Workplace Monitoring and Techno Stress   91
   4.3.4 Online Fraud   92
   4.3.5 Accessing Internet   92

4.3.6   Internet Shutdowns   92
4.3.7   Digital Payments   92
4.3.8   Privacy and Surveillance   93
4.4   Implications for Research   93
4.5   Conclusion   94
References   95

Part II: Plant Pathology   **101**

5   Plant Pathology Detection Using Deep Learning   **103**
Sangeeta V., Appala S. Muttipati and Brahmaji Godi
5.1   Introduction   104
5.2   Plant Leaf Disease   105
5.3   Background Knowledge   109
5.4   Architecture of ResNet 512 V2   111
5.4.1   Working of Residual Network   112
5.5   Methodology   113
5.5.1   Image Resizing   113
5.5.2   Data Augmentation   113
5.5.2.1   Types of Data Augmentation   114
5.5.3   Data Normalization   114
5.5.4   Data Splitting   116
5.6   Result Analysis   116
5.6.1   Data Collection   117
5.6.2   Feature Extractions   117
5.6.3   Plant Leaf Disease Detection   117
5.7   Conclusion   119
References   120

6   Smart Irrigation and Cultivation Recommendation System
for Precision Agriculture Driven by IoT   **123**
N. Marline Joys Kumari, N. Thirupathi Rao
and Debnath Bhattacharyya
6.1   Introduction   124
6.1.1   Background of the Problem   127
6.1.1.1   Need of Water Management   127
6.1.1.2   Importance of Precision Agriculture   127
6.1.1.3   Internet of Things   128
6.1.1.4   Application of IoT in Machine Learning
and Deep Learning   129
6.2   Related Works   131
6.3   Challenges of IoT in Smart Irrigation   133

6.4 Farmers' Challenges in the Current Situation 135
6.5 Data Collection in Precision Agriculture 136
    6.5.1 Algorithm 136
        6.5.1.1 Environmental Consideration on Stage Production of Crop 140
    6.5.2 Implementation Measures 141
        6.5.2.1 Analysis of Relevant Vectors 141
        6.5.2.2 Mean Square Error 141
        6.5.2.3 Potential of IoT in Precision Agriculture 141
    6.5.3 Architecture of the Proposed Model 143
6.6 Conclusion 147
    References 147

7 Machine Learning-Based Hybrid Model for Wheat Yield Prediction 151
    Haneet Kour, Vaishali Pandith, Jatinder Manhas and Vinod Sharma
7.1 Introduction 152
7.2 Related Work 153
7.3 Materials and Methods 155
    7.3.1 Methodology for the Current Work 155
        7.3.1.1 Data Collection for Wheat Crop 155
        7.3.1.2 Data Pre-Processing 156
        7.3.1.3 Implementation of the Proposed Hybrid Model 157
    7.3.2 Techniques Used for Feature Selection 159
        7.3.2.1 ReliefF Algorithm 159
        7.3.2.2 Genetic Algorithm 161
    7.3.3 Implementation of Machine Learning Techniques for Wheat Yield Prediction 162
        7.3.3.1 K-Nearest Neighbor 162
        7.3.3.2 Artificial Neural Network 163
        7.3.3.3 Logistic Regression 164
        7.3.3.4 Naïve Bayes 164
        7.3.3.5 Support Vector Machine 165
        7.3.3.6 Linear Discriminant Analysis 166
7.4 Experimental Result and Analysis 167
7.5 Conclusion 173
    Acknowledgment 173
    References 174

8   **A Status Quo of Machine Learning Algorithms in Smart
Agricultural Systems Employing IoT-Based WSN: Trends,
Challenges and Futuristic Competences**   **177**
Abhishek Bhola, Suraj Srivastava, Ajit Noonia,
Bhisham Sharma and Sushil Kumar Narang
8.1   Introduction   178
8.2   Types of Wireless Sensor for Smart Agriculture   179
8.3   Application of Machine Learning Algorithms
     for Smart Decision Making in Smart Agriculture   179
8.4   ML and WSN-Based Techniques for Smart Agriculture   185
8.5   Future Scope in Smart Agriculture   188
8.6   Conclusion   190
     References   190

Part III: Smart City and Villages   **197**

9   **Impact of Data Pre-Processing in Information Retrieval
for Data Analytics**   **199**
Huma Naz, Sachin Ahuja, Rahul Nijhawan
and Neelu Jyothi Ahuja
9.1   Introduction   200
    9.1.1   Tasks Involved in Data Pre-Processing   200
9.2   Related Work   202
9.3   Experimental Setup and Methodology   205
    9.3.1   Methodology   205
    9.3.2   Application of Various Data Pre-Processing
           Tasks on Datasets   206
    9.3.3   Applied Techniques   207
        9.3.3.1   Decision Tree   207
        9.3.3.2   Naive Bayes   207
        9.3.3.3   Artificial Neural Network   208
    9.3.4   Proposed Work   208
        9.3.4.1   PIMA Diabetes Dataset (PID)   208
    9.3.5   Cleveland Heart Disease Dataset   211
    9.3.6   Framingham Heart Study   215
    9.3.7   Diabetic Dataset   217
9.4   Experimental Result and Discussion   220
9.5   Conclusion and Future Work   222
     References   222

10 Cloud Computing Security, Risk, and Challenges: A Detailed Analysis of Preventive Measures and Applications   **225**
Anurag Sinha, N. K. Singh, Ayushman Srivastava,
Sagorika Sen and Samarth Sinha
10.1   Introduction   226
10.2   Background   228
      10.2.1   History of Cloud Computing   228
            10.2.1.1   Software-as-a-Service Model   230
            10.2.1.2   Infrastructure-as-a-Service Model   230
            10.2.1.3   Platform-as-a-Service Model   232
      10.2.2   Types of Cloud Computing   232
      10.2.3   Cloud Service Model   232
      10.2.4   Characteristics of Cloud Computing   234
      10.2.5   Advantages of Cloud Computing   234
      10.2.6   Challenges in Cloud Computing   235
      10.2.7   Cloud Security   236
            10.2.7.1   Foundation Security   236
            10.2.7.2   SaaS and PaaS Host Security   237
            10.2.7.3   Virtual Server Security   237
            10.2.7.4   Foundation Security: The Application Level   238
            10.2.7.5   Supplier Data and Its Security   238
            10.2.7.6   Need of Security in Cloud   239
      10.2.8   Cloud Computing Applications   239
10.3   Literature Review   241
10.4   Cloud Computing Challenges and Its Solution   242
      10.4.1   Solution and Practices for Cloud Challenges   246
10.5   Cloud Computing Security Issues and Its Preventive Measures   248
      10.5.1   General Security Threats in Cloud   249
      10.5.2   Preventive Measures   254
10.6   Cloud Data Protection and Security Using Steganography   258
      10.6.1   Types of Steganography   259
      10.6.2   Data Steganography in Cloud Environment   260
      10.6.3   Pixel Value Differencing Method   261
10.7   Related Study   263
10.8   Conclusion   263
      References   264

11 Internet of Drone Things: A New Age Invention   269
      Prachi Dahiya
      11.1   Introduction   269
      11.2   Unmanned Aerial Vehicles   271
             11.2.1   UAV Features and Working   274
             11.2.2   IoDT Architecture   275
      11.3   Application Areas   280
             11.3.1   Other Application Areas   284
      11.4   IoDT Attacks   285
             11.4.1   Counter Measures   291
      11.5   Fusion of IoDT With Other Technologies   296
      11.6   Recent Advancements in IoDT   299
      11.7   Conclusion   302
             References   303

12 Computer Vision-Oriented Gesture Recognition System
      for Real-Time ISL Prediction   305
      Mukul Joshi, Gayatri Valluri, Jyoti Rawat and Kriti
      12.1   Introduction   305
      12.2   Literature Review   307
      12.3   System Architecture   309
             12.3.1   Model Development Phase   309
             12.3.2   Development Environment Phase   311
      12.4   Methodology   312
             12.4.1   Image Pre-Processing Phase   312
             12.4.2   Model Building Phase   313
      12.5   Implementation and Results   314
             12.5.1   Performance   314
             12.5.2   Confusion Matrix   318
      12.6   Conclusion and Future Scope   318
             References   319

13 Recent Advances in Intelligent Transportation Systems
      in India: Analysis, Applications, Challenges, and Future Work 323
      Elamurugan Balasundaram, Cailassame Nedunchezhian,
      Mathiazhagan Arumugam and Vinoth Asaikannu
      13.1   Introduction   324
      13.2   A Primer on ITS   325
      13.3   The ITS Stages   326
      13.4   Functions of ITS   327
      13.5   ITS Advantages   328

13.6 ITS Applications 329
13.7 ITS Across the World 331
13.8 India's Status of ITS 333
13.9 Suggestions for Improving India's ITS Position 334
13.10 Conclusion 335
References 335

14 Evolutionary Approaches in Navigation Systems for Road Transportation System 341
Noopur Tyagi, Jaiteg Singh and Saravjeet Singh
14.1 Introduction 342
14.1.1 Navigation System 343
14.1.2 Genetic Algorithm 347
14.1.3 Differential Evolution 348
14.2 Related Studies 349
14.2.1 Related Studies of Evolutionary Algorithms 351
14.3 Navigation Based on Evolutionary Algorithm 352
14.3.1 Operators and Terms Used in Evolutionary Algorithms 353
14.3.2 Operator and Terms Used in Evolutionary Algorithm 357
14.4 Meta-Heuristic Algorithms for Navigation 359
14.4.1 Drawbacks of DE 362
14.5 Conclusion 362
References 363

15 IoT-Based Smart Parking System for Indian Smart Cities 369
E. Fantin Irudaya Raj, M. Appadurai,
M. Chithamabara Thanu and E. Francy Irudaya Rani
15.1 Introduction 370
15.2 Indian Smart Cities Mission 371
15.3 Vehicle Parking and Its Requirements in a Smart City Configuration 373
15.4 Technologies Incorporated in a Vehicle Parking System in Smart Cities 375
15.5 Sensors for Vehicle Parking System 383
15.5.1 Active Sensors 384
15.5.2 Passive Sensors 386
15.6 IoT-Based Vehicle Parking System for Indian Smart Cities 387
15.6.1 Guidance to the Customers Through Smart Devices 389

15.6.2   Smart Parking Reservation System   391
15.7   Advantages of IoT-Based Vehicle Parking System   392
15.8   Conclusion   392
         References   393

**16 Security of Smart Home Solution Based on Secure
Piggybacked Key Exchange Mechanism   399**
Jatin Arora and Saravjeet Singh
16.1   Introduction   400
16.2   IoT Challenges   404
16.3   IoT Vulnerabilities   405
16.4   Layer-Wise Threats in IoT Architecture   406
         16.4.1   Sensing Layer Security Issues   407
         16.4.2   Network Layer Security Issues   408
         16.4.3   Middleware Layer Security Issues   409
         16.4.4   Gateways Security Issues   410
         16.4.5   Application Layer Security Issues   411
16.5   Attack Prevention Techniques   411
         16.5.1   IoT Authentication   412
         16.5.2   Session Establishment   413
16.6   Conclusion   414
         References   414

**17 Machine Learning Models in Prediction of Strength
Parameters of FRP-Wrapped RC Beams   419**
*Aman Kumar, Harish Chandra Arora, Nishant Raj Kapoor
and Ashok Kumar*
17.1   Introduction   420
         17.1.1   Defining Fiber-Reinforced Polymer   421
         17.1.2   Types of FRP Composites   422
                   17.1.2.1   Carbon Fiber–Reinforced Polymer   422
                   17.1.2.2   Glass Fiber   423
                   17.1.2.3   Aramid Fiber   424
                   17.1.2.4   Basalt Fiber   424
17.2   Strengthening of RC Beams With FRP Systems   425
         17.2.1   FRP-to-Concrete Bond   426
         17.2.2   Flexural Strengthening of Beams With FRP
                   Composite   427
         17.2.3   Shear Strengthening of Beams With FRP
                   Composite   427
17.3   Machine Learning Models   428

17.3.1	Prediction of Bond Strength	430
17.3.2	Estimation of Flexural Strength	434
17.3.3	Estimation of Shear Strength	434
17.4	Conclusion	441
	References	441

18 Prediction of Indoor Air Quality Using Artificial Intelligence 447
Nishant Raj Kapoor, Ashok Kumar, Anuj Kumar,
Aman Kumar and Harish Chandra Arora

18.1	Introduction	448
18.2	Indoor Air Quality Parameters	450
	18.2.1 Physical Parameters	453
	18.2.1.1 Humidity	453
	18.2.1.2 Air Changes (Ventilation)	454
	18.2.1.3 Air Velocity	454
	18.2.1.4 Temperature	454
	18.2.2 Particulate Matter	455
	18.2.3 Chemical Parameters	456
	18.2.3.1 Carbon Dioxide	456
	18.2.3.2 Carbon Monoxide	456
	18.2.3.3 Nitrogen Dioxide	456
	18.2.3.4 Sulphur Dioxide	457
	18.2.3.5 Ozone	457
	18.2.3.6 Gaseous Ammonia	458
	18.2.3.7 Volatile Organic Compounds	458
	18.2.4 Biological Parameters	459
18.3	AI in Indoor Air Quality Prediction	459
18.4	Conclusion	464
	References	465

Index 471

Preface

The concepts of machine intelligence, big data analytics and the Internet of Things (IoT) continue to improve our lives through various cutting-edge applications such as disease detection in real time, crop yield prediction, smart parking and so forth. The transformative effects of these technologies are life-changing because they play an important role in demystifying smart healthcare, plant pathology, and smart city/village planning, design and development. This book presents a cross-disciplinary perspective on the practical applications of machine intelligence, big data analytics and IoT by compiling cutting-edge research and insights from researchers, academicians and practitioners worldwide. It identifies and discusses various advanced technologies, such as artificial intelligence, machine learning, IoT, image processing, network security, cloud computing and sensors, to provide effective solutions to the lifestyle challenges faced by humankind.

These practical innovative applications may include navigation systems for road transportation, IoT- and WSN-based smart agriculture, plant pathology through deep learning, cancer detection from medical images and smart home solutions. Moreover, cloud computing has made it possible to access these real-life applications remotely over the internet. The primary concern of this book is to equip those new to this field of application, as well as those with more advanced knowledge related to practical application development, exploit the inherent features of machine intelligence, big data analytics and IoT. For instance, how to harness these advanced technologies to develop practical applications such as drone-based surveillance, smart transportation, healthcare, smart farming solutions, and robotics for automation.

This book is a significant addition to the body of knowledge on practical applications emerging from machine intelligence, big data analytics and IoT. The chapters deal with specific areas of applications of these technologies. This deliberate choice of covering a diversity of fields was to emphasize the applications of these technologies in almost every contemporary aspect of real life to assist working in different sectors by understanding and exploiting the strategic opportunities offered by these technologies. A summary of the main ideas of the work presented in each of the chapters follows:

– Chapter 1 is based on the models used to diagnose Alzheimer's disease (AD). These models utilize CaffeNet, GoogLeNet, VGGNet-16, VGGNet-19, DenseNet with varying depths, Inception-V4, AlexNet, ResNet-18, ResNet-152, or even ensemble transfer-learning, that are pre-trained on generalized images for AD classification to achieve better performance as compared to training a model from scratch.
– Chapter 2 describes how to detect cancerous lung nodules from a lung CT scan image given as input and how to classify the lung cancer along with its severity. A novel deep learning method is used to detect the location of cancerous lung nodules.
– Chapter 3 outlines a classifier used to divide the liver and CT images into normal and abnormal categories based on the main features in terms of shape, texture, and feature statistics. It includes four stages: preprocessing, fuzzy clustering, feature extraction and classification. Furthermore, the grey-level co-occurrence matrix (GLCM) method is used to extract the features.
– Chapter 4 provides some of the major emerging digital technologies which have transformed the lives of individuals by making their future dependent upon the resilience of these technologies. It also highlights some of the major challenges related to these technologies with their suitable implications.
– Chapter 5 describes a model based on ResNet architecture in deep learning to help farmers detect plant leaf diseases at an early stage in order to take precautionary measures against them.
– Chapter 6 discusses an IoT-based smart irrigation system to assist farmers in precision agriculture for increasing crop yield. It uses multiple sensor metrics to help anticipate conditions for irrigation planning by predicting soil moisture, temperature, and humidity.
– Chapter 7 presents a hybrid model for wheat crop yield prediction using machine learning (ML) approaches, namely k-nearest neighbors (KNN), naïve Bayes, artificial neural network, logistic regression, support vector machine and linear discriminant analysis. The model works in two stages: the first stage uses a feature selection strategy to find the best features for wheat crops, and the second stage uses ML to estimate crop yield based on these best features.

- Chapter 8 discusses wireless sensor network (WSN)-based techniques used for smart agriculture and applications of ML for smart decision-making.
- Chapter 9 provides an insight into the applications of data preprocessing techniques and their effects on information retrieval. It covers the major issues that need to be dealt with before beginning any data analysis process.
- Chapter 10 focuses on the security for the latest paradigm shift in cloud and distributed computing. It delineates various risk parameters in the cloud environment and provides some novel methods to be adopted for cloud data security.
- Chapter 11 talks about the internet of drone things (IoDT), its applications in the modern world, research opportunities, and current challenges to be dealt with. Furthermore, it discusses new age inventions, security issues, and attacks that frequently occur in the IoDT.
- Chapter 12 presents an artificial intelligence-based gesture recognition system for the prediction of Indian sign language in real time. It covers different experiments using two-dimensional convolutional neural network-based classification to convert images into text.
- Chapter 13 sets forth applications, challenges, and future developments in the field of intelligent transportation systems (ITS) in India. It explains ITS and evaluates their feasibility in India.
- Chapter 14 provides a survey of evolutionary techniques used in navigation to create opportunities for analysts and researchers seeking to understand the broad pattern of different algorithms used in the navigation system.
- Chapter 15 examines the IoT-based vehicle parking system in Indian cities. Additionally, it discusses vehicle parking and its basic requirements, various technologies incorporated in modern parking systems, different sensors utilized in parking facilities, and the advantages of IoT-based vehicle parking systems in detail.
- Chapter 16 discusses a secure data transmission and key exchange for ensuring the confidentiality of data. Also, a lightweight authentication mechanism for ensuring the integrity and confidentiality of data shared over an unsecured network is presented.
- Chapter 17 delineates machine learning models in the prediction of strength parameters of fiber-reinforced polymer (FRP) wrapped reinforced concrete (RC) beams. It provides a summary of machine learning models in the estimation of bond strength between FRP and concrete surface, and shear and flexural strength of FRP wrapped RC beams.

– Chapter 18 describes existing AI-based studies for forecasting the indoor air quality of buildings and the future of AI-based indoor air quality forecasting. It provides an overview of the important role of machine learning models in the prediction of indoor pollutant concentrations to develop warning systems which help to affect the occupant's health positively.

This book was edited by a team of academicians and experts. It is our hope that readers will draw several benefits from both the theoretical and practical aspects covered in the book to enhance their own practice or research.

The Editors
Dr. Ashok Kumar
Phagwara, India
Dr. Megha Bhushan
Dehradun, India
Dr. José Galindo
Seville, Spain
Dr. Lalit Garg
Valetta, Malta
Dr. Yu-Chen Hu
Tai Chung, Taiwan
January 2023

Part I

DEMYSTIFYING SMART HEALTHCARE

Deep Learning Techniques Using Transfer Learning for Classification of Alzheimer's Disease

Monika Sethi[1], Sachin Ahuja[2]* and Puneet Bawa[1]

1Chitkara University Institute of Engineering & Technology, Chitkara University, Punjab, India
2ED-Engineering at Chandigarh University, Punjab, India

Abstract

Alzheimer's disease (AD) is a severe disorder in which brain cells degenerate, increasing memory loss with treatment choices for AD symptoms varying based on the disease's stage, and as the disease progresses, individuals at certain phases undergo specific healthcare. The majority of existing studies make predictions based on a single data modality either they utilize magnetic resonance imaging (MRI)/positron emission tomography (PET)/diffusion tensor imaging (DTI) or the combination of these modalities. However, a thorough understanding of AD staging assessment can be achieved by integrating these data modalities and performance could be further enhanced using a combination of two or more modalities. However, deep learning techniques trained the network from scratch, which has the following drawbacks: (a) demands an enormous quantity of labeled training dataset that could be a problem for the medical field where physicians annotate the data, further it could be very expensive, (b) requires a huge amount of computational resources. (c) These models also require tedious and careful adjustments of numerous hyper-parameters, which results to under or overfitting and, in turn, to degraded performance. (d) With a limited medical training data set, the cost function might get stuck in a local-minima problem. In this chapter, a study is done based on the models used for AD diagnosis. Many researchers fine-tuned their networks instead of scratch training and utilized CaffeNet, GoogleNet, VGGNet-16, VGGNet-19, DenseNet with varying depths, Inception-V4, AlexNet, ResNet-18, ResNet-152, or even ensemble transfer-learning models pretrained on generalized images for AD classification performed better.

Corresponding author: ed.engineering@cumail.in

Ashok Kumar, Megha Bhushan, José A. Galindo, Lalit Garg and Yu-Chen Hu (eds.) Machine Intelligence, Big Data Analytics, and IoT in Image Processing: Practical Applications, (3–22) © 2023 Scrivener Publishing LLC

Keywords: Alzheimer disease, transfer learning, deep learning, parameter optimization

1.1 Introduction

In the United States, AD is the most widespread neurodegenerative condition and the sixth major cause of fatalities. The global disease burden of AD is expected to exceed $2 trillion by 2030, requiring preventative care [1]. Despite the tremendous study and advancements in clinical practice, nearly half of AD patients are correctly identified for anatomy and progression of the disease based on medical indicators. The existence of neurofibrillary tangles and amyloid plaques in histology is the most definitive evidence for AD. Consequently, the presence of plaque is not associated with the beginning of AD, but rather with sensory and neuron damage. Dr. Alois Alzheimer (a psychiatrist and neuropsychologist) was the origin for the naming of this disease, who studied the brain of a 51-year-old woman who died of severe cognitive impairment in 1906 [2]. Dr. Alois investigated her brain and discovered clumps, which were actually the accumulation of proteins in and around the neurons, resulting in their loss. The key characteristics for identifying or confirming the existence of the illness are shrinkage of the hippocampus and cerebral cortex, as well as growth of the ventricles. The hippocampus is essential in learning and memory, in addition to acting as a connection between the central nervous system of the body's organs. AD eventually destroys the portion of the brain that controls heart and respiratory activity, resulting in death [3].

Unfortunately, AD does not yet have a definitive cure [4]. Instead, the objective is to reduce the illness's development, treat suffering, manage learning disabilities, and enhance the quality of life. Clinical trials, on the other hand, can significantly slow down the progression of psychiatric disorders if the diagnosis is made early. Whereas more psychological therapies and, eventually, prevention or even a cure are essential (long-term) goals, early diagnosis may result in improved treatment outcomes benefits for diseased. Except in a few cases where genetic abnormalities may be identified, the precise cause of AD is still obscure.

The assessment of empirical biomarkers is necessary for the early treatment of disease [5]. A number of noninvasive neuroimaging approaches, including computed tomography (CT) scans, both structural and functional MRI and PET, have been explored for the prediction of AD. To produce cross sectional pictures of the bones, blood arteries, and soft tissues within the body, computer processing is used to integrate a succession of

X-ray images recorded from different angle defined on your body. Plain X-rays do not give as much detail as CT scan imaging. An MRI scan employs a powerful magnet and radio waves to see at structures beneath the brain, according to the National Institute of Health. MRI scans are used by healthcare physicians to examine a variety of diseases, from damaged ligaments to cancer. To see and evaluate changes in cellular metabolism, PET is a functional imaging method, which thus employs radioactive additives termed as radiotracers.

Radiologists and clinicians, who are medical experts, analyze medical imaging data [6]. As a result of the probable tiredness of human specialists while evaluating images manually, a computer-assisted approach has proven to be beneficial for researchers as well as physicians. However, machine learning (ML) approaches are helping to improve the issue. Medical image analysis tasks need the use of ML to discover or learn useful features that characterize the correlations or patterns present in data. Since relevant or task-related characteristics are often created by human specialists on the basis of their domain expertise, it might be difficult for nonexperts to use ML techniques for their own research in the traditional manner. A number of projects are now working on the problem of learning sparse representations from training samples or pre-set dictionaries. Since then, there are attempts to generate sparse representations based on predefined dictionaries, which might be learned from training dataset. As a result of the concept of parsimony, sparse representation is used in many scientific fields. A sparsely inducing penalization and feature learning technique has been shown to be effective in medical image analysis when it comes to determining feature representation and selection [7]. Though data with a shallow architecture are still found to have meaningful patterns or regularities, techniques such as sparse representation or dictionary learning are still limited in their ability to represent them. Feature engineering has been incorporated into a learning phase in deep learning (DL), though, overcoming this issue [8]. Instead of manually extracting features, DL takes simply a collection of data with little preparation, if required, and then learns the valuable interpretations in an automatic method. Due to this shift in responsibility for feature extraction and selection, even nonexperts in ML may now use DL effectively for their own research work, especially in the medical field for imaging analysis [9].

However, DL is afflicted by data dependency, one of the most significant problems. Comparatively to standard ML approaches, DL relies on a significant quantity of training data in order to discover hidden patterns in data. There is an interesting relationship between the size of the model in terms of the numbers of layers and the volume of information required.

Transfer learning (TL) eliminates the dependency of a huge amount of data requirement, which inspires us to utilize this to combat the problem of inadequate training data. This concept is driven by the idea that people may strategically utilize past knowledge to solve new problems or accomplish desirable results. The fundamental reasoning underlying this idea in ML was presented during a Neural Information Processing Systems (NIPS-95) symposium on "Learning to Learn," which emphasized the need of lifelong ML approaches that store and apply previously acquired information [10]. TL approaches have recently shown results in a variety of practical applications. In Verma *et al.* [11], researchers utilized TL methods to transfer text data across domains. For fixing natural language processing issues, structural correspondence learning was presented by an author in Nalavade *et al.* [12]. Researchers employed several Convolutional Neural Network (CNN)-based TL models to detect AD [13].

This chapter presents the results of several TL techniques employed by previous researchers to identify AD.

1.2   Transfer Learning Techniques

TL is an ML research subject that focuses on retaining information received while addressing the problem and adapting it to some other but similar issue. As an instance, knowledge acquired when learning to identify trucks may be used while aiming to classify other four-wheeler vehicles. In CNN, this may be implemented in one of two ways: either the weights of all CNN layers are coupled to some other CNN layer with classification Layer output, as well as just utilizing "off-the-shelf CNN features," whereby CNN serves like a generalized feature extractor to be analyzed later.

Several domains of knowledge engineering, such as classifier, prediction, and segmentation, have already experienced significant results using ML and data mining techniques [14]. Many ML techniques, however, operate successfully with an assumption that training test data are collected from the same dimensional region and variance. Most statistical models must still be redesigned from beginning when the population varies, employing new received data for training. In several practical applications, recollecting the necessary training data and rebuilding the models is either too expensive or not feasible. It would be extremely beneficial if researchers could reduce the need for the time and efforts associated with acquiring training samples. Transferring information or learning across problem contexts would be advantageous in such scenarios.

Researchers face three (1H and 2W) primary research issues in TL: how, when, and what to transfer [15]. What knowledge may be transmitted between areas or tasks is studied under what to transfer? There are certain types of knowledge that are particular to certain domains or tasks, and there are other types of information that are common to several domains and may serve to increase performance in the target task. To answer the "how to transfer" question, learning algorithms must be built after determining which types of knowledge may be transmitted. When to transfer skills is a question that explores when skill knowledge transfer will be performed. In the same way, researchers are curious about when learning should not be revealed. Different TL strategies are used to decide on 1H and 2W as illustrated in Figure 1.1.

In the case of inductive TL, whether the source and target domains are dissimilar or similar, the targeted task is independent of the input space. However, the source and target domains are distinct in a transductive TL situation. In case of unsupervised transfers, the destination task may be distinct from the source [16]. TL allows investigators to apply knowledge gained from previously completed work to related and newer situations. When researchers have a large amount of Learning Task 1 data from Source, they may use TL methodologies to acquire and generalize such Gathered Knowledge (properties, weights) with the goal of Learning Task 2, which has far less data as observed in Figure 1.2.

The shorter training timeframes, improved neural-network efficiency (in most circumstances), as well as the lack of a large quantity of Source Dataset are only a several of the advantages of TL [17]. Similarly, TL in machine learning involves the application of previously learnt models to a new problem. When developing a computational model from the beginning, a considerable quantity data are usually necessary, but access to that knowledge is not always possible—this is when TL gets in handy. Further,

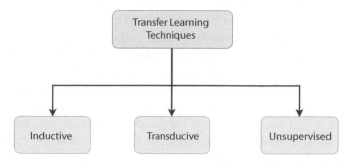

Figure 1.1 Transfer learning techniques.

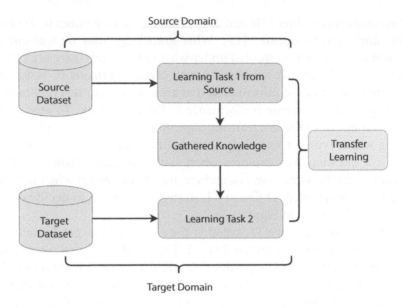

Figure 1.2 TL process.

instead of training the model from start to finish, remove the last fully connected output layer and utilize the pretrained computational model as a feature extractor [18, 19]. Similarly, we may utilize a new dataset to solve a different problem using similar strategy. One may develop a convenient and rapid linear model to change the output based on the new dataset since the pretrained advanced artificial neural network is used as a characteristic for the new task. Whenever the target task data is limited, the feature extraction technique is the best option. Fine-tuning is accomplished by unfreezing the underlying model (or a portion of it) and retraining the entire model on the full dataset at a low learning speed [20]. Effectiveness of the model in terms of accuracy performance somewhat on the given new dataset will be improved because to the modest learning rate, which also helps in minimizing over fitting in limited data scenarios. The learning rate has to be low since the model is large and the dataset is small. This is an overfitting equation that explains why and how the learning rate is very low. In such scenario, one would need to recompile the model that they have made these adjustments for them to take effect [21]. This is because the compile function locks the behaviour every time it is invoked. This means that if you want to change the model's behavior, you will have to recompile it. The model will be retrained and monitored by call back to ensure that it does not over fit.

1.3   AD Classification Using Conventional Training Methods

As a result of technological systems, e-health has gained remarkable progress. In addition to clinical predictive analytics, tele-health and patient monitoring tools are also examples of this technology. It is crucial to be able to predict and diagnose disease before it occurs. Psychiatric evaluation and smart watch data can be used by doctors to provide a diagnosis for the patient or forecast the possibility of future disease to aid patients postpone and avoid illnesses. Several diseases, including AD, are extremely challenging to diagnose in the initial phase because of their mild symptoms [22]. The usage of artificial intelligence systems centred on neural networks in patient care has steadily increased over the last decade [23]. These systems are now utilized in a variety of fields, such as medical assessment, categorization, and forecasting domains. In an artificial neural network, the computation is performed in a distributed and parallel manner using many integer cores. Patterns are learnt through training samples and depending on the gained knowledge throughout training then applied to new unseen data. Table 1.1 shows the performance of several AD classification models utilized by the scientific community.

Table 1.1 AD classification via deep learning models.

Article reference	Objective (binary/ ternary classification)	Dataset	Architecture	Classification (accuracy in %)
[24]	Binary	ADNI	Sparse Autoencoder with CNN	AD v/s HC (93.8) AD v/s. MCI (86.3) MCI v/s HC (83.3)
[25]	Binary	ADNI	Autoencoder stack with softmax Regression	AD v/s NC (88) NC v/s MCI (77)
[26]	Binary	ADNI	Deep Boltzmann Machine	AD v/s NC (93.5) MCI v/s NC (85) cMCI v/s ncMCI (74.5)

(*Continued*)

Table 1.1 AD classification via deep learning models. (*Continued*)

Article reference	Objective (binary/ ternary classification)	Dataset	Architecture	Classification (accuracy in %)
[27]	Binary and Ternary	ADNI	2D-CNN 3D- CNN	2D-CNN AD v/s HC (95.39) HC v/s MCI (90) AD v/s MCI (82) AD v/s MCI vs. HC (86) 3D-CNN AD v/s HC (95.39) HC v/s MCI (92) AD v/s MCI (87) AD v/s MCI vs. HC (89)
[28]	Binary	-	CNN	AD vs. NC (86.7)
[29]	Binary	ADNI	Ensemble DBM	AD v/s NC (90) MCI v/s AD (84) MCI v/s NC (83)
[30]	Binary	ADNI	Cascaded 3D-CNN	AD v/s HC (92)
[31]	Binary	ADNI	CNN	AD v/s NC (Sensitivity-.69, Specificity-.98)
[32]	Binary	ADNI	ensemble CNN	AD v/s NC (89.60)
[33]	Binary	ADNI	3D CNN	AD v/s NC (80) lMCI v/s eMCI (52)
[34]	Binary	ADNI	Multiple 3D CNN	AD v/s NC (92.26)
[35]	Binary	ADNI	CNN	AD v/s NC (79.9)
[36]	Binary	ADNI	2D CNN with RNN	AD v/s NC (95.3) NC v/s MCI (83.9)

(*Continued*)

Table 1.1 AD classification via deep learning models. (*Continued*)

Article reference	Objective (binary/ ternary classification)	Dataset	Architecture	Classification (accuracy in %)
[37]	Multiclass Classification	ADNI	SAE and SVM	With SAE AD v/s cMCI v/s ncMCI v/s NC (53) With SVM AD v/s cMCI v/s ncMCI v/s NC (47)
[38]	Binary	ADNI	CNN	AD v/s CN (90.1) CN v/s pMCI (87.46) pMCI v/s sMCI (76.9)
[39]	Binary	ADNI	3D CNN with LSTM	AD v/s NC (94.8) pMCI v/s NC (86.3) sMCI v/s NC (65.3)
[40]	Binary	ADNI	CNN	AD v/s NC (92.3) AD vs. MCI (85.6) MCI v/s NC (78.1)
[41]	Binary	ADNI	2D CNN, 3D CNN, 3D CNN SVM	NC v/s MCI (98.8) NC v/s AD (99) MCI v/s AD (89.4)
[42]	Binary	ADNI	3D CNN	AD v/s MCI (89.3) MCI v/s NC (87.5)
[43]	Multiclass Classification	OASIS	CNN	NonD v/s VMild v/s Mild v/s Mod (71)

For binary classification, in 2014, the researchers employed SAEs with a regression layer soft max as activation function and reached an accuracy of 88% for AD v/s NC and 77% for MCI v/s NC [2]. With DBM architecture [26], researchers attained an accuracy of approximately 93% for AD vs. NC on the ADNI. There was a comparison between 2D CNN and 3D CNN for both binary as well as ternary AD classification, and 3D CNN model outperformed 2D CNN for both binary and ternary classification [3].

The accuracy of AD vs. CN for multiple 3D CNN was 92% in [8], while 2D CNN performed better than 3D CNN utilising RNN with an accuracy of 95.3% in [10]. For multiclass classification, only 71% accuracy was achieved [16] using the CNN architecture on OASIS dataset. Overall, 3D CNN performed better than 2D CNN, and the best accuracy was attained in the AD vs. NC classification, when compared to other binary and ternary classes.

1.4   AD Classification Using Transfer Learning

Unlike traditional ML, DL enables for automatic feature extraction from low to high level. In contrast, the network is trained from scratch using DL methods, which has certain disadvantages. In the medical area where clinicians annotate the data, this may be an issue and highly expensive. It also requires a significant amount of computing resources. In addition, several hyper-parameters in these models must be carefully adjusted, which could cause to under or over fitting, and degradation in performance. Due to the lack of data, the cost function may become stuck in 'local minima'. TL may be used to fine-tune a deep network (such as CNN) rather than learning from scratch. A neural network model is initially trained on data from a source domain that is similar to the problem being targeted. It then uses the top few layers from the trained model to create a new one that uses the target dataset as its training data set. In order to perform TL, weight initialization and feature extraction are the two most important approaches. As a result, this approach is far quicker and produces better results than training a network from scratch. As a result of its huge computational capacity, it is mostly used for computer vision applications, such as emotional analysis and classification issues. Many research teams fine-tuned their networks instead of starting from scratch and used CaffeNet, GoogLeNet, VGGNet-11, VGGNet-16, DenseNet with varying depths (121-161-169), Inception-V4, AlexNets, ResNet-18, ResNet-152, or even ensemble TL models that are pretrained on generalized images for the AD classification and improved performance. Table 1.2 illustrates the success of various designs via TL techniques utilized by researchers to classify AD.

Ternary AD classification using VGG-16 architecture was reported to be 92% accurate in [44]. On the other hand, VGG-based AD binary classification was shown to be more accurate by the authors [50, 55, 58, 61]. For ternary classification, GoogleNet outperformed ResNet-18 and ResNet-152 in [45] on ADNI dataset. There was a comparison of four different TL models

Table 1.2 AD classification via transfer learning techniques.

Article reference	Objective (binary/ ternary/ multi-class classification)	Dataset	Architecture	Classification (accuracy in %)
[44]	Ternary	ADNI	VGGNet	AD v/s MCI v/s HC (92)
[45]	Binary	ADNI	LeNet	AD v/s HC (98.84)
[46]	4-Way	ADNI	GoogleNet ResNet-18 ResNet-152	AD v/s MCI v/s LMCI v/s HC 98.9 98.01 98.14
[47]	Multiclass Classification	OASIS	CNN (Hyperparameters of the Inception-V4 model)	NonD v/s VMild v/s Mild v/s
[48]	Ternary	ADNI	GoogleNet CaffeNet	sMCI v/s cMCI v/s HC (83.23) 87.78
[49]	Multiclass Classification	OASIS	Inception-V4 ResNet	NonD v/s VMild v/s Mild v/s Mod (95)
[50]	Binary	ADNI	VGG	AD v/s NC (96.81) MCI v/s NC (92.62)
[51]	Binary	ADNI	VGG-16	AD v/s CN vs. MCI (95.73)
[52]	Ternary	OASIS	Ensemble ResNet50 DenseNet	AD v/s MCI vs. CN (95.23)

(*Continued*)

Table 1.2 AD classification via transfer learning techniques. (*Continued*)

Article reference	Objective (binary/ ternary/ multi-class classification)	Dataset	Architecture	Classification (accuracy in %)
[53]	Binary	OASIS	Siamese CNN (ResNet-34)	AD v/s NC (98.72)
[54]	Binary	ADNI	AlexNet GoogLeNet ResNet	Highest Accuracy achieved in AlexNet then ResNet.
[55]	Binary	ADNI	VGG	AD v/s NC (98.7) EMCI v/s LMCI (83.7)
[56]	Ternary	ADNI	AlexNet	AD v/s NC (97.2)
[57]	Binary	ADNI	ResNet-18	AD v/s NC (96.9)
[58]	Binary	ADNI	VGG-16	NC v/s AD (99.01) MCI v/s AD (98.71)s

(*Continued*)

Table 1.2 AD classification via transfer learning techniques. (*Continued*)

Article reference	Objective (binary/ ternary/ multi-class classification)	Dataset	Architecture	Classification (accuracy in %)
[59]	Multiclass Classification	OASIS	VGG DenseNet ResNet EfficientNet	VGG NonD v/s VMild v/s Mild v/s Mod (79) DenseNet NonD v/s VMild v/s Mild v/s Mod (92) ResNet NonD v/s VMild v/s Mild v/s Mod (93) EfficientNet NonD v/s VMild v/s Mild v/s Mod (96)
[60]	Multiclass Classification	OASIS	AlexNet	NonD v/s VMild v/s Mild v/s Mod (95)
[61]	Binary	ADNI	VGG-16	AD v/s CN (99.95)

(VGG, DenseNet, ResNet, and EfficientNet) in [54] using the same OASIS dataset and EfficientNet had the greatest accuracy of 96% out of four.

As a whole, it is suggested to combine pretrained CNNs for initialization and retrain them utilising just fine-tuning of the CNNs layers for better performance for AD classification.

1.5   Conclusion

This chapter discussed binary and multiclass classification approaches using conventional and DL models using TL for AD classification. AD is a leading risk factor for death in developed nations. From a scientific perspective, computer-aided algorithms have yielded excellent findings, but practically, there is still no viable diagnostic technique usable. In the past few years, DL models have grown increasingly popular, especially when it comes to AD classification. A deep model was trained from scratch in the majority of the experiments, although this is sometimes impractical because the training procedure is time-consuming and a significant bit of training dataset is necessary to make it work efficiently. For general face recognition, there are billions of images in a dataset, while for neuroimaging there are just a thousand. As initialization, CNNs that have been pretrained on a dataset can be utilized for classification of neuroimaging data to classify AD. The CNNs may then be trained again on neuroscans using TL. Thus, researchers are able to classify AD more efficiently using pre-trained models, such as VGG (16, 19), ResNet, DenseNet (121,161,169), LeNet, and Inception as compared to building a model from scratch. Researchers have achieved the highest accuracy for AD vs. NC (approx 99.5%) in comparison to other binary (AD v/s MCI, MCI v/s NC, cMCI v/s ncMCI), ternary (AD v/s MCI v/s NC, AD vs. cMCI v/s ncMCI) and multiclass classification (NonD v/s VMild vs. Mild vs. Mod).

References

1. Shen, D., Wu, G., Suk, H.I., Deep learning in medical image analysis. *Annu. Rev. Biomed. Eng.*, 19, 221–248, 2017.
2. Taqi, A.M., Awad, A., Al-Azzo, F., Milanova, M., The impact of multi-optimizers and data augmentation on TensorFlow convolutional neural network performance, in: *2018 IEEE Conference on Multimedia Information Processing and Retrieval (MIPR)*, April 2018, IEEE, pp. 140–145.
3. Sethi, M., Ahuja, S., Rani, S., Bawa, P., Zaguia, A., Classification of Alzheimer's disease using gaussian-based bayesian parameter optimization for deep convolutional LSTM network. *Comput. Math. Methods Med.*, 2021, 2021.
4. Liu, M., Li, F., Yan, H., Wang, K., Ma, Y., Shen, L., Xu, M., Alzheimer's disease neuroimaging initiative, a multi-model deep convolutional neural network for automatic hippocampus segmentation and classification in Alzheimer's disease. *Neuroimage*, 208, 116459, 2020.

5. Ebrahimighahnavieh, M.A., Luo, S., Chiong, R., Deep learning to detect Alzheimer's disease from neuroimaging: A systematic literature review. *Comput. Methods Programs Biomed.*, 187, 105242, 2020.

6. Kadyan, V., Bala, S., Bawa, P., Mittal, M., Developing in-vehicular noise robust children ASR system using Tandem-NN-based acoustic modelling. *Int. J. Veh. Auton. Syst.*, 15, 3-4, 296–306, 2020.

7. Suk, H.I., Lee, S.W., Shen, D., Deep sparse multi-task learning for feature selection in Alzheimer's disease diagnosis. *Brain Struct. Funct.*, 221, 5, 2569–2587, 2016.

8. Sethi, M., Ahuja, S., Bawa, P., Classification of Alzheimer's disease using neuroimaging data by convolution neural network, in: *2021 6th International Conference on Signal Processing, Computing and Control (ISPCC)*, 2021, October, IEEE, pp. 402–406.

9. Sethi, M., Ahuja, S., Kukreja, V., An empirical study for the deep learning models. *J. Phys.: Conf. Ser.*, 1950, 1, 012071, IOP Publishing, 2021.

10. Sethi, M., Ahuja, S., Rani, S., Koundal, D., Zaguia, A., Enbeyle, W., An exploration: Alzheimer's disease classification based on convolutional neural network. *BioMed. Res. Int.*, 2022, 2022.

11. Verma, K., Bhardwaj, S., Arya, R., Islam, M.S.U., Bhushan, M., Kumar, A., Samant, P., Latest tools for data mining and machine learning. *Int. J. Innov. Technol. Exploring Eng.*, 8, 9S, 18–23, July 2019, Available:https://doi.org/10.35940/ijitee.I1003.0789S19.

12. Nalavade, A., Bai, A., Bhushan, M., Deep learning techniques and models for improving machine reading comprehension system. *IJAST*, 29, 04, 9692–9710, 2020.

13. Ashraf, A., Naz, S., Shirazi, S.H., Razzak, I., Parsad, M., Deep transfer learning for Alzheimer neurological disorder detection. *Multimed. Tools Appl.*, 1–26, 2021.

14. Yang, Q. and Wu, X., 10 Challenging Problems in Data Mining Research. *Int. J. Inf. Technol. Decis. Mak.*, 5, 04, 597–604, 2006.

15. Pan, S.J. and Yang, Q., A survey on transfer learning. *IEEE Trans. Knowl. Data Eng.*, 22, 10, 1345–1359, 2009.

16. Weiss, K., Khoshgoftaar, T.M., Wang, D., A survey of transfer learning. *J. Big Data*, 3, 1, 1–40, 2016.

17. Kholiya, P.S., Kapoor, A., Rana, M., Bhushan, M., Intelligent process automation: The future of digital transformation, in: *2021 10th International Conference on System Modeling& Advancement in Research Trends (SMART)*, IEEE, pp. 185–190, 2021.

18. Wu, H., Luo, J., Lu, X., Zeng, Y., 3D transfer learning network for classification of Alzheimer's disease with MRI. *Int. J. Mach. Learn. Cybern.*, 13, 2, 1–15, 2022.

19. Sayeedakhanum, P., Megha, B., Anita, B., A study on healthcare using data mining techniques. *J. Crit. Rev.*, 7, 19, 7877–7890, 2020.

20. Bawa, P. and Kadyan, V., Noise robust in-domain children speech enhancement for automatic Punjabi recognition system under mismatched conditions. *Appl. Acoust.*, 175, 107810, 2021.

21. Singh, V.J., Bhushan, M., Kumar, V., Bansal, K.L., Optimization of segment size assuring application perceived QoS in healthcare, in: *Proceedings of the World Congress on Engineering*, vol. 1, 2015.

22. Hu, C., Ju, R., Shen, Y., Zhou, P., Li, Q., Clinical decision support for Alzheimer's disease based on deep learning and brain network, in: *2016 IEEE International Conference on Communications (ICC)*, May 2016, IEEE, pp. 1–6.

23. Shafi, I., Ansari, S., Din, S., Jeon, G., Paul, A., Artificial neural networks as clinical decision support systems. *Concurr. Comput.: Pract. Exp.*, 33, 22, e6342, 2021.

24. Gupta, A., Ayhan, M., Maida, A., Natural image bases to represent neuroimaging data, in: *International Conference on Machine Learning*, May 2013, PMLR, pp. 987–994.

25. Liu, S., Liu, S., Cai, W., Pujol, S., Kikinis, R., Feng, D., Early diagnosis of Alzheimer's disease with deep learning, in: *2014 IEEE 11th International Symposium on Biomedical Imaging (ISBI)*, April 2014, IEEE, pp. 1015–1018.

26. Suk, H.I., Lee, S.W., Shen, D., Alzheimer's disease neuroimaging initiative, hierarchical feature representation and multimodal fusion with deep learning for AD/MCI diagnosis. *NeuroImage*, 101, 569–582, 2014.

27. Payan, A. and Montana, G., Predicting Alzheimer's disease: A neuroimaging study with 3D convolutional neural networks. in *Proc. Int. Conf. Pattern Recogn. Appl. Meth.*, 2015.

28. Gao, X.W. and Hui, R., A deep learning based approach to classification of CT brain images, in: *2016 SAI Computing Conference (SAI)*, July 2016, IEEE, pp. 28–31.

29. Ortiz, A., Munilla, J., Gorriz, J.M., Ramirez, J., Ensembles of deep learning architectures for the early diagnosis of the Alzheimer's disease. *Int. J. Neural Syst.*, 26, 07, 1650025, 2016.

30. Cheng, D. and Liu, M., Classification of Alzheimer's disease by cascaded convolutional neural networks using PET images, in: *International Workshop on Machine Learning in Medical Imaging*, September 2017, Springer, Cham, pp. 106–113.

31. Luo, S., Li, X., Li, J., Automatic Alzheimer's disease recognition from MRI data using deep learning method. *J. Appl. Math. Phys.*, 5, 9, 1892–1898, 2017.

32. Cheng, D. and Liu, M., CNNs based multi-modality classification forAD diagnosis, in: *2017 10th International Congress on Image and Signal Processing, Biomedical Engineering and Informatics (CISP-BMEI)*, October 2017, IEEE, pp. 1–5.

33. Korolev, S., Safiullin, A., Belyaev, M., Dodonova, Y., Residual and plain convolutional neural networks for 3D brain MRI classification, in: *2017 IEEE*

14th International Symposium on Biomedical Imaging (ISBI 2017), April 2017, IEEE, pp. 835–838.

34. Cheng, D., Liu, M., Fu, J., Wang, Y., Classification of MR brain images by combination of multi-CNNs for AD diagnosis, in: *Ninth International Conference on Digital Image Processing (ICDIP 2017)*, July 2017, vol. 10420, International Society for Optics and Photonics, p. 1042042.

35. Lin, W., Tong, T., Gao, Q., Guo, D., Du, X., Yang, Y., Guo, G., Xiao, M., Du, M., Qu, X., Alzheimer's disease neuroimaging initiative, convolutional neural networks-based MRI image analysis for the Alzheimer's disease prediction from mild cognitive impairment. *Front. Neurosci.*, 12, 777, 2018.

36. Liu, M., Cheng, D., Yan, W., Alzheimer's Disease Neuroimaging Initiative, Classification of Alzheimer's disease by combination of convolutional and recurrent neural networks using FDG-PET images. *Front. Neuroinform.*, 12, 35, 2018.

37. Shi, J., Zheng, X., Li, Y., Zhang, Q., Ying, S., Multimodal neuroimaging feature learning with multimodal stacked deep polynomial networks for diagnosis of Alzheimer's disease. *IEEE J. Biomed. Health Inform.*, 22, 1, 173–183, 2017.

38. Huang, Y., Xu, J., Zhou, Y., Tong, T., Zhuang, X., Alzheimer's disease neuroimaging initiative (ADNI, diagnosis of Alzheimer's disease via multi-modality 3D convolutional neural network. *Front. Neurosci.*, 13, 509, 2019.

39. Feng, C., Elazab, A., Yang, P., Wang, T., Zhou, F., Hu, H., Xiao, X., Lei, B., Deep learning framework for Alzheimer's disease diagnosis via 3D-CNN and FSBi-LSTM. *IEEE Access*, 7, 63605–63618, 2019.

40. Choi, B.K., Madusanka, N., Choi, H.K., So, J.H., Kim, C.H., Park, H.G., Bhattacharjee, S., Prakash, D., Convolutional neural network-based MR image analysis for Alzheimer's disease classification. *Curr. Med. Imaging*, 16, 1, 27–35, 2020.

41. Feng, W., Halm-Lutterodt, N.V., Tang, H., Mecum, A., Mesregah, M.K., Ma, Y., Li, H., Zhang, F., Wu, Z., Yao, E., Guo, X., Automated MRI-based deep learning model for detection of Alzheimer's disease process. *Int. J. Neural Syst.*, 30, 06, 2050032, 2020.

42. Pei, Z., Gou, Y., Ma, M., Guo, M., Leng, C., Chen, Y., Li, J., Alzheimer's disease diagnosis based on long-range dependency mechanism using convolutional neural network. *Multimed. Tools Appl.*, 1–16, 2021.

43. Ajagbe, S.A., Amuda, K.A., Oladipupo, M.A., AFE, O.F., Okesola, K.I., Multi-classification of alzheimer disease on magnetic resonance images (MRI) using deep convolutional neural network (DCNN) approaches. *Int. J. Adv. Comput. Res.*, 11, 53, 2021.

44. Billones, C.D., Demetria, O.J.L.D., Hostallero, D.E.D., Naval, P.C., DemNet: A convolutional neural network for the detection of Alzheimer's disease and mild cognitive impairment, in: *2016 IEEE Region 10 Conference (TENCON)*, November 2016, IEEE, pp. 3724–3727.

45. Sarraf, S. and Tofighi, G., Deep learning-based pipeline to recognize Alzheimer's disease using fMRI data, in: *2016 Future Technologies Conference (FTC)*, December 2016, IEEE, pp. 816–820.

46. Farooq, A., Anwar, S., Awais, M., Rehman, S., A deep CNN based multi-class classification of Alzheimer's disease using MRI, in: *2017 IEEE International Conference on Imaging Systems and Techniques (IST)*, October 2017, IEEE, pp. 1–6, 2017.

47. Islam, J. and Zhang, Y., A novel deep learning based multi-class classification method for Alzheimer's disease detection using brain MRI data, in: *International Conference on Brain Informatics*, November 2017, Springer, Cham, pp. 213–222.

48. Wu, C., Guo, S., Hong, Y., Xiao, B., Wu, Y., Zhang, Q., Alzheimer's disease neuroimaging initiative, discrimination and conversion prediction of mild cognitive impairment using convolutional neural networks. *Quant. Imaging Med. Surg.*, 8, 10, 992, 2018.

49. Islam, J. and Zhang, Y., 36Early diagnosis of Alzheimer's disease: A neuroimaging study with deep learning architectures, in: *Proceedings of the IEEE Conference on Computer Vision and Pattern Recognition Workshops*, pp. 1881–1883, 2018.

50. Tang, H., Yao, E., Tan, G., Guo, X., A fast and accurate 3D fine-tuning convolutional neural network for Alzheimer's disease diagnosis, in: *International CCF Conference on Artificial Intelligence*, August 2018, Springer, Singapore, pp. 115–126.

51. Jain, R., Jain, N., Aggarwal, A., Hemanth, D.J., Convolutional neural network based Alzheimer's disease classification from magnetic resonance brain images. *Cognit. Syst. Res.*, 57, 147–159, 2019.

52. Jabason, E., Ahmad, M.O., Swamy, M.N.S., Classification of Alzheimer's disease from MRI data using an ensemble of hybrid deep convolutional neural networks, in: *2019 IEEE 62nd International Midwest Symposium on Circuits and Systems (MWSCAS)*, August 2019, IEEE, pp. 481–484.

53. Amin-Naji, M., Mahdavinataj, H., Aghagolzadeh, A., Alzheimer's disease diagnosis from structural MRI using Siamese convolutional neural network, in: *2019 4th International Conference on Pattern Recognition and Image Analysis (IPRIA)*, March 2019, IEEE, pp. 75–79.

54. Simon, B.C., Baskar, D., Jayanthi, V.S., Alzheimer's disease classification using deep convolutional neural network, in: *2019 9th International Conference on Advances in Computing and Communication (ICACC)*, November 2019, IEEE, pp. 204–208.

55. Mehmood, A., Yang, S., Feng, Z., Wang, M., Ahmad, A.S., Khan, R., Maqsood, M., Yaqub, M., A transfer learning approach for early diagnosis of alzheimer's disease on MRI images. *Neuroscience*, 460, 43–52, 2021.

56. Sambath Kumar, S. and Nandhini, M., Automated classification of alzheimer's disease using MRI and transfer learning, in: *Mobile Computing and Sustainable Informatics*, pp. 663–686, Springer, Singapore, 2022.

57. Ebrahimi, A., Luo, S., Chiong, R., Introducing transfer learning to 3D ResNet-18 for Alzheimer's disease detection on MRI images, in: *2020 35th International Conference on Image and Vision Computing New Zealand (IVCNZ)*, November 2020, IEEE, pp. 1–6.

58. Tanveer, M., Rashid, A.H., Ganaie, M.A., Reza, M., Razzak, I., Hua, K.L., Classification of Alzheimer's disease using ensemble of deep neural networks trained through transfer learning. *IEEE J. Biomed. Health Inf.*, 26, 4, 1453–1463, 2021.

59. Kadri, R., Tmar, M., Bouaziz, B., Alzheimer's disease prediction using EfficientNet and Fastai, in: *International Conference on Knowledge Science, Engineering and Management*, 2021, August, Springer, Cham, pp. 452–463.

60. Fu'adah, Y.N., Wijayanto, I., Pratiwi, N.K.C., Taliningsih, F.F., Rizal, S., Pramudito, M.A., Automated classification of Alzheimer's disease based on MRI image processing using convolutional neural network (CNN) with AlexNet architecture. *J. Phys.: Conf. Ser.*, 1844, 1, 012020, IOP Publishing, 2021.

61. Janghel, R.R. and Rathore, Y.K., Deep convolution neural network based system for early diagnosis of Alzheimer's disease. *IRBM*, 42, 4, 258–267, 2021.

Medical Image Analysis of Lung Cancer CT Scans Using Deep Learning with Swarm Optimization Techniques

Debnath Bhattacharyya[1]*, E. Stephen Neal Joshua[2] and N. Thirupathi Rao[2]

[1]Department of Computer Science and Engineering, Koneru Lakshmaiah Education Foundation, Vaddeswaram, Guntur, Andhra Pradesh, India
[2]Department of Computer Science and Engineering, Vignan's Institute of Information Technology, Visakhapatnam, AP, India

Abstract

Image processing is the process of extracting information from a photograph. It is required to make visual data more understandable by humans and machines. In image processing, you start with a CT scan, modify it, and create a new image or report. Processing digital images is becoming increasingly crucial in healthcare as more individuals utilize digital imaging for diagnosis. Medical image processing is vital because it can detect dangerous diseases and other healthcare issues. It spreads through the blood vessels and lymphatic system. This comprises the brain, bones, adrenal glands, and liver. Patients with lung cancer have a slim probability of survival, no matter what stage it is detected at. It is initially assumed to be a lung tumor, a non-normal lump of tissue that can be solid or fluid-filled. Benign tumors grow slowly and cause no complications. Malignant tumors are more harmful than benign tumors because they can spread to other regions of the body and become cancerous. As a result, finding and identifying the malignant tumor quickly is critical to predicting the patient's prognosis. Because tumors generally show no symptoms when first discovered, the odds of a cure are poor.

Keywords: Medical image analysis, machine learning, deep learning, multimodal fusion, principal component analysis

*Corresponding author: debnathb@gmail.com

Ashok Kumar, Megha Bhushan, José A. Galindo, Lalit Garg and Yu-Chen Hu (eds.) Machine Intelligence, Big Data Analytics, and IoT in Image Processing: Practical Applications, (23–50) © 2023 Scrivener Publishing LLC

2.1   Introduction

When a person has lung cancer, they have an excessive number of abnormal cells [1] developing in one or both of their lungs, most often in the cells that line their airways. This is referred to as "uncontrolled cell proliferation." In contrast to normal lung tissue, the cells that are not normal expand rapidly and develop into tumors. When tumors grow in size and become more numerous, the number of tumors increases, making it more difficult for the lung to give adequate oxygen to the body. People who have "benign tumors" do not have malignant tumors since their tumors remain in one location and do not seem to be spreading. Tumors have the ability to spread to other sections of the body via the bloodstream or lymphatic system. In cancer, the word "metastasis" refers to cancer that has migrated from its initial site to other sections of the body. Following the spread of cancer, it becomes considerably more difficult to effectively treat it. Secondary lung cancer develops in another part of the body and spreads to the lungs as a result. Primary lung cancer starts in the lungs [2] and spreads throughout the body. As a result of the diverse forms of cancer that they have, everyone has a unique treatment strategy. Lung cancer is classified into two types: small cell lung cancer and non-small cell lung cancer. Small cell lung cancer is the most common kind of lung cancer. In addition to coughing up blood and experiencing chest discomfort, people with lung cancer experience weight loss and decreased appetite, as well as shortness of breath and feeling weak. When cancer is discovered early, the chance of survival increases from 15% to 50%, according to the American Cancer Society [2]. A significant increase in its survival rate is required in contrast to the current situation. Images, such as X-rays, computed tomography (CT) scans, magnetic resonance imaging (MRI), and others may aid in the detection of lung cancer in its early stages without the need for surgery beforehand. When a person has a CT scan, they are provided with three-dimensional images of their lungs as a result. It has been shown that early identification and treatment of illness reduces mortality rates. Detection of cancer [3] cells at an early stage is critical for halting the progression and spread of cancer cells. At the moment, the methods available for identifying lung cancer are inadequate. As a consequence, it is critical to develop novel methods of detecting lung cancer at an early stage. Is it possible to forecast the performance of multilayer and neural network classifiers that have been trained using independent component analysis feature extraction and eleven training techniques? Is this conceivable? A MATLAB-based software application may be used to clean up cancer images before sending them. It has been shown that an image processing technology is very effective in detecting tumor cells. They developed a method of searching for lung cancer that is based on common data properties and image processing methods. According to the authors, the most significant challenge in diagnosing lung cancer is the time it takes for physical tests to be completed. As a consequence, they devised a method of eliminating the need for the null hypothesis test, which they implemented using a standard statistical model. CT scans may be processed in a variety of methods, including image pre-processing, image erosion, median filtering, thresholding, and feature extraction, among others. Researchers at the University of California, Los Angeles [4] focused on creating a CT-based [5] image processing approach for identifying lung cancer. A technique for detecting lung nodules in chest X-rays that makes use of neural networks. They have developed a method of detecting lung cancer based on a CT-based Artificial Neural Network (ANN). Lung cancer was detected using an image processing technology that was developed and tested. In this research, the color characteristic was used in the feature extraction approach to predict lung cancer in its early stages while it was in its early stages. Making sense of the usage of a Computer-Aided Identification (CAD) system, which utilizes pictures from a CT scan to aid in the early diagnosis of lung cancer, is difficult when you do not know what to say (to differentiate between benign and malignant tumors). The National Cancer Registry Program of the Indian Council of Medical Research examined data from six different regions of the nation, including both rural and urban areas, as part of its research. The study found there were significant disparities in the number of cancer cases in various sections of the nation. In 1989, the most common kinds of cancer among men in India's main cities were prostate cancer and testicular cancer. The three most prevalent kinds of cancer in males were trachea cancer, bronchus cancer, and lung cancer. By the end of 2012, there will be 226,160 new lung cancer diagnoses and 160,340 lung cancer-related deaths in the United States. According to the World Health Organization (WHO), cancer claims the lives of 7.6 million people worldwide each year, accounting for about 13% of all fatalities worldwide in that year WHO [1, 2]. As seen in the graph below, lung cancer is the leading cause of mortality in the United States. The majority of persons who get lung cancer pass away within the first three months of the year. Stomach cancer takes the lives of 736,000 people, while liver cancer claims the lives of 690,000. Breast cancer claims the lives of 458,000 people [3, 4]. A recent study by the American Cancer Society found that lung cancer accounts for 14% of all new cancer cases diagnosed in the United States today [5]. According to research, lung cancer claims the lives of more people each year than breast, prostate, and colon cancers combined, which is why it is so vital to maintain a healthy lifestyle (in the USA). Lung cancer accounts for the vast majority of lung cancer cases. Under a microscope, the cells are examined in order to reach a determination on the condition. In 2012, small cell lung cancer and small cell lung cancer claimed the lives of 226.160 people in the United States [6]. Some CAD programs use algorithms such as Support Vector Machines (SVM), fuzzy logic, and neural networks to ensure that the designs they generate are valid. They have a number of disadvantages, including a lengthy training time and a large amount of data. Because of this, they created the Hidden Markov model in order to be better at what they do than their competition.

2.2   The Major Contributions of the Proposed Model

The major contributions of the proposed method are as follows:

The convolution neural network (CNN) may be used to weight the output gradients pixel by pixel in terms of how much weight they should have by using the final convolutional feature map that the network produced. When the CNN algorithm reaches its final conclusion, this approach is applied to each feature map pixel in order to determine its value in relation to the final conclusion reached by the CNN algorithm. A book may be used to get the mathematical formulae for the SoftMax and exponential output activation functions, as well as for pixel-wise weighting. Even while our technology is comparable in operation to previous gradient-based systems, the aesthetic impact produced by it is far greater, as seen in Figure 2.1.

When evaluating the quality of a visualization, many methods, such as deconvolution, guided backpropagation, Computer Aided Mechanism (CAM), and Gradient Weighted Class Activation (Grad-CAM) [7] may be used. It is common practice to employ human or auxiliary assessments instead, such as the location error relative to bounding boxes (ground truth). Moreover, the decision-making rules of the network should be thrown out in this instance. According to the findings of research, new criteria have been developed to assess the degree to which theories and explanations reflect the models on which they are founded (objectively). The results of the research show that swarm optimization outperforms [8] the other alternatives.

Like previous attempts to explain CNNs, the model has been tested on simulated work. Grad-CAM shots do not instil the same amount of trust in the model as Graduated CAM++ photos (among human users). Swarm Optimization outperformed Grad-CAM in finding weakly directed object

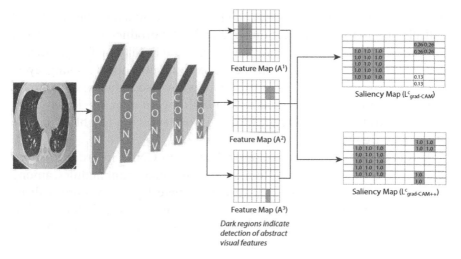

Figure 2.1 Basic block diagram of CNN with PSO.

classes in photographs. Logic should be able to summarize any set of data properly. The explanation of artificial intelligence [9] has received much attention. Grad-explanation Its maps let students construct their own loss functions, which may help them perform better in one-on-one sessions with professors. Statistical sampling has shown success [10], and a method was established to help achieve goals. Swarm optimization [11] may be used in a variety of applications for data synthesis, including image classification and object recognition. When it comes to photos, 3D-CNNs, which are often employed to assist people in understanding films, are not commonly seen. Figure 2.2 illustrates how, until recently, CNNs were only capable of displaying two-dimensional imagery.

It is possible to have input and output layers in each layer of a CNN. There are also convolutional and Max pooling layers on top of the top layer. Incorporating convolutional layers enables for the most efficient possible

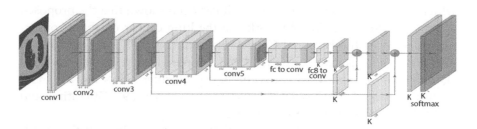

Figure 2.2 Proposed 3D CNN architecture.

use of the available input pools. Each layer will be linked to the others and will all point to the same location. Taking the dot product of the kernel and the data results in this kind of map being created. Making things smaller in size is best accomplished by using a maximum number of pooling layers available. This improves processing speed while also reducing the amount of overfitting that is performed. When building an item, the lower layers are taught to be more difficult, while the top layers are taught to include pieces that are simpler.

It is necessary to use convolutional layers in CNNs in order to simulate characteristics, such as benign nodules and malignant nodules, among many other things. Rather than using a single layer, a volume will be employed instead, and the kernels or filters will be three-dimensional. It is hypothesized that, instead of seeing 2D qualities, such as limits and corners, it would notice 3D traits, such as the ones listed above.

2.3   Related Works

Tumors form in the lungs as a result of the uncontrolled development of abnormal cells [12]. Early detection of lung cancer helps patients to live longer and get more effective treatment. A variety of medical imaging methods may be used to aid doctors in the diagnosis of disorders. A slew of research initiatives to aid in the early detection of lung tumors have been proposed. The most often encountered issues are significant computation delays and misidentification of tumors. In order to overcome these difficulties, this study offers a hybrid classifier that is based on the Atrous Spatial Pyramid Pooling (ASPP)-Unet architecture and the Whale Optimization Algorithm (ASPP-Unet-WOA). This model necessitates the use of the Gabor filter during pre-processing in order to achieve more accurate tumor detection in CT images of the lung. Second, to segment features, guaranteed convergence particle swarm optimization is applied. Last but not the least, the binary grasshopper optimization algorithm is used to identify which characteristics should be used. It has been shown that the proposed approach (ASPPUnet-WOA) works on the lung cancer database consortium dataset of the National Cancer Institute (NCI). A large number of performance measures are analyzed and compared to previously produced classifiers to determine which is the best performer. CNN has a 93.45% accuracy rating, CNN has a 91.67% rating, UNet has a 95.75% rating, and ASPP-UNet-WOA has a 98.68% rating.

Lung cancer is a dreadful disease that destroys the lives of thousands of individuals every year all around the globe. Early identification of lung cancer may improve the chances of a patient's survival. CT scans are used to detect the size and extent of a tumor's spread throughout the body. Problems with CT imaging, on the other hand, result in lower tumor visibility regions and higher unconstructive rates in tumor sites. An optimization-based strategy for lung cancer diagnosis is described in this chapter. The CT scan is used to determine the location of the tumor. The Deep Joint model is used to segment the CT picture at this point. Additionally, local ternary pattern-based features, HoG features, and statistical parameters, such as variance, mean, kurtosis, energy, entropy, and skewness, are extracted. A hierarchical attention network (HAN) is used to classify lung cancer. To train HAN, the firefly competitive swarm optimization (FCSO) technique is employed. It combines the firefly algorithm (FA) with the competitive swarm optimization (CSO) to provide optimal results. The accuracy, sensitivity, and specificity of the proposed FCSO-based HAN were all high, with 91.3% accuracy, 88% sensitivity, and 89.1% specificity for the HAN [13].

A review of the literature on frequent item set mining algorithms; the impact of segmentation techniques for condition monitoring of electrical equipment from thermal images; a comprehensive survey study on various issues and techniques for healthcare record security and privacy; and performance analyses of different classification algorithms for the bank loan sector are among the topics covered [14]. A multiplexer was utilized to produce a universal shift register for low supply voltages in 20-nm FinFETs, which was designed for low supply voltages. Facebook data mining predicts the graphical personality of its users. For classification of class-imbalanced diabetic retinopathy images, precision medicine necessitates deep learning using generative models in order to generate synthetic data. To improve stress prediction, an SVM classifier based on differential boosting particle swarm optimization is used. Controlling a solar-powered high gain converter in a closed loop drive application using an optimized algorithm; using a denoising autoencoder for effective ensemble dimensionality reduction in intrusion detection systems; smart speed control system for highway transportation vehicles using the Internet of Things is being developed. Lung cancer detection using CT images with computer assistance; quote prediction using LSTM and greedy algorithms.

In the realm of medical informatics, classification of medical data is a challenging task to do [15]. The classification algorithms derived from medical datasets must be employed in order to fulfil the goal of reducing human stress via medical data categorization. The major purpose of this study is to categorize medical data using a Decision Tree-based Salp Swarm Optimization (DT-SWO) approach. Following pre-processing, the hybrid feature selection technique chooses the qualities of medical data. Discrete independent component analysis (DICA) is used to lower the number of high-dimensional features, whereas DT-SWO is used to select the most significant class of medical data. The UCI machine learning repository includes four datasets connected with four illnesses of the heart, liver, cancer, and lungs: leukaemia, diffuse larger B-cell lymphoma (DLBCL), lung cancer, and colon cancer. Finally, the study results indicated that the recommended DT-SWO approach outperforms current algorithms for medical data classification.

The purpose of this study is to construct an ideal back propagation network (BPN) model by decreasing the amount of input characteristics and optimizing the number of neurons in each layer of the BPN classifier while keeping accuracy [16]. This work provides a hybrid AFSO-EA strategy for feature set reduction that blends Artificial Fish Swarm Optimization (AFSO) with evolutionary algorithms (EA). The number of neurons in each hidden layer is optimized in this research using the same hybrid strategy that is used to create the BPN model. According to the results, the recommended hybrid AFSO-EA strategy provides a BPN model with a classification accuracy of 97.5% and a computational overhead substantially lower than that of existing techniques.

Recently identified genes that indicate cancer-causing disorders have become increasingly relevant in the analysis of microarray data [17]. Due to the frequent changes in situation, a considerable volume of data was essential. Traditional data mining methodologies are inefficient in terms of geographical and temporal complexity. The specified aim is fulfilled by the exploitation of big data. In this study, feature extraction is achieved utilizing the improved supervised principal component analysis (ISPCA). A covariance matrix is constructed for gene expression, and ISPCA is used to classify cancer through feature selection. Modified particle swarm optimization (PSO) is then done utilizing a unique wrapper model method and a boundary tuned support vector machines (BT-SVM) classifier. The research makes use of a range of datasets from the University of California, Irvine collection, including those on leukaemia, breast cancer, brain cancer, colon cancer, and lung carcinoma. The suggested technique is assessed in terms of accuracy, recall, and precision on six benchmark datasets for deoxyribonucleic acid (DNA) microarray data. It excelled in a range of frequently used measures for judging the effectiveness of planned work, obtaining the best levels of accuracy, recall, precision, and training length with and without feature selection.

Lung cancer is no longer a side effect [18]. Following the end of various researches, it was discovered that the mortality rate had risen. As a consequence, patients should be diagnosed more promptly to limit fatalities. Lung cancer is commonly discovered via a range of procedures from a variety of professions. Various approaches for identifying lung cancer exist now in the domain of machine learning for feature extraction and selection. Segmentation using super-pixels is a frequent method. We shall present an algorithm that will produce accurate results in the detection of lung cancer in this study, develop or refine current algorithms, such as genetic optimization, PSO, and SVM, and compare them to existing algorithms before agreeing on a standard strategy that delivers accuracy in the design of a lung cancer detection system.

Lung cancer is a primary cause of death in men and women worldwide, killing an estimated five million people each year [19]. A CT scan may be highly effective in diagnosing lung disorders. This study's principal purpose is to detect malignant lung nodules in a lung image and to diagnose lung cancer according to its severity. This work applies new deep learning methods to localize cancerous lung nodules. The best feature extraction algorithms are employed in this work, including the histogram of oriented gradients (HoG), wavelet transform-based features, local binary pattern (LBP), scale invariant feature transform (SIFT), and Zernike Moment. After accumulating textural, geometric, volumetric, and intensity data, the best feature is identified using the fuzzy particle swarm optimization (FPSO) approach. Following that, deep learning is employed to classify these traits. The computational complexity of CNN has been lowered as a consequence of the new fuzzy particle swarm optimization convolutional neural network (FPSOCNN).

Under the area of biomedical research, it is vital to check diseased tissue under a microscope and assess the severity of the illness using a procedure called histopathology [20], in which laboratories generate tissue slides for digital viewing. These histology photos have been scanned in their entirety. After digitizing the slides, pathologists may be able to view them on a computer rather than a microscope. As a consequence, it is crucial for cancer detection to be able to distinguish and characterize cell nuclei. This process becomes considerably more difficult owing to the high degree of noise and modest fluctuation in the size of cell nuclei in histopathological photographs. To tackle this difficulty, we designed a method for automatic nucleus cell segmentation based on optimization-based superpixel clustering. The histopathological image collection is first gathered from an acceptable source. After that, a normalizing method is employed to remove noise from the images. Following denoising, an optimum clustering algorithm is utilized to segment cells into malignant and benign cells. The recommended strategy's major purpose is to employ an effective segmentation technique to handle the issue of benign tumors. The recommended study produced remarkable performance in segmenting histological images of lung/breast/liver/brain cancer, thereby supporting the early diagnosis of cancer by retrieving crucial parts for categorizing the presented CT scans as tumors or non-tumors.

Due to the various benefits of automating medical diagnosis, lung cancer detection has been a prominent focus of investigation [21]. In recent years, automated lung cancer screening utilizing CT scans has been identified as a crucial tool. Despite the fact that several approaches for lung cancer detection have been published in the literature, establishing an effective system capable of automatically recognizing lung cancer is challenging. As a consequence, the purpose of this project is to create a deep learning-based automated lung cancer detection system employing a hybrid optimization strategy. The CT images from the lung cancer database are preprocessed here before being submitted to the active contour lung segmentation. Then, using the grid-based technique, the nodules in the segmented image are discovered. Several image components, including intensity, wavelet, and scattering transformations, are input into the proposed Elephant Herding Optimization Algorithm-Deep Belief Network (SEOA-DBN) for categorization. SEOA is a newly discovered approach that combines the Salp Swarm Algorithm (SSA) with optimization for Elephant Herding Optimization (EHO). The experiment compares lung CT images from a conventional database to a number of cutting-edge techniques. According to the statistics, the planned SEOA-based DBN functioned beautifully, obtaining an accuracy rate of 96%.

2.4   Problem Statement

Medical pictures are becoming more significant as more people rely on them to diagnose and cure ailments. These images are critical as medical professionals seek a better understanding of human anatomy and function. For a long time, medical image processing was considered a niche field, but it has now become ubiquitous. Among the techniques used in medical surgery are image segmentation, image registration, and image-guided surgery, just to name a few. A major concern with cancer is the late discovery and diagnosis of the disease. Thus, few patients get therapy, treatment costs are prohibitively expensive, and death seems to be practically certain. A diagnosis and classification of the illness must be determined at the earliest opportunity. This is an attribute that can, without a doubt, save lives.

2.5   Proposed Model

2.5.1   Swarm Optimization in Lung Cancer Medical Image Analysis

PSO is a computer-based approach for identifying the most optimal solution to a given issue. It makes use of a quality measure in order to attempt to increase the quality of a potential solution. This is how PSO refers to each of their solutions. Particles make up the swarm's structure. Because of the speed at which individuals walk and the distance between them, PSO determines where they should be assigned to work. Particles migrate in the direction of the most well-known locations in the search space since each particle has a unique local best-known location.

Changes are made to the particle placements in order to better steer the swarm to the locations where the best possibilities have been identified. In the search space, particles that are close to one another in proximity might represent a solution to a certain optimization issue. A consequence of the fact that the search procedure is random is that the particle may come up with a solution that is less satisfactory than the one it came up with before. When it comes to the PSO algorithm, there are a lot of particles, and each one has a unique current goal number and position, as well as a personal best goal number that is the particle's best goal number to date. The individual who is performing at their peak is considered the most valuable to them. PSO also retains the world's best value and position, which is the greatest value and position that any particle has ever had in the universe. At this moment, the most optimal global value has been identified. During each cycle, the best values, pbest and gbest, are utilized to alter the values of all particles.

These two ideal values are what allow particles to shift to a more advantageous location. The particle's speed and position are changed once it has selected the two optimum values. When it comes to computer science, PSO is a technique that is used to repeatedly attempt to enhance the quality of a proposed solution. PSO solves issues by generating a population of potential solutions (referred to as particles) and moving them about in the search space according to their position and speed. If a particle travels, the best-known position of the particle has an impact on where the particle goes. It is also propelled to the best-known search locations, which are updated as other particles discover better locations. It is also anticipated to travel in the direction with the greatest likelihood of success. Initially, so was thought to be a model of social behaviour since it seemed to be a flock of birds or fish moving together in a herd-like manner.

Once it became evident that the system was effective at what it was intended to do, it was refined to be more efficient. The book by Kennedy and Eberhart discusses a wide range of philosophical issues related to PSO and swarm intelligence. Piercingly spent a significant amount of time researching PSO applications. Bonyad and Michaelis provided an excellent summary of both theoretical and practical PSO research in their presentation. A PSO is a metaheuristic in that it does not make many assumptions about the issue at hand and is capable of generating a large number of potential solutions to the problem.

As an added advantage, since PSO does not make use of the gradient of the optimization issue, it does not need the optimization problem to be differentiable in the same way as gradient descent and quasi-newton do. It is not certain that the optimal solution will be discovered at some point throughout a procedure, including metaheuristics, such as PSO.

2.5.2   Deep Learning with PSO

In the computer industry, deep learning is a sort of machine learning that is effective at identifying patterns, but it requires a large amount of data to function properly. It is an important process to check the quality of data [22–25] used for machine learning and deep learning algorithms. A broad variety of domains, including image classification, natural language processing, computer vision, and bioinformatics, are addressed using deep neural network topologies, which are utilized to solve issues in the approach. Several recent research studies have shown that deep learning can yield cutting-edge outcomes in a broad variety of applications, prompting its resurgence.

Making neural networks with more than one layer of neurons buried away from the surface is the process of deep learning, which is also known as convolutional neural network training. Additionally, these "deep" designs varied widely, with implementations that were customized to certain activities or ends in mind. Due to the rapidity with which people are doing research, they are always developing new and exciting methods of learning about deep learning. As a result of having three or more layers of ANNs, each of which extracts one or more visual features, deep learning is more effective at determining what objects are in a picture. Machine learning algorithms become less effective as the quantity of data increases.

Deep learning, which is one of the best machine learning approaches, may be able to speed up the feature extraction process by using automated feature extraction methods. Deep learning is one of the greatest machine learning techniques since it is very fast. In the area of medical image processing, deep learning is becoming more popular. Approximately $300 million in revenue will be generated by 2021, which is a significant amount of money.

The company will have raised more money for medical imaging than the whole analytical sector did in 2016, according to the figures. Using machine learning in this manner is the most efficient and supervised method available. A sort of neural network known as a "deep neural network" is used in this procedure.

In recent years, this sort of neural network has gained popularity for applications such as image classification and computer vision. DCNN-based architecture: This study demonstrates how to detect and classify cancers from CT scan pictures of the lungs. Thus, this piece of work demonstrates how this architectural structure is put together.

CNN has mainly three important layers:

    i. Input Image layer
    ii. Convolution pooling Layer
    iii. Classification layer

2.5.3   Proposed CNN Architectures

The CNN model makes use of local receptive fields, shared weights, and subsampling to make it more resistant to shifting, scaling, and distorting than a single model would be able to do. It is possible for this model to learn how to map a large number of distinct characteristics to one another by using a linear filter, bias, and nonlinear function. The convolutional layer is referred to as Conv, the pooling layer is referred to as Pool, and the fully connected layer is referred to as FC, which is an abbreviation for completely connected.

A SoftMax classifier with 32 filters was included, as were three convolutional layers, two max-pooling layers, a fully connected layer, and a fully connected layer. The picture is delivered to the first convolutional layer, which separates it using 2424 patches created by the first convolutional layer. With a one-pixel gap between each of the 32 feature maps in C1, they are connected to a patched picture of 24 24 pixels by five filters, with a one-pixel gap between each of the feature maps in C1. They were 20 pixels wide by 20 pixels high. This is the second layer of the structure. A total of two digits are used in the kernel size and two in the stride. There was a second convolutional layer, which was a 10 10 convolutional layer, in addition to the first one. The output of P1 was routed via this layer. Following that, the convolutional and pooling layers are processed in exactly the same way as before. At the end, there are layers that serve to connect multiple separate units with one another. After that, the SoftMax classifier is employed to determine if the image is normal or cancerous. DCNN architecture 1 was created with the fewest number of layers feasible in order to be able to cope with images of up to 1,000 pixels in size. The accuracy of the dataset increases in direct proportion to the number of photographs in the collection. As a consequence, this work employs two alternative DCNN topologies, each of which is described below. Compared to architecture 1, architecture 2 features a greater number of convolutional and pooling layers. This is done in order to increase accuracy. These classifiers were applied to around ten thousand photos and performed well. Figure 2.3 is a representation of the architecture 2 that was proposed by the audience.

CNN architecture in Figure 2.4 is composed of thirteen layers. Seven of them are convolutional neural networks, four are pooling neural networks, one is connected neural networks, and one is a SoftMax classifier. However, some layers have a filter size in Figure 2.5 of 5 5, while other layers have a filter size of 2 2, and the two sizes are not the same. Following the introduction of convolutional layers, the number of filters used to feature maps increases from 64, to 96, to 128, to 192, to 256, as seen below.

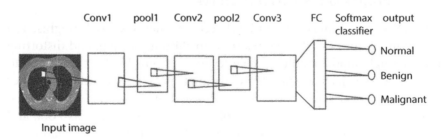

Figure 2.3 Linear filter with bias.

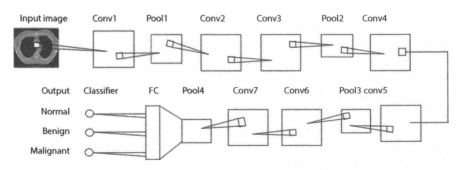

Figure 2.4 Fully connected CNN.

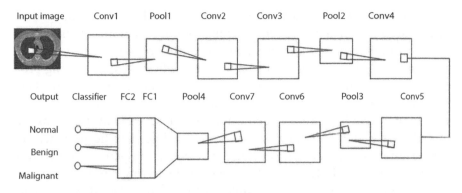

Figure 2.5 Showing the CNN with filter and stride.

Seven convolutional layers, four pooling layers, two fully connected layers, and the SoftMax classifier are included in this algorithm. It adds 64 filters to the 256 256 patch size input pictures in the 256-patch size mode. Approximately the same size as the patch, each filter has a similar footprint. Max pooling layer has two 2 filters, which work together to reduce the output of the preceding convolutional layer to the smallest possible size. 126126 is generated by the first pooling layer, which is then sent on to the second and third convolutional layers, which apply 98 and 128 filters to the resulting image, respectively. After that, the maximum pooling is employed, resulting in an image of 5959 pixels in size. Similarly, with the fourth and fifth convolutional layers, which feature 128 and 192 filters, this is likewise the case. They apply the down-sampled photos from the third max pooling layer to these layers, resulting in a 2525-dimensional feature map with a total of 2525 dimensions. Method: The process is repeated until all neurons in each layer are connected to all neurons in the layer preceding them, and so on. Last but not the least, the SoftMax classifier is utilized to categorize the images into three categories: normal, benign, and malignant. The second architecture outperforms the previous one in every way. Architecturally speaking, the third layer is totally interconnected by layer descriptions and hyperparameters that are quite similar to the previous layers.

2.6   Dataset Description

Lung cancer is the most frequent kind of cancer in the world, and it is also the leading cause of cancer-related mortality. In the United States, low-dose CT scans are being used to screen for lung cancer to identify people who are at high risk of developing the disease. It is predicted that more nations would follow suit in the near future. Millions of CT pictures will need to be reviewed for CT lung cancer screening, putting a significant amount of strain on radiologists. Therefore, there is considerable interest in creating computer algorithms that might improve the efficiency of screening procedures. Lung nodules, which may or may not be indicators of early-stage lung cancer, are the first thing to check for on CT imaging for lung cancer screening. There has already been evidence that several CAD systems are capable of achieving this aim. An automated nodule identification method will be tested on a broad scale during the LUNA16 competition, which will leverage the LIDC/IDRI data set to conduct the testing. Four radiologists provide information regarding nodules as part of the LIDC/IDRI data collection. The information is accessible to the general public, and it contains descriptions of the individuals. To make matters even better, the LUNA16 challenge is now open to anybody who wants to participate. The tracks for systems that can locate nodules in their totality, as well as tracks for systems that work with a list of locations that may have nodules in some form are offered. The list was compiled so that teams may assist one another by utilizing an algorithm to determine where a lung nodule could be seen on a CT scan and where it might not be visible. The Lung Image Database Consortium image collection contains thoracic CT pictures with annotated lesions, which may be found here (LIDC-IDRI). These photos are used to diagnose lung cancer and to look for symptoms of the disease. It is possible to get assistance with CAD processes for identifying and diagnosing lung cancer on the internet, which is among the greatest resources available. The National Cancer Institute (NCI), the Foundation for the National Institutes of Health (FNIH), and the Food and Drug Administration (FDA) are all participating in this public-private cooperation to demonstrate how a consensus-based consortium operates.

Together with eight medical imaging firms, seven academic institutions developed this data collection, which has 1018 occurrences. For each individual, images from a thoracic CT scan are shown alongside an XML file containing the findings of a two-phase image annotation system created by four thoracic radiologists over a two-year period are also displayed with the images. This is exactly what occurred during the initial blinded-reading phase. Each CT image was reviewed by a radiology specialist who categorized the lesions as "nodule > or =3 mm," "nodule 3 mm," or "non-nodule > or =3 mm." Each radiologist reviewed their own markings, as well as the marks of the other three radiologists, before reaching a final judgment during the unblinded-read part of the procedure. The goal of this method was to discover as many lung nodules as feasible on each CT scan without requiring consensus from the team.

2.7   Results and Discussions

2.7.1   Parameters for Performance Evaluation

The basic performance metrics for classification of lung cancer are:

- Mean Absolute Error (MAE)
- Classification Accuracy
- Mean Squared Error (MSE)
- F1 Score
- Logarithmic Loss
- Confusion Matrix
- Area Under Curve (AUC)

MAE can be calculated in two different ways. It demonstrates that the forecasts were incorrect. If data is forecasted incorrectly, the data does not show the direction of the error. It can be expressed mathematically as follows:

$$Mean\_Absolute\_Error = \frac{1}{N}\sum_{J=1}^{N} |\, actual - predicted$$

In classification, frequently, the term "accuracy" is misspelled as "precision." The correct predictions are divided by the total number of input samples as follows:

$$Accuracy\_of\_Model = \frac{Number\_of\_Correct\_Predictions}{Total\_Number\_of\_Predictions\_Made}$$

This method works best when both classes have the same number of samples, because each class does not always have the same number of samples. Take into account that the training set includes samples from 98% of the classes A samples and only 2% of the classes B samples. As a result, by simply predicting all class A training samples, the model achieves 98 % training accuracy. According to the findings, when the same model is tested with 60% class A samples and 40% class B samples, the accuracy of the test drops to 60%. As a result, the accuracy of the classification in this case appears to be quite high. The real issue arises when the cost of incorrectly classifying minor class samples becomes prohibitively expensive.

The cost of failing to diagnose rare but fatal diseases outweighs the cost of additional testing on a healthy individual.

The only difference between MSE and MAE is that MSE averages the square of the difference between the original and predicted values. MSE is easier to compute than MAE, which requires sophisticated linear programming tools. Because larger errors have a greater impact when squared, the model can now emphasize larger errors.

$$Mean\_Squared\_Error = \frac{1}{N} \sum_{J=1}^{N} (actual - predicted)^2$$

The Harmonic Mean is used to calculate the F1 Score, which is a precision and recall measure. The F1 Score ranges from 0 to 1. It displays the precision and strength of the classifier (number of times it correctly classifies) (number of times it correctly classifies) (it does not miss a significant number of instances). With a high precision but a low recall, the result is extremely accurate, but many difficult-to-classify cases are missed. The F1 Score indicates how accurate the model is. In terms of math, it is as follows:

$$F1 = 2 * \frac{1}{\frac{1}{Precision} + \frac{1}{recall}}$$

The property of the F1 score is to find the proper balance between the recall and precision. Precision is the number of correctly predicted positive results divided by the classifier's prediction accuracy.

$$Precision = \frac{True\_Positives}{True\_Positives + False\_Positives}$$

The recall is the proportion of correctly identified positive samples to total relevant samples (all samples that should have been identified as positive) (all samples that should have been identified as positive).

$$Precision = \frac{True\_Positives}{True\_Positives + False\_Negatives}$$

Logarithmic loss is a penalty imposed on false classifications. It excels at categorizing several classes. The classifier must give probabilities to each class when using log loss. The log loss is: If there are N samples from M classes, the log loss is:

$$Logarithmic\_Loss = \frac{-1}{N} \sum_{i=1}^{N} \sum_{j=1}^{M} y_{i,j} * log(p_{i,j})$$

where y ij indicates whether a sample was taken. I am a jth class member.

P ij signifies the chance that sample I is of class j.

Log loss has no upper limit; it happens between [0,1]. Precision is illustrated by a log loss that is close to zero, but not exactly. Lowering log loss generally increases classifier accuracy.

As an output, the confusion matrix summarizes the model's overall performance. Assume there is a binary classification issue; confusion matrix was shown in Table 2.1.

The samples that may be classified as yes or no and also have a classifier that guesses the class of an input sample. The table below shows the findings of the model on 165 samples.

Four terms to remember for confusion Matrix:

TP: It occurs when predicted YES and the result was YES.

TN: When correctly anticipated NO.

FP: It occurs when anticipating YES but get NO.

FN: When anticipated NO but received YES.

Averaging along the "principal diagonal,"

$$Accuracy = \frac{truepositives + trueNegatives}{Total\_Sample}$$

Confusion matrices are the main fundamental basic block for the other metrics.

The AUC is an often-used evaluation metric. It is used to solve problems that need binary classification. The AUC value of a classifier is the probability that a randomly selected positive example will be ranked higher than a randomly selected negative example. Let us define two concepts before moving on to AUC: sensitivity is measured using the TPR (TP/(FN+TP)). The TPR is the % of positive data points that are correctly categorized as positive after all positive data points have been evaluated.

Table 2.1 Confusion matrix.

Total population		True condition	
		Condition positive	**Condition negative**
Predicted Condition	Positive Predicted Situation	Actually positive	Type I error (false positive)
	Negative Expected Situation	False Negative (Type II error)	ACTUALLY NEGATIVE \
		Sensitivity = True Positive/ Condition Positive, True Positive Rate (TPR), Recall	FPR = False Positive/ Condition Negative.
		Accuracy (ACC) = (Total Population/ True Positive + True Negative) ((1/Recall + 1/ Precision)/2) F1 score	Precision = Predicted Condition Positive/ True Positive
			Selectivity=True negative/Condition Negative Specificity (SPC)

$$True\_Positive\_Rate = \frac{True\_Positive}{False\_Negative + True\_Positive}$$

True negative rate is defined as TN/(FP+TN). The FPR is the proportion of negative data points that are correctly recognized as negative when all negative data points are included.

$$True\_Negative\_Rate = \frac{True\_Negative}{True\_Negative + False\_Positive}$$

(FP/(FP+TN)) False Positive Rate (FP/(FP+TN))

FPR (FP/(FP+TN)) is the % of negative data items that are incorrectly assessed positive when all negative data points are included.

$$False\_positive\_Rate = \frac{False\_Negative}{True\_Negative + False\_Positive}$$

TPR and FPR are both in the range of 0 to 1.

A graph is created when the FPR and TPR are computed for a variety of threshold values (0.00, 0.02, 0.04, 1.00).

The AUC of a plot of FPR versus TPR at different points in the range [0, 1] is defined in Figure 2.6.

AUC has a range of [0, 1], as previously stated. The proposed model performs better when the value is higher as shown in Table 2.4 compared with Tables 2.2 and 2.3. Performance metrics were shown in Tables 2.5 and 2.6 and the last CT image was shown in Figure 2.6.

Figure 2.7 deals about the training technique for the LUNA 16 dataset. The picture X was used as a stimulus in the training of the model. This comparison results in the model suffering a loss, and new parameters are introduced into the model as a result of this. It was necessary to learn estimates before they could be utilized to construct the loss function, which was accomplished via the application of the SoftMax activation

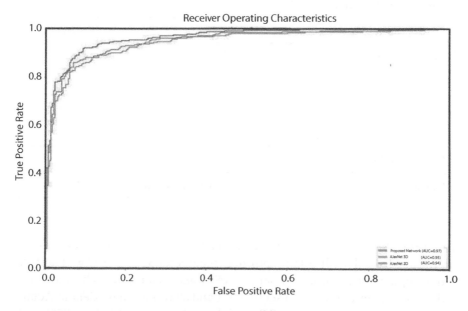

Figure 2.6 AUC-ROC curve of the proposed algorithm.

Table 2.2 Network model of the 2D CNN AlexNet.

Layer	Output shape	Param (#)
Convolutional Layer_2D	32*32*16	273
Mini__Batch__Normalization_V1	32*32*16	65
Kernal__Layer__2D	32*32*16	1
Convolutional Layer_2D _1	32*32*32	4641
Mini__Batch__Normalization _v1_1	32*32*32	129
Kernal__Layer__2D_1	32*32*32	1
Convolutional Layer_2D _2	32*32*64	18495
Mini__Batch__Normalization _v1_2	32*32*32	257
Convolutional Layer_2D _3	32*32*32	36927
Mini__Batch__Normalization _v1_3	32*32*32	257
Convolutional Layer_2D _4 (Conv 2D)	8*8*32	18455
Mini__Batch__Normalization _v1_4	8*8*32	127
Kernal__Layer__2d_2	8*8*32	1
Layers_Panel_Flatten	512	1
Dense__layer_V1	200	102601
Mini__Batch__Normalization _v1_5	200	856
Drop_Out_Layer	200	1
Dense__layer_V1	75	15071
Mini__Batch__Normalization _v1_6	75	356
Drop_Out_Layer	75	1
Dense__layer_V1	Benign or Malignant	151

function. Use of cross entropy was crucial in this situation. For example, all models have default Adam optimizer parameter 3 set to 0.99 and default Adam optimizer parameter 1 set to 0.09, whereas all models have default Adam optimizer parameter 3 set to 0.001 and all models have default Adam optimizer parameter 3 set to 0.99. With an Accuracy (AEC) of 95% and

Table 2.3 Network model of the 3D CNN AlexNet.

Layer	Output shape	Param (#)
Convolutional Layer_3D_1 (Conv 3D)	32*32*32*16	8207
Mini__Batch__Normalization _v1	32*32*32*16	66
Kernal__Layer__3D_1	16*16*16*16	1
Convolutional Layer_3D _1(Conv 3D)	16*16*16*32	13857
Mini__Batch__Normalization _v1_2	16*16*16*32	129
Kernal__Layer___2	8*8*8*312	1
Convolutional Layer_3D _3	8*8*8*64	55361
Mini__Batch__Normalization _v1_3	8*8*8*64	257
Convolutional Layer_3D _4	8*8*8*64	1106562
Mini__Batch__Normalization _v1_4	8*8*8*64	252
Convolutional Layer_3D _5 (Conv 2D)	8*8*8*32	53329
Mini__Batch__Normalization _v1_5	8*8*8*32	121
Kernal__Layer__2d_3	4*4*4*32	1
Layers_Panel_Flatten	204	1
Dense__layer_V1	200	409801
Mini__Batch__Normalization _v1_6	200	801
Drop_Out_Layer	200	1
Dense__layer_V1	75	15072
Mini__Batch__Normalization _v1_7	75	301
Drop_Out_Layer	75	1
Dense__layer_V1	Benign or Malignant	153

an AUC of 97%, AlexNet CNN outperformed the CNN, highlighting the value of incremental advancement in machine learning algorithms. The AEC of the CNN was 94%, whereas Alex's 3D CNN had an AEC of 95% as shown in Table 2.4 to 2.6.

Table 2.4 Proposed CNN model.

Layer	Output shape	Param (#)
Convolutional Layer_3D _1 (Conv 3D)	27*27*27*16	3472
Mini__Batch__Normalization _v1	27*27*27*16	64
Convolutional Layer_3D _1(Conv 3D)	27*27*27*16	2064
Mini__Batch__Normalization _v1_2	27*27*27*16	64
Convolutional Layer_3D _3	27*27*27*16	16400
Kernal__Layer__3d_2	8*8*8*312	0
Mini__Batch__Normalization _v1_3	23*23*23*16	64
Kernal__Layer__3d_1	11*11*11*16	0
Convolutional Layer_3D _4	10*10*10*32	4128
Mini__Batch__Normalization _v1_4	10*10*10*32	129
Convolutional Layer_3D _5 (Conv 2D)	9*9*9*32	8225
Mini__Batch__Normalization _v1_5	9*9*9*32	125
Kernal__Layer__3d_2	4*4*4*32	1
Convolutional Layer_3D _6 (Conv 2D)	3*3*3*64	16441
Dense__layer_V1	200	409801
Mini__Batch__Normalization _v1_6	3*3*3*64	251
Kernal__Layer__3d_3	1*1*1*64	1
Layers_Panel_Flatten	64	1
Dense__layer_V1	256	163240
Mini__Batch__Normalization _v1_7	256	1021
Drop_Out_Layer	256	1
Dense__layer_V1	Benign or Malignant	524

Because this recall value is only 0.3% better than the Alex Net CNN, every 0.1% improvement in memory saves the life of another patient, making this recall value very useful, especially if it can be increased without compromising accuracy. Visual insights with AUC-ROC curve shown in Figure 2.7.

Table 2.5 Performance metrics of the 3D-CNN classifier.

S. no.	CNN	AUC	Accuracy	F-score	Precision	Recall
1	AlexNet2D-CNN	0.94	87.67	0.89	0.85	0.91
2	AlexNet3D-CNN	0.95	89.17	0.91	0.91	0.88
3	Proposed CNN	0.97	97.17	0.92	0.87	0.94

Table 2.6 Results comparisons with other classifiers.

Results of experiments								
Dataset information for LUNA 16				Existing technique			Methodology proposed	
S. no.	No. of samples	Training	Testing	Authors	Accuracy	2D AlexNet	3D AlexNet	Proposed 3D AlexNet
1	888	90%	10%	[19]	88.17%	89.45%	90.23%	97.17%
2	1018	90%	10%	[20]	89.67%	88.78%	91.13%	97.17%

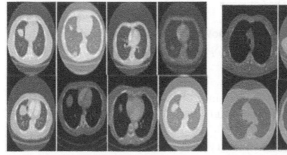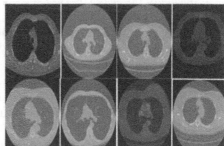

Figure 2.7 Training image of the Luna-16 dataset.

2.8 Conclusion

Various optimization algorithms were employed to locate the tumor in this study. Medical images are frequently cleaned up before statistical analysis. The adaptive median filter outperforms the median and mean filters in terms of speckle suppression, speckle preservation, and mean preservation. The proposed algorithm is the most accurate of the five algorithms for removing tumors, with a maximum accuracy of 95.8079% across all images. It outperforms the previous approach, which was 90% accurate in 4 of 10 datasets before being upgraded. In order to enhance accuracy, future study will use more optimization approaches, and hence employ more. The level of diagnostic certainty increases when the complexity of the relationships decreases. Using a hidden Markov model, CT scans can look for lung cancer even if it is not present at the time of the scan. This method improves the accuracy of the findings while speeding up the process. The data are cleaned up by removing noise. The images of the lungs are then segmented. A doctor can tell if a patient has cancer by looking at their symptoms and age. The Updated model creates the image, and the image is used to diagnose. The ultimate goal is to locate and categorise 4-D CT scans. The work carried in this are working on automated medical image processing systems that can detect cancer cells in real time.

References

1. Neal Joshua, E.S., Bhattacharyya, D., Thirupathi Rao, N., The use of digital technologies in the response to SARS-2 CoV2-19 in the public health sector, in: *Digital Innovation for Healthcare in COVID-19 Pandemic: Strategies and Solutions*, pp. 391–418, 2022.
2. Joshua, E.S.N., Battacharyya, D., Doppala, B.P., Chakkravarthy, M., Extensive statistical analysis on novel coronavirus: Towards worldwide health using apache spark, in: *EAI/Springer Innovations in Communication and Computing*, pp. 155–178, 2021.
3. Bhattacharyya, D., Kumari, N.M.J., Joshua, E.S.N., Rao, N.T., Advanced empirical studies on group governance of the novel corona virus, MERS, SARS and EBOLA: A systematic study. *Int. J. Curr. Res. Rev.*, 12, 18, 35–41, 2020.
4. Marline Joys Kumari, N., Bhattacharyya, D., Thirupathi Rao, N., Improving the diagnostic accuracy using amplification and sequencing of the SARS-CoV-2 genome, in: *Digital Innovation for Healthcare in COVID-19 Pandemic: Strategies and Solutions*, pp. 331–350, 2022.
5. Eali, S.N.J., Rao, N.T., Swathi, K., Satyanarayana, K.V., Bhattacharyya, D., Kim, T.-H., Simulated studies on the performance of intelligent transportation system using vehicular networks. *Int. J. Grid Distrib. Comput.*, 11, 4, 27–36, 2018.
6. Bhattacharyya, D., Long term prediction of rainfall in Andhra Pradesh with deep learning. *J. Med. Pharm. Allied Sci.*, 10, 4, 3132–3137, 2021.

7. Maran, P., S., S., K., T., D., H., K., N., A novel deep learning method for identification of cancer genes from gene expression dataset, in: *Advances in Computer and Electrical Engineering*, pp. 129–144, 2020.

8. Bhattacharyya, D., Comprehensive analysis on comparison of machine learning and deep learning applications on cardiac arrest. *J. Med. Pharm. Allied Sci.*, 10, 4, 3125–3131, 2021.

9. Joshua, E.S.N., Bhattacharyya, D., Chakkravarthy, M., Kim, H.J., Lung cancer classification using squeeze and excitation convolutional neural networks with Grad Cam++ class activation function. *Trait. du Signal*, 38, 4, 1103–1112, 2021.

10. Kim, E. and Chung, Y., Comparison and optimization of deep learning-based radiosensitivity prediction models using gene-expression profiling in National Cancer Institute-60 cancer cell lines. *Nucl. Eng. Technol.*, 54, 8, 2022.

11. Doppala, B.P., NagaMallik Raj, S., Stephen Neal Joshua, E., Thirupathi Rao, N., Automatic determination of harassment in social network using machine learning, Springer, India, 2021. Retrieved from www.scopus.com.

12. Altan, G., A deep learning architecture for identification of breast cancer on mammography by learning various representations of cancerous mass, in: *Deep Learning for Cancer Diagnosis*, pp. 169–187, 2020.

13. Lata, K. and Saini, S., A review of deep learning-based methods for cancer detection and classification, in: *Deep Learning for Biomedical Applications*, pp. 229–253, 2021.

14. Kotani, D., Fujii, S., Yamada, T., Suzuki, M., Yoshino, T., O12-1 A novel gene-prediction model, virtual sequencing with deep learning to predict gene alterations in colorectal cancer. *Ann. Oncol.*, 32, S290, 2021.

15. Mahanty, M., Bhattacharyya, D., Midhunchakkaravarthy, D., Kim, T., Detection of colorectal cancer by deep learning: An extensive review. *Int. J. Curr. Res. Rev.*, 12, 22, 150–157, 2020.

16. Passi, K., Shi, Z., Jain, C.K., Improved prediction of gene expression of epigenomics data of lung cancer using machine learning and deep learning models, in: *Knowledge Modelling and Big Data Analytics in Healthcare*, pp. 165–182, 2021.

17. Gene expression data based deep learning model for accurate prediction of drug-induced liver injury in advance. *J. Chem. Inf. Model*, 59, 7, 3240–3250, 2019.

18. Joshua, E.S.N., Chakkravarthy, M., Bhattacharyya, D., Lung cancer detection using improvised grad-cam++ with 3D CNN class activation. *Lecture Notes in Networks and Systems*, pp. 55–69, 2021.

19. Choi, H. and Na, K.J., A risk stratification model for lung cancer based on gene coexpression network and deep learning. *BioMed. Res. Int.*, 2018, 1–11, 2018.

20. Joshua, E.S.N., Chakkravarthy, M., Bhattacharyya, D., An extensive review on lung cancer detection using machine learning techniques: A systematic study. *Rev. Intell. Artif.*, 34, 3, 351–359, 2020, https://doi.org/10.18280/ria.340314.

21. Choi, H. and Na, K.J., A risk stratification model for lung cancer based on gene coexpression network and deep learning. *BioMed. Res. Int.*, 2018, 1–11, 2018.

22. Bhushan, M., Goel, S., Kumar, A., Improving quality of software product line by analysing inconsistencies in feature models using an ontological rule-based approach. *Expert Syst.*, 35, 3, e12256, 2018.

23. Megha, Negi, A., Kaur, K., Method to resolve software product line errors, in: *International Conference on Information, Communication and Computing Technology*, Springer, pp. 258–268, 2017, https://doi.org/10.1007/978-981-10-6544-6_24.

24. Bhushan, M., Goel, S., Kumar, A., Negi, A., Managing software product line using an ontological rule-based framework, in: *2017 International Conference on Infocom Technologies and Unmanned Systems (Trends and Future Directions) (ICTUS)*, IEEE, pp. 376–382, 2017.

25. Bhushan, M. and Goel, S., Improving software product line using an ontological approach. *Sādhanā*, 41, 12, 1381–1391, 2016.

Liver Cancer Classification With Using Gray-Level Co-Occurrence Matrix Using Deep Learning Techniques

Debnath Bhattacharyya[1]*, E. Stephen Neal Joshua[2] and N. Thirupathi Rao[2]

[1]Department of Computer Science and Engineering, Koneru Lakshmaiah Education Foundation, Vaddeswaram, Guntur, Andhra Pradesh, India
[2]Department of Computer Science and Engineering, Vignan's Institute of Information Technology, Visakhapatnam, AP, India

Abstract

Liver cancer is the most common cause of death in the world. In order to find liver cancer, you need to figure out what medical images mean adaptive thresholding with a watershed transform is used to make the liver stand out from the other parts of the body. Optimal strategies and the swarm optimization model are some of the methods used to separate the malignant area of the liver. The data included 225 images from patients with different types of liver cancer. In order to build a big dataset, the Gray Level Matrix, the Local Binary Pattern was used to find features that are important. It then gets broken down into different types of cancer using neural network, support vector machine, random forest, and deep neural network classifiers. These include hemangioma, hepatocellular carcinoma. The suggested methods are judged on their sensitivity, specificity, accuracy, and Jaccard index. People who use watershed Gaussian-based deep learning algorithms have found that they can be used to diagnose liver cancer. DNN classifiers were found to be the best, with an accuracy of 99.25% and a Jaccard index of 0.97 across 300 epochs with a low validation loss of 65.73% of the accuracy.

Keyword: Machine learning, deep learning, cancer classification, SVM, random forest, deep neural network, semantic image segmentation

---

*Corresponding author: debnathb@gmail.com

Ashok Kumar, Megha Bhushan, José A. Galindo, Lalit Garg and Yu-Chen Hu (eds.) Machine Intelligence, Big Data Analytics, and IoT in Image Processing: Practical Applications, (51–80) © 2023 Scrivener Publishing LLC

3.1   Introduction

Cancer diagnosis and therapy are very significant medical [1] topics to address since there are so many distinct forms of cancer and because of the high mortality and recurrence rates associated with cancer treatment and diagnosis. Infections, drug or alcohol withdrawal, cancer, hereditary disorders, immune system difficulties, and other problems with the body are examples of ailments that may be acquired. It is one of the most prevalent causes of death in this species. Tumorous liver disease is a life-threatening disorder that necessitates the collaboration of a number of different doctors. The majority of the time, items like radiography, medical treatments, gastrointestinal, endoscopy [2], hepatology, and other restorative procedures are the ones that make themselves heard. Nonmedical areas, on the other hand, may be able to assist with diagnostics, drug development, and other tasks. It is one of the methods that individuals use to obtain knowledge about themselves. Its primary function is to examine the photographs that have been taken. Image processing is a computer method that is becoming more popular [3] these days due to the fact that it is simple to apply. This technology may be utilized for a variety of various tasks, including managing engineering output, ensuring and safeguarding people's safety, and evaluating biometric data from individuals. MedTech includes medical

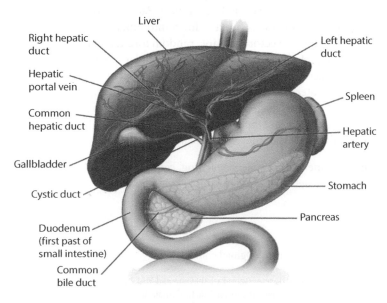

Figure 3.1 Basic anatomy of human liver [1].

imaging as a significant component of its work, with image processing and Artificial Intelligence (AI) technologies being utilized to tackle medical challenges such as determining what kind of liver tumors are there, identifying them, and diagnosing them. Ultimately, the purpose of this study is to determine the best methods for diagnosing and categorizing liver cancers, as well as to provide a framework for medical imaging.

Cancer is fourth on the list of leading causes of mortality across the globe. Figure 3.1 shows the anatomy of the human liver. The liver, one of the most essential organs in the body, is located in the upper right region of the stomach, as illustrated. It performs a variety of critical functions, including eliminating toxins from the bloodstream, digesting medications, and supplying blood proteins and bile for digestion. What is more, the liver may be harmed by a variety of different factors, some of which are potentially deadly.

3.1.1   Liver Roles in Human Body

The liver performs four essential functions listed below.

Albumin, coagulation factors, very low-density lipoprotein, and a variety of other blood proteins, including those produced by the liver, are all produced by this organ.

It stores supplements on hand in case of emergencies and splits [4] the supply of supplements according to the amount of interest earned. The body stores glucose as glycogen and converts it back to glucose when it is required. Aside from that, the liver uses a range of lipids, such as unsaturated fatty acids, cholesterol, amino acids, and others. The liver transforms excess glucose and amino acids to fats that may be stored when the circulatory system is overwhelmed with glucose. Both cholesterol absorption and disposal from the body are controlled by the liver. Finally, it may increase the creation of amino acids that the body requires.

It helps to remove harmful substances and contaminants from the environment. Because of the liver's capacity to digest fat-soluble compounds, this is the case. It secretes bile, which aids in the digestion of lipids in the small intestine by breaking them down into smaller pieces. Bile also contributes to the clearance of toxins, cholesterol, hematoidin, and other waste products from the liver.

3.1.2   Liver Diseases

If the liver is not performing properly, it may get ill, and if not treated properly, cancer can develop. The liver is responsible for a number of important activities in the body, and if it develops cancer, these functions may be lost, causing serious harm to the body. Seventy-five percent or more of the liver's tissues are usually destroyed before its function deteriorates. All liver function is destroyed in a couple of minutes. Cirrhosis, alcoholic liver disease, viral hepatitis, hemochromatosis, and liver cancer are just a few of the conditions that may affect the liver. Chronic hepatitis is a comprehensive and dynamic stable hepatic disease in which hepatocyte activity is hampered by the existence of stiff blood tissue that obstructs [5] blood flow. Individuals with this blood flow-preventive gateway may have high blood pressure, which may lead to further problems, such as nerve damage around the liver and other complications. Cirrhosis may cause drainage issues, renal problems, osteoporosis, and liver and kidney abnormalities, among other things. Excessive alcohol use may lead to fatty liver disease, alcoholic liver disease, or liver cirrhosis.

The most common substance present in the liver is fat. All three of these characteristics may be present in the same individual, either alone or in close proximity. Nobody is permitted to get the kind of alcohol treatment that they need. The two types of liver are smooth and firm. Smooth livers are the most prevalent and suffer from the least amount of damage. When individuals consume a large amount of alcohol over a lengthy period, it is known as binge drinking. Fat accumulates in the cytoplasm of liver cells, causing the organ to enlarge and disseminate a large amount of fat over time. Acute or chronic alcoholic hepatitis may occur when you consume too much alcohol. Acquired hepatitis, a chronic illness that often precedes cirrhosis, is very dangerous. Fluffy cirrhosis, like other types of cirrhosis, may be very harmful if it develops often. People with this condition have expanding knobs of liver tissue surrounded by totally tiled tissues, as seen in the image. In this scenario, scar tissue develops faster than normal liver cells, and its regulatory system continues to govern blood flow in the same manner it did before. Even if the patient is kept unprotected for a long time, the chances of developing a sluggish liver increase considerably as the cirrhosis progresses.

Hepatitis is the medical term for liver inflammation. Hepatitis may be brought on by a variety of factors, the most prevalent of which being infection and alcohol consumption [6]. Hepatitis produced by a virus may manifest in a variety of ways. Three of the most well-known are hepatitis B, hepatitis A, and hepatitis C. Viral hepatitis may be transmitted via the bloodstream. It may manifest itself in a variety of ways, some of which might progress to cirrhosis and hence be dangerous. Hemochromatosis is a condition in which the iron levels in the body are abnormally high (iron overload). Hemochromatosis may develop to cirrhosis, cancer, and cardiac issues if left untreated. The initial stage of hepatic malignancy occurs when a tumor in the liver grows out of control. The great majority of liver tumors [7] fall into one of two categories: essential or nonimportant to the body. The most prevalent kind of malignant development is when a benign tumor forms on the liver, causing liver disease. Noncancerous tumors are the reason. Because hemangiomas and central nodular hyperplasia do not obstruct the liver's capacity to function, they are seldom damaging to the liver. Hepatitis B or C, as well as numerous poisons and viral diseases of the liver, may all lead to the establishment of dangerous hepatic cancer cells. Nonliver carcinogenic cells spread to other organs, causing supplementary malignant development to increase. This illness has the potential to spread to other organs of the gastrointestinal system. The liver receives blood from various organs, including the stomach. The term "metastatic sickness" refers to another kind of hepatic malignant growth in addition to liver malignant growth. It is often a side effect of a growing liver tumor. It might also be caused by advanced breast cancer, colorectal cancer, lung cancer, or kidney cancer, among other cancers.

3.1.3   Types of Liver Tumors

Liver tumors may be caused by liver tumors or growths, as well as other factors. This suggests that liver tumors may be a contributing factor to liver tumors. Normally, cells proliferate [8] in a predictable and controlled manner in a healthy liver, producing minimal disturbance to neighboring cells. Tumors arise in the liver when the proliferation of hepatocytes (liver cells), bile duct cells, or blood vessels inside the liver becomes too rapid. Tumors are classified as either benign or malignant in nature. It is in this instance that the benign tumors are shielded by a cell wall, which prevents them from interacting with surrounding tissues. On the other hand, tumors have the potential to be fatal. They have the potential to spread to other regions of the body, causing harm to the tissues in their immediate vicinity. In general, males have a higher risk of developing this kind of tumor than women do. They may be discovered by medical imaging, or patients may present with symptoms such as an abdominal mass, discomfort, jaundice, or other signs and symptoms of liver failure.

3.1.3.1   Benign Tumors

The greater the number of patients who have stomach imaging exams, the greater the number of benign malignancies discovered. Many individuals are misdiagnosed with liver illness despite the fact that they do not exhibit any symptoms and do not have a family history of liver disease. In certain cases, noninvasive liver tumor tests and diagnostic imaging examinations may make it difficult to distinguish between those who are healthy [9] and those who are unwell. Some benign tumors, on the other hand, need a thorough examination to rule out any impurities, making it critical to have a comprehensive picture of the disease. There are many benign tumors in the liver that cannot be removed due to the danger of causing hemorrhage (such as hemangioma and hepatic adenoma). Most prevalent benign tumors include hemangioma, hepatocellular carcinoma (HCC), and Focal nodular hyperplasia, which are all benign tumors. A mass in the liver, on the other hand, might be caused by a variety of conditions. The most common benign liver tumor is a hemangioma, which has the appearance of a blood artery but is not cancerous. The majority of the time, it is discovered by chance and has no clinical significance.

This is based on a general prevalence estimate of 20% for the condition in question. Hemangiomas may occur in people of any age, but they are most frequent in children between the ages of three and five because they are more prone to developing them than other children. The greater the number of women who have them, the more probable it is that they are located in the right hepatic lobe of the body. There may be a significant amount of blood during a biopsy for a diagnosis. FNH is the second most frequent benign concrete tumor in the liver, and it is also the most prevalent kind of benign concrete tumor in the liver. Despite the fact that HPV may affect both men and women, young women are more prone to getting the disease. This may happen to anybody at any age. It becomes most obvious in the third and fifth decades [10] of life. H. adenoma, commonly known as HCC, is an uncommon solid tumor that affects women of reproductive age more than any other group of people. HCC is the term used to describe it. Oestrogens in general, and oral contraceptives in particular, have a significant impact on the development of breast cancer. People who have diabetes, as well as those who have difficulties with glycogen storage, have all been related to the ailment. In addition, men and children have been discovered who do not have any recognized risk factors for the disease.

Women in their forties and fifties constitute the majority of those who develop angiomyolipoma. This is a relatively unusual form of liver tumor to be diagnosed with. This is an extremely uncommon kind of liver tumor that mainly affects adults in their forties and fifties, although it may strike anybody at any time in their lives. In the majority of instances, there are no noticeable symptoms associated with it. During a laparotomy or autopsy, it is often discovered by chance. Pseudolipoma is a term used to describe mature fat that develops outside of the liver but inside the Glisson's capsule. This is a relatively uncommon medical issue.

3.1.3.2   Malignant Tumors

The two forms of liver cancer that may be seen in patients are primary liver tumors and cancers that have spread to other regions of the body. Primary liver tumors are the most common type of liver cancer. The most prevalent reason for liver metastasis is malignant tumors in the digestive system, as well as in the lungs and breasts. Various kinds of tissues, including hepatocytes, bile duct epithelium, and other types of tissues, may be used to form primary hepatic tumors. Hepatocellular Carcinoma (HCC) is the most prevalent kind of primary lymphatic tumor in the liver in people over the age of 50. For people over the age of 50, it may account for as much as 85% of all primary liver cancers (2011). HCC, often known as "deadly hepatoma," is a kind of liver cancer that has the potential to spread to other regions of the body if left untreated. HCC [11] is most often caused by cirrhosis or by a virus that causes hepatitis B or C infection (alcoholism is the most common cause of liver cirrhosis). HCC is now the sixth most frequent kind of cancer in the world, according to the World Cancer Research Fund.

Asia was compelled to pay attention as a result of the catastrophe. In most cases, only approximately 10% to 20% of HCC can be removed entirely with surgery. Patients normally die within 3 to 6 months if their tumor is not entirely removed, but this is not always the case. A large part of this is due to the tumor being discovered late in the process. Fibro lamellar HCC is a form of HCC that is distinguished by its distinct clinical and pathological characteristics. HCC differs from normal HCC in that it is more common in young people with healthy livers than in older adults with unhealthy livers. It affects around 1% of those who have HCC, and it affects both men and women equally in this regard. Cholangiocellular carcinoma (CCC) is the medical term for this condition.

It is a form of bile duct epithelial carcinoma that accounts for around 10% of all liver malignancies that begin in the liver, according to the American Cancer Society. The practice is less popular in Western nations, but it is more prevalent in Asian ones. CCC is found in fewer than 10% of the population, with the majority of those affected being over the age of 50. It is the most prevalent kind of malignant liver tumor in children, and it is also the most dangerous type of liver tumor in children. Angiosperms are found in just a small percentage of the population. They are suffering from liver mesenchymal neoplasm. Endothelial cells are waiting to attack you as soon as your blood vessels open. Hemangioendothelioma epithelioid is a rare kind of liver cancer that affects the endothelial cells of the liver.

3.1.4   Characteristics of a Medical Imaging Procedure

When it comes to medical image information, it is often employed to enhance things like transmission, compression, and function. It is essential to review this information to ensure that you have a better grasp of the program; there is a potential to enhance transmission, storage, and archiving by making use of image-related information. Achieving a natural language explanation of how a picture was captured, why the process was completed, and the fact that no one has stored the semantics of the processes described in the guide is difficult, despite the fact that many systems can manage a large amount of information. An archival responsibility may be used for a variety of different tasks, such as compressing large amounts of data. When developing novel designs, database designs, and compression technologies, it is important to consider these possibilities. This kind of material appeals to a large number of authors. Using the list of important criteria provided below, for example, it is possible to generate a diverse list of distinct imaging processes. The list of different imaging processes may then be generated in any number of ways.

The therapeutic purpose of a procedure is to perform a satisfactory diagnostic or therapeutic function.

It is necessary to understand the connection between series in order to extract information from them. Image processing is used to extract valuable information from a large amount of data in order to accomplish a process objective. As a consequence, it is necessary to document the procedures and factors that have an impact on the overall success of the processing. The majority of digital tools make use of some kind of image processing. Simple processing techniques like windowing, zooming, and measuring are often used. In the medical area, nuclear medicine is the only one that employs cutting-edge technology. Processing time needed by the technique of processing, on the other hand, has no influence on the quantity of storage space that is required during the processing process. To determine how tumor pathology evolves over time, the most frequent medical rationale for retaining pictures is so that they may be compared to future treatments. Liver cancer and the cells that produce it are both examples of cancer. The majority of malignancies grow more aggressive as they go through the phases of development. The Barcelona Clinic Liver Cancer Center says that the American Association for the Study of Liver Diseases

(AASLD) recommendations are based on how patients with liver cancer are treated in the real world, not on theoretical models.

It assesses the extent to which the tumor has spread, the extent to which liver function has been maintained, and the general health of the patient. From the liver, the obtained gray co-occurrence matrices, gray level run length matrices, and gray level gradient co-occurrence matrices were analyzed. Using computed tomography (CT) scans, it is possible to distinguish between two forms of liver illness in the location where Computer Aided Diagnosis (CAD) has discovered liver cancer. There are few therapeutic choices available for those who have liver cancer, and their prognosis is not very promising. As a consequence, selecting the most appropriate hepatic cancer treatment strategy might be critical to the patient's overall health and even survival after diagnosis. It is necessary to initially divide the liver into two halves in order for computer-assisted techniques of identifying liver disease to be effective. In accordance with studies, the liver may be divided into a variety of distinct parts. Because of the amount of overlap (equalization) between the ligaments that link the liver and the kidneys, as well as the fact that the liver is not particularly stiff, they are very difficult to comprehend. When liver tumors can be clearly detected and segmented in CT scans, the likelihood of early cancer identification and diagnosis increases significantly.

Malignancies of the liver that have migrated from other regions of the body include secondary hepatic malignancies and HCCs. Additionally, the liver may develop a primary fatal tumor such as HCC or Schweich, which is the most prevalent kind of cancer and the leading cause of mortality in the world. A noninvasive method for searching for organs and soft tissues within a person's belly is Ultrasound Diagnostics (US), which does not need any invasive procedures. This enables the operator to demonstrate both the image plane choices and the anatomy of the illness in the most accurate manner possible. It may be necessary to examine the liver using Diagnostic Ultrasound in this situation. It is a noninvasive, safe method of checking on the patient that is simple to do, inexpensive, and likely to be repeated. In the case of CT or MRI, you are unable to utilize them since they are either too radiation-intensive or too costly to be used in this situation. To assist the expert doctor while also ensuring that the outcomes seem excellent, the CAD system makes use of image processing and AI. There are occasions when the doctor is the one who makes the ultimate choice, but this is not the case in every instance.

3.1.5   Problems Related to Liver Cancer Classification

Liver cancer, which is the third most prevalent kind of cancer in the United States, is also the third most common type of cancer in the globe. When it comes to detecting liver cancer, a variety of diagnostics, including blood tests, imaging investigations, and biopsy, are performed. Early identification of liver cancer will be achievable because of a more straightforward and quicker method of searching for it. A blood test is the first stage in determining a patient's health and whether or not their liver enzymes are abnormally elevated. People who have hepatitis viruses will have a blood test done to determine whether they are infected. This is due to the fact that both hepatitis B virus (HBV) and hepatitis C virus (HCV) are known to induce liver cancer. If any issues are discovered, the patient may be subjected to further procedures, such as tumor marker tests. An ultrasound scan of the liver is performed to determine if it is larger or thicker than usual, or whether it contains a solid or fluid-filled tumor. Based on the results of an ultrasound examination, CT scans are used to get cross-sectional images of the liver.

A CT scan is performed four times in order to detect HCC. Using a different contrast material for each of the four scans, the results are different. If this approach reveals that the patient has HCC, there is no need for a liver biopsy. Furthermore, it may be used to get cross-sectional images of organs, tissues, and bones in order to determine whether or not liver cancer has spread to neighboring lymph nodes, tissues, or organs. A biopsy is performed if a CT scan does not reveal the presence of HCC. Because liver cancer is located near the rib cage, it is more difficult to detect early indications of the disease. The tumor may have grown to such a It is so small that it is no longer detectable by the time it is discovered. CT-based liver tumor identification has been around for a long time, but it is more costly and less safe than ultrasound imaging, which provides more information about soft tissue and is less expensive and less hazardous. A variety of applications for ultrasound is possible, including the ones listed below:

- It is critical to consider the size, location, and structure of the soft tissue organs in the body.
- It is important to understand the distinction between liquid-filled cysts and tumors.
- Contribute in some way to the doctor's success during a needle aspiration, biopsy, or other cancer-treatment operation.
- Determine the extent to which the cancer has spread (staging).

Blood flows through arteries may be shown to demonstrate how rapidly and where they pass through them (a Doppler ultrasound). In the past, CAD-based approaches for identifying liver cancer relied on features extracted from a single or multiple feature space from a collection of liver pictures to make their determination. When faced with a large number of feature options, vectors are picked by themselves. The computational method is really tough. It is possible to reduce the amount of computation time required while simultaneously improving diagnostic extraction in the detection and classification of liver tumors by using a specified best feature value selection from a collection of feature vectors. These are the most important characteristics to look for in an ultrasound picture to aid in the diagnosis and categorization of liver cancer. Because of this, the researchers came up with a strategy to utilize ultrasound images to screen for liver tumors in patients. This strategy is focused on identifying the most prominent aspects of the photos.

3.1.6   Purpose of the Systematic Study

One of the primary objectives of this research is to determine ways to identify and categorize liver tumors utilizing ultrasonic imaging techniques. Pre-processing, feature extraction and selection, optimal feature fusion, and classification are all processes that must be completed at the subtask level in order for the task to be completed successfully. The significance of processing liver ultrasound pictures before they may be utilized to identify and categorize cancer is discussed in the article. When this information is employed in this way, it aids in the pre-processing and the building of a filtering algorithm that is both efficient and effective. This information is used to determine how fast tumors can be classified based on their characteristics. A new approach for filtering away speckles from ultrasound pictures has been discovered, which makes it simpler to get rid of them than the previous method.

This approach combines two forms of spatial filtering, nonlinear filtering and diffusion filtering, into a single operation. In order to examine the texture region, nonlinear filtering is used, and a diffusion filter is employed in order to maintain the picture borders. The results of the studies demonstrate that the suggested strategy is very effective in reducing speckle noise while maintaining edges, resulting in a better ultrasound image. It is possible to construct a higher-level approach by creating and selecting the greatest feature value from photographs that have been denoised. For the purpose of identifying and selecting the most essential characteristics, this technique employs the dynamic vector warping (DVW) distance method. In order to forecast the optimal value from a fused feature space, a novel metaheuristic approach known as the Penguin Search Optimization (PeSO) method is applied. Specifically, it is determined by the value that is used to categorize the classifier.

3.2   Related Works

One of the objectives of this research is to examine all of the novel methods of detecting liver cancer tumors that have been mentioned in the literature. Image processing methods are used to study cancer processes, and subsequently risk assessments and evaluations are carried out utilizing a range of imaging modalities. Additionally, research into existing cancer prevention measures is carried out as part of this process. Aside from that, there are certain technological concepts that researchers might use in order to develop a new method of diagnosing HCC.

Metagenomic sequencing technologies have provided us with a wealth of genomic information on the human microbiome, which has assisted us in better understanding and diagnosing microbial disorders [12]. Patients and healthy people have significantly different compositions, which may be utilized to assist physicians determine what is wrong with them. Despite significant advancements, the diagnostic and therapeutic accuracy of these instruments still need improvement. It is possible to predict whether or not someone is unwell with great accuracy using an approach called MDL4 Microbiome, which combines several characteristics from metagenome sequencing with a multimodal deep learning model. The combination of three diverse criteria, such as standard taxonomic profiles, genome-level relative abundance, and metabolic functional traits, might help you increase your classification abilities. Using this deep learning approach, the researchers were able to create a classifier that included all of the various sorts of things that were discovered in the human microbiome. Predictive accuracy for patients with IBD, type 2 diabetes (including insulin resistance), liver cirrhosis (including liver cancer), and colorectal cancer (including colorectal cancer) was 0.98, 0.76, 0.84, and 0.97, respectively. Compared to classic machine learning approaches, these results are comparable to or better than those obtained with these new methods. The features that were picked were examined even more extensively in order to have a deeper understanding of how their distinct characteristics functioned. MDL4 Microbiome is a classifier that outperforms or is on par with other machine learning algorithms in terms of accuracy, depending on the situation. Specifically, it provides insight into how features are generated in deep learning models based on metagenome sequences, how they might be utilized to diagnose the host's illness condition, and the mechanisms by which they operate.

The medical industry for data to be accumulated over time, progressing from basic to more complicated categories as time goes on. In the case of medical data collection, the data and structures that make up the data may be broken down into their original forms and the process by which they arrived at the current state of the data [13]. Take, for example, the cancer dataset, which is divided between early stages of benign and malignant cancers. A binary structure is replaced by a multiclass structure as medical knowledge evolves and new data becomes available. More labels for sub-categories of sickness are added to the dataset as it progresses from a binary structure to a multiclass structure. Making a multiclass model in machine learning necessitates the use of greater computer resources. Model optimization is used in order to improve the accuracy of multiclass models. Making life-or-death choices is very essential, and model optimization is critical in this situation. The models must be trained for a lengthy period of time before they can be used in this model optimization assignment. In this research, a novel strategy known as Group-of-Single-Class Prediction (GOSC) will be employed to get the best possible outcomes while spending the least amount of time possible training the model. It goes hand in hand with majority voting and model transfer, as well. The most significant advantage is that it enables you to create an ideal multiclass classification model with the highest possible accuracy that is near to the absolute maximum while reducing training time by up to 70%. Machine learning was used in order to categorize liver data. The classification of COVID-19 lung CT data was accomplished via the application of deep learning. According on early statistics, it seems that this new procedure may be viable.

It is possible to learn about many different illnesses via the examination of histological pictures, including cancer, brain tumors, fatty liver, and congenital heart problems, by looking at histopathological images [14]. In the actual world, field expertise is required in order for the time-consuming diagnostic approach to be effective. During each phase of the decision-making process, pathologists may come to different conclusions. When it comes to solving these challenges at the beginning of the diagnostic process, machine learning may be employed in many ways. When pathologists are examining histopathological pictures, deep learning algorithms may be able to assist them in working more rapidly and arriving at more correct conclusions. Deep learning algorithms that learn features on their own do better in analysing histopathology pictures than other machine learning approaches that rely on characteristics that have been pre-programmed into the system. Using patch-based, lightweight CNN, This research aims to categorize histopathology pictures in order to detect radiation-induced liver damage in this research. The proposed technique was used to categorize 555 histological pictures from a dataset of radiation-induced liver disease, and the results were published in Radiology. Testers were not given access to the test set, which resulted in accuracy of 100% for binary classification and 87.577% for multiclass classification being attained on the test set. Models such as ResNet-50, Vgg16, and Google Net are not good enough for this purpose. To have a better model than they do. This is the first time that deep learning has been used to detect data from radiation-induced liver illness, which is a first in the field. The approach may be valuable to pathologists in the future. This research is significant in the world of medicine since it identifies a risk factor for impaired liver function.

TILs have piqued the attention of translational cancer researchers because they have the potential to be employed as a biomarker for disease progression and clinical outcomes [15]. The refined and enlarged deep learning algorithm that may be used to determine if 50×50 µm tiled picture patches that are 100×100 pixels at 20× magnification are positive or negative for TIL, or whether they are positive or negative for a specific gene. It is possible to create TIL maps using this technique to determine how many and where TILs are present in 23 distinct forms of cancer. The utilized pathologist annotations and computer-generated labels from the first-generation TIL model to train three popular CNN architectures for 13 different forms of cancer using pathologist annotations and computer-generated labels from The first-generation TIL model. This was accomplished with the use of a large amount of training data (model-generated annotations). This training dataset contains both TIL positive and negative patches from tumors in various areas of the body, as well as data that has been reviewed and approved by specialists in the field. All of this contributes to the improvement of the algorithm's performance by lowering the number of known false positives and negatives. In order to create TIL maps, The developed a novel TIL approach that leverages automated use thresholding to turn model predictions into binary classifications. Everyone who made modifications to TIL models performed better than the originals, with advances in accuracy and F-score of up to 13% and 15%, respectively, in comparison to the original models. In addition to the evaluation code, the public will have access to these one-of-a-kind TIL models, as well as TIL-Maps-23, which has TIL maps for 7983 WSIs from 23 distinct forms of cancer compiled into a single collection.

Liver illness is a significant contributor to the increase in the global mortality rate [16]. If liver illness is detected early, it may be possible to reduce the number of people who die from it. In a doctor's office, CT is a procedure that allows you to see images of the liver. A large number of liver images might lead radiologists to believe they understand what is wrong with the liver. This may result in potentially life-threatening illnesses such as liver cancer. A machine-learning algorithm is required to determine the nature of these challenges based on the texture of their solutions. According to this study, there are two alternative techniques to screen for liver issues. In the first technique, which is based on classical machine learning, the categorize objects based on the texture that they possess. By merging automated texture analysis with supervised machine learning approaches, this technology enhances the capabilities of machine learning algorithms. This topic was answered using over 3000 CT imaging samples taken from 71 different people. All of the CT pictures were clinically confirmed prior to being used. It was necessary to employ supervised learning approaches to create image classes that represented the same condition in order to ensure that liver tissues were not normal. To detect asymmetric patterns in CT images by using machine learning techniques. It was determined how effectively the feature vector operated using the KNN, Naive Bayes (NB), Support Vector Machine (SVM), and Random Forest (RF) classifiers, among other methods. When looking for liver illness, the second step is to divide the term down into several categories of terms. The use of a CNN is accomplished via the use of a safety injection system (SIS). High-density maps are encoded using this approach, which takes advantage of focused attention. The algorithm has been taught to distinguish between five distinct kinds of diseases based on the results of CT scans. The model's remarkable findings demonstrate that it is capable of doing what it claims to be capable of doing. Evidence suggests that texture analysis approaches may be used to discriminate against and detect anomalies in the human liver, which is a promising development. This might be useful to radiologists and medical physicists in predicting the severity and spread of liver disease.

An automated cancer screening system that assists radiologists is becoming more important as the number of cancer cases and the number of persons utilizing Positron Emission Tomography/Computed Tomography (PET/CT) increases [17]. Today's image-based PET/CT approaches are frequently restricted to a particular body part, which is a significant limitation. This article suggests a deep learning-based strategy for recognizing distinct forms of organ-specific malignancies in order to aid radiologists in their job. In the classification model, this study includes a module that incorporates PET pictures, CT images, and multiorgan segmentation data from several sources. The output of the segmentation provides a map for the network, which instructs it on where it should concentrate its efforts. It informs the network that it needs to learn about organ-related aspects from the PET/CT image of the whole body in order to function properly. Lung Data Computed Tomography (LDCT) images, which are becoming more popular in PET/CT scans, are shown here. A two-level V-net is used to segment a large number of distinct organs in LDCT images, which is the author's recommended grayscale transformation approach tackles the issue of unlabeled data, while the double-level V-net adds context information to the image by transforming it into grayscale. PET/CT scans may be classified into six distinct categories of screening depending on the technology used to perform the scan (health, oesophageal cancer, gastric cancer, liver cancer, pancreatic cancer, and lung cancer). The F-score of the classifier is 82.3%, which shows that it might be useful in radiologists' cancer screening efforts.

3.3   Proposed Methodology

According to current medical imaging studies, the incidence of liver cancer is increasing in Asian nations. Invasive procedures such as biopsies, which may be risky and unpleasant, are one of the most effective methods to determine what is wrong with the body. Therefore, unless there is a compelling need to do so, it is not advised in the vast majority of circumstances. Image processing techniques and procedures, as an alternative to this paradigm, may be utilized to assist clinicians in determining what is wrong with a patient. In order to be effective with these tools and procedures, however, a high level of competence and practice is required. The texture of liver images demonstrates the differences between healthy and sick livers in a manner that is not immediately apparent. Get the material and examine it with the help of cutting-edge image processing technologies. Generally speaking, when it comes to providing the most realistic texture information possible to the viewer, there are two sorts of methodologies to choose from: gradient-based methods and gray level co-occurrence analyses. It is possible to look at images of the liver using a computer-assisted imaging equipment known as CT scans. This instrument may be used in conjunction with a CT scan [17, 18], an MRI, an X-ray, and an ultrasound. They provide information on the tumor's location and size, as well as how it is related to other organs and tissues in the body, using imaging modalities. The purpose of this research is to determine how the classification rate of the classifier evolves during the course of the liver tumor detection procedure. It is possible to divide the process of identifying a liver tumor into four major parts.

i) Image acquisition
ii) Pre-processing
iii) Feature extraction
iv) Classification

There are two approaches to look at the classification rate of liver tumor images: first, without pre-processing, and then with pre-processing and further processing. Pre-processing is accomplished via the use of neigh-bour shrink, and feature extraction is accomplished through the use of the pooling layer approach. When it comes to classification, the CNN [19, 20] approach is applied.

The block Diagram of the proposed research is shown in Figure 3.2.

In Figure 3.2, showing the flow of the Pre-Processing, the methods and content descriptions for this research went into considerable depth about how these investigations were carried out and what they looked at in great detail. However, the majority of the studies reviewed did not include any information on the number of patients who were engaged, the number of photographs that were utilized, or the number of clinical procedures that were performed. According to some researchers, this variance in reported approaches may be due to differences in author backgrounds and the out-lets that disseminate their results. Each of the eleven articles was prepared by a team of 58 persons, with five of them working in radiology depart-ments, one in pathology, and two working in the medical profession.

The other 50 persons that contributed came from a variety of back-grounds, including engineering, computer science, and a medical image processing facility. In certain cases, such as neurocomputing, IEEE Journal of Biomedical Health Informatics, Computers in Biology and Medicine, and the International Journal of Computer-Assisted Radiology [20] was

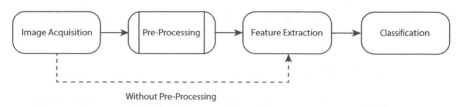

Without Pre-Processing

Figure 3.2 Basic flow of the pre-processing.

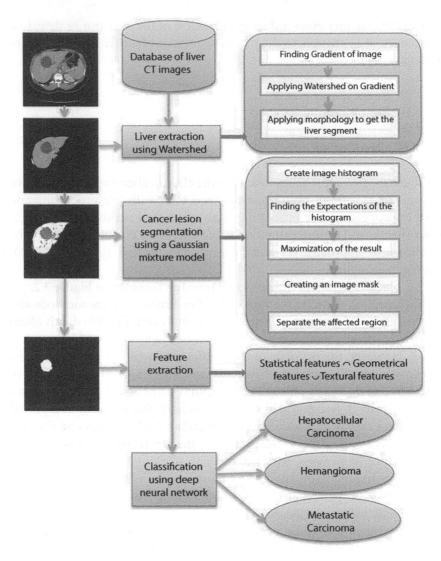

Figure 3.3 Proposed flow diagram for liver cancer classification.

clearly drawn in Figure 3.3 [21], and Surgery, there were exceptions. The majority of these publications were published in journals devoted to computer science or biomedical informatics.

3.3.1   Gaussian Mixture Model

The Watershed transition is a method of dividing property depending on its appearance. Grayscale pictures are seen as "space-alleviation" in this manner, and a near least is viewed as a "catchment cup." A watershed is formed when water overflows and establishes a boundary. As a result, the image was entirely divided in half. The morphological approach is used to construct the image structure. Often, structural noise and other undesirable elements are removed from the grayscale image during this procedure. The altered the inclination picture to make it seem as though it were a watershed. It became much simpler to construct a smooth limit structure as a result of this. The image is a component of a larger system, and each component of the larger system communicates with a single pixel. In the Gaussian mixture model, picture pixels are regarded as random variables in Gaussian Mixture Modelling (GMM). To represent the image pixels, the variable x is a three-dimensional variable with RGB values. When determining how probable an image is, weighted Gaussian dispersion is utilized. The total number of districts is k, and the set of loads that fulfil the criteria is the set of all the loads that meet the criterion. In that portion of the nation, there is a term that refers to the mean and standard deviation of the Gaussian transmission in that area. To determine the model's parameters before determining how good GMM. The most popular method of determining the GMM parameter is to seek for the most probable result. The primary objective of the test is to increase the likelihood of the GMM dataset. The expectation-maximization (EM) approach is used in the testing process.

3.3.2 Dataset Description

A large number of persons were observed and imaged throughout the inquiry (http://medicaldecathlon.com/) [22]. The study included 20 individuals who had a total of 68 lesions. It was done using a testing set that consisted of 14 persons who had a total of 55 lesions and who also had CT scans done. Another CT imaging investigation revealed the findings of a small study that looked at 182 liver lesions, which was conducted in 2010. (53 cysts, 64 metastases, and 65 hemangiomas). Although medical databases are accessible online, the authors claim that they are restricted in breadth and only applicable to a select group of medical diseases. They also claim that obtaining datasets is difficult. 75 participants who received 3D multiphase contrast-enhanced liver CT scans were studied by the researchers. The doctor examined cysts, FNH, HCC, hemangioma, and metastases, among other things.

A CT scan of the liver with enhanced contrast has been performed for the fourth time can be seen in Figure 3.4. In a previous investigation, images of liver tumors at three distinct stages were discovered (noncontrast-agent

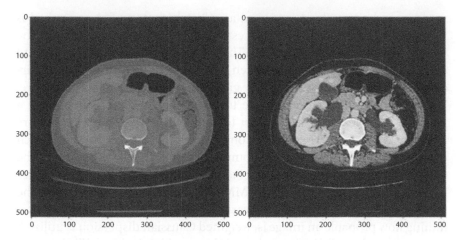

Figure 3.4 Basic CT scan of the liver with enhanced multiphase.

enhanced, arterial, and delayed). As previously stated, the spectators were divided into five groups. 134 patients, on the other hand, had MRI scans of their upper abdomens performed (37.7% primary liver mass and 62.3% metastatic lesion). There were 1700 different areas in the MRI scan images of twenty patients that did not contact each other. Only one researcher used ultrasonic imaging to examine the livers of 94 participants. There were 48 normal livers, 50 chronic livers, 50 cirrhosis, and 41 HCCs that had formed over cirrhosis in the photo files, thus they were all represented. Other files included experimental data, such as 127 photos of liver illness.

3.3.3   Performance Metrics

3.3.3.1   Accuracy Measures

A broad variety of accuracy metrics were utilized in this research, as a starting point, the system operates and how accurate it is in segmenting the liver into parts and identifying abnormalities. After that, how well the system performs in comparison to other CNNs. Finally, the characteristics of the deep CNN method's sensitivity, specificity, and accuracy, among other things. Sixth, the deep CNN approach is compared to the Bayesian model and the benchmark method to see which is more accurate. A trained radiology professional uses their eyes to compare two pieces of data (precision and recall rates) A CNN-based system is being used to determine how well existing approaches for diagnosing liver lesions that are not network-based perform in this study. Many experiments did not assess the sensitivity or specificity of their results.

CT images were used to segment a liver tumor shown in Figure 3.5, which had been discovered earlier. In particular, liver segmentation, the formation of a multiscale candidate tumor, the building of an active contour model, and the selection of a tumor candidate are all covered in this section. Machine learning algorithms have been used to divide liver tumors into multiple segments, according to the researchers. The quality of data [23–26] is of utmost importance for the implementation of machine learning algorithms. A large number of CNNs have been created with the goal of separating the liver from the lesions. In this data collection, you will find that there are not many hand masks that can be used to help doctors view where lesions are located in the body more easily. With limited training samples, it is critical to teach the network the invariances and efficient qualities that it will need in the future. When the size of the lesion varies in this scenario, scales ranging from 0.8 to 1.2 are generated.

After obtaining consistent measurements, fresh pictures are re-sampled using the near-neighbour approach to provide more accurate results. The CNN segmentation technique produced the following findings, which are shown below. Four distinct scales were used to determine the size of each photo in the collection. With the Confusion Matrix in Figure 3.6, the aim

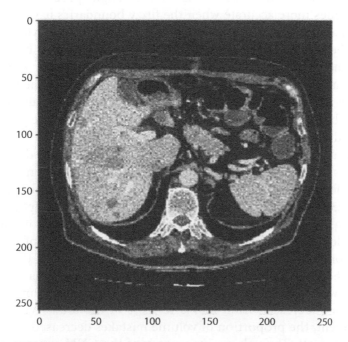

Figure 3.5 CNN-based classified image.

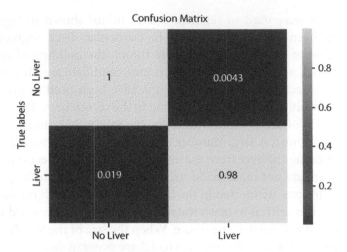

Figure 3.6 Confusion matrix of the ROC curve.

is to enhance the accuracy of lesion segmentation while simultaneously reducing the value of the pixels that are being examined. The precision with which the liver and tumor can be distinguished demonstrates the effectiveness of the hybrid learning strategy. High-speed data segmentation becomes more accurate when the fuzzy boundaries have been effectively separated. Large margins help to achieve this. The hybrid feature, despite the fact that it aids in tumor segmentation, can only be employed for tiny malignancies since they are detected in fewer slices. The training and validation accuracy of the proposed CNN in terms of the number of periods it must learn and test throughout the training and validation process (the blue curve reveals the accuracy of training and the green curve suggests the accuracy of validation).

In order to determine how well the process performed with various kinds of liver, the volume error was calculated. When around 14 training forms are employed, the inaccuracy remains consistent. When around 23 training forms are used, the error rate is not optimal. Even in the most severe cancer cases, when the amount of noise in the image and the location of the patient are altered, the accuracy of predicting the tumor load increases dramatically. A comparison of the findings is done between them before and after the type has been corrected. In cases where there were significant errors in the segmentation, volume errors decreased from 28.9% to 6.6% of the total. When tumors were present, the proportion of volume mistakes decreased from 17.0% to 5.2% as well. The volume inaccuracy of the CNN approach is seen in Figure 3.7. Retrospective clinical research including the utilization of CT scans of the liver was conducted three times (enhanced, arterial, and delayed, noncontrast agent). A mass that occurs seldom cannot be utilized to determine the form of a lesion since it does not have a consistent pattern. As a consequence, it is necessary to validate a diagnosis by comparing the rate of contrast injections over time to ensure that it was made correctly. There have been many distinct traditions utilized in the extraction process throughout history. Many of these traditions may be found in the extraction functions used in computer vision. Because the number of nodes in a deep-learning network has a significant influence on how well the network can categorize and detect lung CT nodules, the conventional method of building the network is to make it as deep as possible before training the network. As a result of this method, the number of false positives for lung CT nodule classification is reduced. A method of categorizing items that makes use of CNNs. This study employed a two-dimensional CNN-based method to categorize lung nodule candidates as either positive lung nodules or negative nonnodules. This was

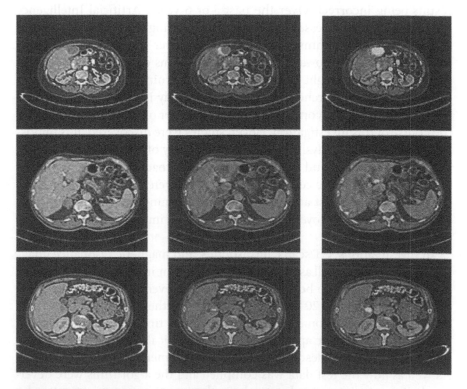

Figure 3.7 Classified output after applying proposed algorithm.

done in order to eliminate false positives from recently discovered lung nodule candidates.

The network output is the likelihood of the central pixel in the input patch (target pixel). The network has two output neurons. One is the likelihood of a nontumor (0~1); the other is the probability of a tumor (0~1 was shown in Figure 3.7. The output will be (0, 1 if the target pixel is tumor labelled, while the target pixel is tumor-free (1, 0) when the objective pixel is tumor-free. A cross-entropy function is used to measure the precision of the learning as a loss function. The proposed method has less complexity when compared to other existing methods. This function has been utilized to measure weight and weight loss error.

3.3.3.2   Key Findings

They discovered that the mean accuracy for discriminating liver masses was 0.84 and that the slope of the under receiver operating characteristic curve was 92 when they conducted the testing. During three cross-validation studies, 94.6% of the findings were correct, with just 2.9% of the results being incorrect. Over the past 5 or 6 years, Artificial Intelligence (AI) has grown more essential in image processing research, with the use of hand-made algorithms taking precedence over the use of deep learning frameworks and toward the use of algorithms created by humans. To determine if CNN might be utilized in deep learning to analyze images of liver cancer, the researchers conducted this study in order to answer that question (HCC and secondary cancers in the liver or liver masses). During the course of this investigation, eleven studies were discovered. It is impossible to do a meta-analysis due to the vast variety of data available, gaps in the reported results, and the many diverse approaches to accomplish the same tasks. There were certain guidelines that were established as a result of these inquiries. First and foremost, deep learning architectures such as CNN have been employed to enhance medical imaging via the usage of CNN.

In investigations, CT scans, ultrasounds, and magnetic resonance imaging (MRI) scans, as well as histological and cellular imaging of liver tumors, have all been found to be quite accurate. All eleven research papers were published in 2017 and 2018, despite the fact that there were just a few thorough original studies on each malignancy at the time of their publication. Thus, CNNs and their applications in the medical sciences demonstrate that deep learning as described by CNNs is an emerging topic of research. As well as how the test data was compiled and the algorithms were developed, it is necessary to discuss how the image sources, patients who participated in the study, and clinical data were obtained and evaluated. When this examination was completed, it was discovered that there was no consistency or uniformity among the methodologies used. More studies with equal numbers of authors from medical, computer science, and engineering sectors must be conducted in order to ensure that research may be utilized by individuals from a variety of different professions. Compare the outcomes of research articles that were tested with the same datasets would be an interesting exercise in comparison.

Unfortunately, most studies did not specify where the datasets were from, making it difficult to determine whether or not the tests were performed in the same way. This information is necessary in order to compare and assess. When authors write about deep learning, we believe that publications that wish to publish their work will follow the same set of principles as well. They will be interested in knowing what datasets were utilized and how the results were evaluated. However, it is true that there are several problems in this study. Because there are so many various types of lesions, when interpreting the data and determining their significance. Because of a "publication bias," it is possible that research that was not excellent was kept from the public's view. It is likely that there were excellent pieces written about this topic in other languages, but only those that were published in English were included in this collection. According to the findings of this study, this demonstrates what has been done in the English-language literature on this issue to date. Each one was unique, depending on who conducted the study, what sort of research was conducted, how it was conducted, and what type of methodology was used.

3.3.3.3   Key Issues Addressed

Several various aspects of CNN were investigated in this research, including how they functioned and how they were trained (CNNs). For starters, CNNs are capable of distinguishing between HCC and other kinds of liver cancer in a short period of time. Second, CNNs may be utilized to perform a variety of tasks in the area of liver cancer, including the identification of lesions, the classification of lesions, and the segmentation of lesions. Third, CNN is not simply for seeing radiological pictures of liver cancer; it may also be utilized for other purposes. The technique may also be utilized to investigate pathogenic and cellular research issues. But there are several issues with the current literature in this field that need to be addressed. Unfortunately, there are not many long-term liver Computed Tomography (CT) scan studies available that can be used to compare the outcomes of standalone and follow-up approaches with the outcomes of long-term CT scans.

In addition, there are just a few research on the issue, and there are not many multicentre studies either. Researchers may be able to compare changes over time to baseline scans, which may allow them to detect new tiny liver cancers more promptly in long-term trials.

CNN, a kind of AI utilized in biomedical research, and other deep learning techniques [27, 28] have renewed interest in the issue. Hepatocellular carcinoma (HCC) and other liver cancers are discussed in this article, which examines some current studies on how to utilize computer networks (CNNs) to look at these tumors. To also consider how this cutting-edge technology may assist physicians in making more accurate diagnoses in the real world. While the study is centered on a promising new field in gastroenterology and oncology, some researchers encountered difficulties in obtaining images, which made it difficult to apply machine learning algorithms to medical datasets and obtain high-accuracy image classification. Ultimately, the study was unsuccessful. Consequently, the CNN design must be revised to include a much greater number of patients with cirrhosis for a variety of causes, as well as patients with HCC on top of cirrhosis or patients with liver secondary disorders such as fibrosis. Investigate HCC or liver masses using CNNs to search PubMed, EMBASE, and Web of Science for publications that detail the pathological architecture, cell composition, and radiographic pictures of the tumor or mass. Using a particular extraction method, the data was analysed to see if the CNNs were accurate and effective in diagnosing cancer or cancer in its early stages. Principal results of the research were the kind of cancer or liver tumor that had the highest accuracy for cancer diagnosis and the sorts of photographs with the highest accuracy for cancer detection, both of which were discovered in the study. There was just a little research that was discovered. When the tests were performed, it was discovered that it was capable of distinguishing liver masses from other liver issues, as well as distinguishing HCC from cirrhosis or tumor growth. The nuclei of HCC tumors were investigated in two distinct approaches. A number of experiments revealed that CNNs were capable of being accurate in their predictions.

The researchers attempted to locate, characterize, and dissect lesions as part of their investigation. Many different approaches were used to determine how accurate CNN models were. Most current research has concentrated on liver tumors; however, the number of clinical trials has been small and restricted, and more study is needed to determine what is wrong with the algorithms that CNNs use. When it came to segmenting, classifying, and recognizing lesions in pictures of common cancers from radiology and pathology, CNNs performed well. Because of all of these issues, it is also difficult to study these days. A multicentre investigation is required in order to provide a comprehensive picture of how CNN may be employed in the treatment of HCC and other hepatic disorders. Because of the limitations of radiological imaging, it might be difficult to distinguish between primary and secondary liver cancers (liver metastases) or other liver tumors. The use of CT-based deep learning algorithms to distinguish liver metastases from primary liver tumors has been shown in this research to be effective. A greater emphasis should be placed on the accuracy and sensitivity of the CNNs that make up a system's performance when assessing system performance and computing positive prediction values.

3.4   Conclusion

In light of the increasing number of studies that are using CNNs to interpret pictures of liver cancer, it is critical to ensure that these studies are correct and to establish objectives for further study. Because large liver tumors have historically been the focus of study, the number of papers that can be seen by the general public is quite restricted. The majority of the studies had an imbalance in the substance of their procedures between the description of the patients, the medical element, and the technical computer-related part of the processes. In addition, CNNs should be compared to other models, particularly when it comes to how accurate and sensitive each model is when it comes to a set of photographs that are the same size and form as the other pictures. Future research in these areas should thus be carried out collaboratively and with the participation of a large number of patients suffering from various illnesses. It is becoming more vital for CNNs to be used in liver oncology.

References

1. Neal Joshua, E.S., Bhattacharyya, D., Thirupathi Rao, N., The use of digital technologies in the response to SARS-2 CoV2-19 in the public health sector, in: *Digital Innovation for Healthcare in COVID-19 Pandemic: Strategies and Solutions*, pp. 391–418, 2022.
2. Joshua, E.S.N., Battacharyya, D., Doppala, B.P., Chakkravarthy, M., Extensive statistical analysis on novel coronavirus: Towards worldwide health using apache spark, in: *EAI/Springer Innovations in Communication and Computing*, pp. 155–178, 2021.
3. Bhattacharyya, D., Kumari, N.M.J., Joshua, E.S.N., Rao, N.T., Advanced empirical studies on group governance of the novel corona virus, MERS,

SARS and EBOLA: A systematic study. *Int. J. Curr. Res. Rev.*, 12, 18, 35–41, 2020.

4. Marline Joys Kumari, N., Bhattacharyya, D., Thirupathi Rao, N., Improving the diagnostic accuracy using amplification and sequencing of the SARS-CoV-2 genome, in: *Digital Innovation for Healthcare in COVID-19 Pandemic: Strategies and Solutions*, pp. 331–350, 2022.

5. Eali, S.N.J., Rao, N.T., Swathi, K., Satyanarayana, K.V., Bhattacharyya, D., Kim, T.-H., Simulated studies on the performance of intelligent transportation system using vehicular networks. *Int. J. Grid Distrib. Comput.*, 11, 4, 27–36, 2018.

6. Bhattacharyya, D., Long term prediction of rainfall in Andhra Pradesh with deep learning. *J. Med. Pharm. Allied Sci.*, 10, 4, 3132–3137, 2021.

7. Maran, P., S., S., K., T., D., H., K., N., A novel deep learning method for identification of cancer genes from gene expression dataset. *Adv. Comput. Electr. Eng.*, 3, 129–144, 2020.

8. Bhattacharyya, D., Comprehensive analysis on comparison of machine learning and deep learning applications on cardiac arrest. *J. Med. Pharm. Allied Sci.*, 10, 4, 3125–3131, 2021.

9. Joshua, E.S.N., Bhattacharyya, D., Chakkravarthy, M., Kim, H.J., Lung cancer classification using squeeze and excitation convolutional neural networks with Grad Cam++ class activation function. *Trait. du Signal*, 38, 4, 1103–1112, 2021.

10. Kim, E. and Chung, Y., Comparison and optimization of deep learning-based radiosensitivity prediction models using gene-expression profiling in National Cancer Institute-60 cancer cell lines. *Nucl. Eng. Technol.*, 54, 8, 3027–3033, 2022.

11. Eali, S.N.J., Rao, N.T., Swathi, K., Satyanarayana, K.V., Bhattacharyya, D., Kim, T., Simulated studies on the performance of intelligent transportation system using vehicular networks. *Int. J. Grid Distrib. Comput.*, 11, 4, 27–36, 2018.

12. Altan, G., A deep learning architecture for identification of breast cancer on mammography by learning various representations of cancerous mass, in: *Deep Learning for Cancer Diagnosis*, pp. 169–187, 2020.

13. Lata, K. and Saini, S., A review of deep learning-based methods for cancer detection and classification, in: *Deep Learning for Biomedical Applications*, pp. 229–253, 2021.

14. Kotani, D., Fujii, S., Yamada, T., Suzuki, M., Yoshino, T., O12-1 A novel gene-prediction model, virtual sequencing with deep learning to predict gene alterations in colorectal cancer. *Ann. Oncol.*, 32, S290, 2021.

15. Lin, P.-Y. and Chen, P.-Y., Review of: "Deep-BGCpred: A unified deep learning genome-mining framework for biosynthetic gene cluster prediction, *Qeios*, 1, 1–4, 2021.

16. Passi, K., Shi, Z., Jain, C.K., Improved prediction of gene expression of epigenomics data of lung cancer using machine learning and deep learning

models, in: *Knowledge Modelling and Big Data Analytics in Healthcare*, pp. 165–182, 2021.

17. N, T.R., Bhattacharyya, D., Sk, M., Hong, S., Performance evaluation of a distributed energy model with compound poisson arrivals on an improvised forked network: A detailed analysis. *Sustain. Comput.: Inform. Syst.*, 3, 33, 120–124, 2022.

18. Joshua, E.S.N., Chakkravarthy, M., Bhattacharyya, D., Lung cancer detection using improvised grad-cam++ with 3D CNN class activation, in: *Lecture Notes in Networks and Systems*, pp. 55–69, 2021.

19. Choi, H. and Na, K.J., A risk stratification model for lung cancer based on gene coexpression network and deep learning. *BioMed. Res. Int.*, 2018, 1–11, 2018.

20. Joshua, E.S.N., Chakkravarthy, M., Bhattacharyya, D., An extensive review on lung cancer detection using machine learning techniques: A systematic study. *Rev. Intell. Artif.*, 34, 3, 351–359, 2020, https://doi.org/10.18280/ria.340314.

21. Das, A., Rajendra Acharya, U., Panda, S.S., Sabut, S., Deep learning based liver cancer detection using watershed transform and Gaussian mixture model techniques. *Cognit. Syst. Res.*, 54, 165–175, 2019, https://doi.org/10.1016/j.cogsys.2018.12.009.

22. Medical Segmentation Decathlon, APE1, 2022, http://medicaldecathlon.com.

23. Bhushan, M., Goel, S., Kumar, A., Improving quality of software product line by analysing inconsistencies in feature models using an ontological rule-based approach. *Expert Syst.*, 35, 3, e12256, 2018.

24. Megha, Negi, A., Kaur, K., Method to resolve software product line errors, in: *International Conference on Information, Communication and Computing Technology*, Springer, pp. 258–268, 2017, https://doi.org/10.1007/978-981-10-6544-6_24.

25. Bhushan, M., Goel, S., Kumar, A., Negi, A., Managing software product line using an ontological rule-based framework, in: *2017 International Conference on Infocom Technologies and Unmanned Systems (Trends and Future Directions) (ICTUS)*, IEEE, pp. 376–382, 2017.

26. Bhushan, M. and Goel, S., Improving software product line using an ontological approach. *Sādhanā*, 41, 12, 1381–1391, 2016.

27. Pal, S., Mishra, N., Bhushan, M., Kholiya, P.S., Rana, M., Negi, A., Deep learning techniques for prediction and diagnosis of diabetes mellitus. *2022 International Mobile and Embedded Technology Conference (MECON)*, 10-11 March, 2022, pp. 588–593.

28. Rana, M. and Bhushan, M., Advancements in healthcare services using deep learning techniques. *2022 International Mobile and Embedded Technology Conference (MECON)*, 10-11 March, 2022, pp. 157–161.

Transforming the Technologies for Resilient and Digital Future During COVID-19 Pandemic

Garima Kohli[1]* and Kumar Gourav[2]

[1]The Business School, University of Jammu, Jammu, India
[2]UIC, Chandigarh University, Mohali, India

Abstract

The emergence of the COVID-19 pandemic has transformed society, which was initiated through widespread, impulsive, and vivid digitalization. This disaster has brought an astonishing digital hop in various sectors of the economy. This digital leap in different sectors gave rise to the emergence of digital technologies. These technologies provide solutions and open the doors for the difficulties being faced during a crisis. Different digital technologies like smart healthcare systems, blockchain-based monitoring platforms, the Internet of Things, Artificial Intelligence, etc. came into existence. Robotics also came into the life of an individual to deliver food, medications, screen temperature and also manage social distancing. E-learning has changed the approach of education with the advancement of these digital tools. The support of these digital techniques can be a boon and has helped society for a resilient future in various sectors of the economy. This situation of disaster has forced every member of society to such a position where digitalization has become a requirement rather than a choice that will help the business and government units in reshaping the future. Thus, this chapter highlights the impact of digital transformation during this pandemic, major challenges related to it and its implications.

Keywords: COVID-19, digital transformation, economy, society, technology

*Corresponding author: garima.kohli5@gmail.com

Ashok Kumar, Megha Bhushan, José A. Galindo, Lalit Garg and Yu-Chen Hu (eds.) Machine Intelligence, Big Data Analytics, and IoT in Image Processing: Practical Applications, (81–100) © 2023 Scrivener Publishing LLC

4.1   Introduction

COVID-19 pandemic has drastically changed and hindered the regular flow of economic activities globally in such an extraordinary way that will have a remarkable impression on the history of humanity. This disaster has put the entire globe into the most horrible economic emergency since World War II, wherein employees have lost their jobs, daily wagers also suffered, and these circumstances gave rise to poverty all around. With the stretch of this pandemic worldwide, the lockdown was imposed everywhere, all the activities like malls, educational institutions, corporates, government offices, temples, airports, and railway stations that involve huge crowds were halted. When all the activities were shut down, then Internet services came as a boon to the society wherein people can exchange messages, ideas can interact, shop, work from home and many more. The technological development in the past few years has led to the beginning of the new era of digitalization of services at each level of the business and processes. With the advancement in technology, automation has become an important aspect of digitalization [1]. This gave rise to the increase in the 100% usage of the Internet. Digital technologies have played a crucial part with the help of Internet and broadband networks in addressing the effects of this pandemic. There was a tremendous rise in using video-conferencing services like Zoom and delivery services, like Akamai have also shown a steep rise of 30% increase in content usage [2]. With the widespread of this pandemic everywhere, the majority of the population across the globe have shifted to the digital platform which was in progress for the past few decades. The change into this digital transformation is the need of the hour which gave rise to e-shopping, e-banking, e-learning, work from home culture, etc. To maintain and preserve the regular flow of revenue many business houses used digital business models, many mobile applications were also launched to keep and trace a fair record for the overall improvement of this pandemic. This pandemic forced us to take an astonishing digital leap in all the sectors of an economy [3].

As the silhouette of 2020 is progressively elaborated by this dreadful pandemic, digitalization has become the modern way that is encumbered with the latest innovations and has vast scope for changing the battle against this contagious disease. Individuals and organizations across the globe have to fiddle with the new ways of work and life and thus transform society.

The omnipresence of Big Data and the arrival of emergent digital technologies like AI, Blockchain, the Internet of Things (IoT) and robotics are expected to have deep effects on business [4–6]. Intelligent process automation is the blend of robotics process automation with Artificial Intelligence (AI) with a vision to produce end-to-end processes that can learn, think and adapt on their own [7]. Technologies adopted during the pandemic were IoT, smart healthcare systems, AI, Blockchain based epidemic monitoring platforms, fifth-generation (5G) mobile networks, cloud computing, big data analysis, robotics and many more. To avoid human interaction during pandemics robots came into being for sending food parcels to patients, for medication purposes, to monitor the temperature, to maintain social distancing in manufacturing plants and various other applications. Online learning platforms like Google meet, zoom, etc and work from home culture are the major digital transformations that took place during this pandemic. For the social welfare programs, government organizations have employed various digital platforms for payment purposes. With the advancement and implementation of these digital technologies into day to day activities of the individual, it has been observed that society is moving to the world of technology. This society was based on the traditional economy consisting of directorial, industrious and authoritative systems, which collides with the digital platforms consisting of its pioneering attributes in terms of business models, production, business groups and supremacy. All these gave rise to the new culture, which is digitally interwoven with the multifaceted ecosystem that is presently undergoing organizational, institutional and regulatory transformation [8].

For maintaining a resilient, sustainable and secure future science, technology and advancement play a significant role in the present era. To accomplish resilience and sturdiness in future networks, there is an urgency to keep an eye on the system to check the errors, immediate reply to a solution and immediate trigger of remedial actions to refurbish the network back to normal functionality. In plummeting the harmful radiations of the COVID-19 pandemic both wired and wireless mediums have played a crucial role in different sectors of the economy such as health, education, transportation, manufacturing, etc. With the emergence of these technologies and mediums of communication and by taking the correct actions, during the crisis, these sectors were given a new life and continue to survive and can secure their future. Investments in these technologies have enormously supported the prospect resilience of different sectors of the economy and the public services delivery. Thus, with the emergence of these technologies into the daily lifestyle of individuals that will escort the next wave and make their life more expedient, more pleasant, and more vigorous.

The study further reveals how society is benefited during this pandemic by adopting these smart technologies. The main contribution of this chapter is to study the different digital technologies by highlighting their challenges and their implications. The rest of this chapter is organized as follows: section 4.2 discusses the various digital technologies used during the pandemic. Section 4.3 highlights the challenges caused by using these digital technologies. Section 4.4 depicts the implications of highlighted challenges. Section 4.5 concludes the chapter.

4.2   Digital Technologies Used

Technologies played a vital role in justifying the impact of the pandemic. Digital technology allows businesses and provides a platform to get products and services in different areas. Moreover, looking forward to building a resilient and digital future requires balancing short-term goals with achieving long-term opportunities. The emergence of the World Wide Web and its global adaption have increased the use of digital technologies. Some of them are shown in Figure 4.1 below.

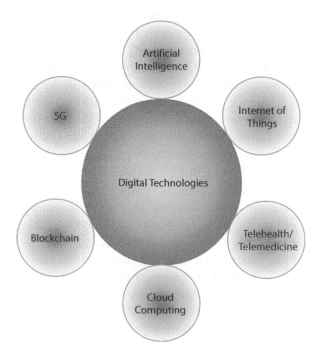

Figure 4.1 Digital technologies.

4.2.1   Artificial Intelligence

AI is the technique that makes a computer think and work like humans. It is the developing technology that is used for numerous intellectual applications in different fields like healthcare and medical diagnosis, image processing, cyber security, banking sector, telemedicine, robotics and driverless cars. AI techniques have various approaches but Deep Learning (DL) [9] and Machine Learning (ML) [10] are the two significant approaches. The DL approach is used to deal with composite difficulties and represent them simply and effectively, whereas ML is used to extract the significant patterns from the data [8]. According to researchers [11], the use of AI in banks facilitates the client experience by providing effortless communications. By using AI banking service was not restricted to retail banking services but also uses three-channel service in which informal banking, anti-scam and endorsing are the main parts. The education sector is also affected by AI, researcher [12] recommended in his study that by using AI, there is an impressive learning experience for students offered by the teachers using different tools that also minimizes the excessive load of the teacher. These days, healthcare sector is one of the most auspicious and fast-rising sectors.

AI and ML help medical consultants to easily diagnose and treat that disease [13, 14]. There are several applications of AI in the field of healthcare. By using this technique, doctors can easily monitor the patient's health, manage patient data, prescribe medicine, perform surgery, treat patients in remote areas and also provide personalized treatment to the patients [15]. AI can offer health workers supporting diagnoses and proper medication that reduces the burden of health workers and increases medical services and efficiency [16, 17]. The robot is one of the main achievements of AI in healthcare. Now in the current situation robots are used to perform surgeries and also deal with difficult medical situations [18]. AI also has a significant impact to deal with threats that emerged during this situation of the pandemic. AI also helps us to deal with coronavirus by allowing population screening, medical assistance, announcements and providing recommendations related to the prevention of this disease [19–21]. This automation can lay out better treatment for the infected patients with the best possible outcomes.

4.2.2   Internet of Things

IoT is the novel approach that has significantly transformed the conventional living standard into a modern lifestyle. IoT is defined as the technology in which every device is connected to the Internet. Smart cities, smart homes, smart transportation and smart industries are the major transformations of IoT. A commonly used application of IoT that includes smart homes too is the smart city. The idea of developing smart cities is to enhance the quality of well-being by using smart technologies [22]. Smart homes become the latest trend of IoT that makes the life of people more comfortable with smart home appliances like smart television, smartphone, smart lights, air-conditioners, audio/video streaming devices and security systems. All this communication has been done by using an IoT based central device that is connected to the Internet [23]. In the study of [24], the IoT also benefited the agriculture sector in which soil moisture examination and monitoring the trunk diameter become the major concerns. It also controls and protects the number of products for agricultural use and standardizes the weather conditions to make massive production with good quality. IoT also has different benefits like tracking of goods, processing of payment according to their geographical limits, notifying the storage circumstances during the supply chain process, theme-based parks and fitness centers in the field of retail management. For retail IoT can be used for instant payments by enabling biometrics, identifying the irritant products and also managing the rotation of goods and automating restocking of warehouses [25]. IoT can be used in weather forecasting that gives accurate data and good resolution for observing the weather information by allocating and exchanging information.

Smart health sensing system (SHSS) is another achievement of IoT. This technology uses sensors and actuators to manage the health scenario's well-being [26]. To monitor the various indoors and outdoors aspects like a heartbeat, pulse rate, blood pressure and glucose level for good health and fitness of the patient's smart devices like sensors and actuators are used [27, 28]. Therefore, this technology has changed the whole situation of the healthcare industry by easing with this latest innovation and smart gadgets [29, 30]. IoT applications also become the emerging technology during this pandemic by tracing the pattern of infection transmission, diagnosing patients and providing telemedicine services and also progress the efficiency, accuracy and pace for the analysis and cure of infected individuals [31, 32]. During quarantine of patients, this technology was also beneficial for medical experts to observe them with ease. All the patients with high risk are followed without any difficulty through the Internet.

4.2.3   Telehealth/Telemedicine

The advancement in the information and communication sector becomes important for the enhancement of services related to healthcare [33]. Healthcare-related services provided remotely by using the means of telecommunication technologies are the novel concept that is known as telehealth [34]. It includes various services associated with healthcare, i.e.. guidance, administration, education, remote clinical assistance, public health and also elaborates these services extensively. According to Hau [35], telemedicine is defined as the medical services that offer consultation, diagnosis, treatment, and delivery of medicines to the patients by the healthcare professional that uses telecommunication set-up to provide medical help at a remote site. It allows patients to receive medical services at their doorstep [36]. During the COVID-19 pandemic, this digital technology prevents many people from getting infected. They get medical assistance from the experts at their respective homes without visiting the overcrowded emergency or waiting rooms [37]. This service also benefits the patients by getting advice from doctors whenever required and trying to stabilize their condition at home keeping them away from overloaded hospitals [38]. Digital technologies, like medical support apps, also play a vital role in the current scenario, which helps to face this pandemic and also addresses the day-to-day health problems related to isolation or quarantine, like mental health conditions, physical inactivity and psychological needs [39–41]. Robotic telemedicine carts were used for observing the situation of isolated patients in which the physical existence of the medical staff is not required. Many hospitals used two-way audio-video communication to monitor patients' conditions in intensive care units [42].

4.2.4   Cloud Computing

Cloud computing is the digital computing platform that plays a significant aspect in the era of digital technology [43]. In the study of [44], cloud computing is the essential component of technology and business models that has changed the concept of businesses by adopting new strategies. It offers the concept of pay and use service. Many organizations used this third-party service to save costs, time and space. The education sector plays a major role in handling the monetary development of the country. In the existing situation, teaching students in the classroom has become more innovative and is getting involved with this online education system i.e. Cloud Computing. This technology enhances the level of teaching by providing better services and tools that are not easily available to the students [45]. It also reduces the cost of purchasing licensed software for educational institutes. Now, they use the third-party service by opting for pay and use service. Online courses are available for teachers as well as for students in different fields to enhance their skills with the latest knowledge and technology. The concept of cloud computing has been used in many areas like agriculture, entertainment, healthcare, etc. In agriculture, there are innovative applications that use cloud computing and create a complete network with the help of sensors that manage and monitor techniques for gathering data related to the soil from agricultural field images and interpretations by individuals on the ground precisely by using the Global Positioning Service (GPS) coordinates. Information related to demand by area/markets, product prediction, and production-related data is also accessed using cloud computing [46, 47].

Cloud computing can be used for entertainment. Cloud facilitates the users with a single device that is used and provides access to the various entertainment platforms like on demand gaming, online media stores, Over-the-Top (OTT) platforms and online music apps. The Healthcare cloud also has an important place, by using this technology all the data related to the medicine becomes digitized and stored in the central location. This centrally stored data is easily accessed and shared by healthcare experts, insurance companies and also by the patients too. During the COVID-19 Pandemic, the clouds have enabled us to meet one another digitally by using applications like Google Meet, Zoom Video, Skype, YouTube, etc. Several cloud service providers offer cloud services to the users some of them are Microsoft Azure, Amazon Web Services, and Google cloud. By using these facilities, people can survive their lockdown period during the pandemic successively.

4.2.5   Blockchain

Blockchain technology is the technology that transformed the industry, operating modes of commerce, healthcare, and even education too. This technology also promotes the economy globally and has the potential to deal with all types of applications. All the transactions executed in blockchain become transparent, immutable, and trustworthy [48]. Presently most educational institutions have made blockchain technology a part of the curriculum and some of them consider academic degrees for management [49, 50]. This technology allows financial services without involving financial organizations like banks or other intermediaries. It can also be used to manage various services like e-payments, digital assets and reimbursement of funds [51]. The use of blockchain technology in the aviation industry is beneficial to many by providing secure and strong mutual partnerships between product and service providers. To deal with important data like a patient's health records blockchain is used by converting these records in encoded form and can be accessed by using a secured key that would permit authorized persons to get access to the information [52]. Likewise, a similar procedure can be used to examine the information that is required by Health Insurance Portability and Accountability Act (HIPAA) laws to certify the privacy of the data. A patient's health information is stored in the blockchain, then it is transferred to the insurance providers automatically in some cases doctors also transfer the information to the authorized persons [53]. Blockchain contributes to building a transparent and efficient system that deals with the COVID-19 pandemic by using trusted, verified, distributed and tamper-resilient ledger technology. According to [54] blockchain technology can permanently store the information related to different stages, phases and events of the COVID-19 vaccine, like development, production, certification and allotment to the authorized administrations for immunization purposes. A generic blockchain-based system is used to keep eye on COVID-19–related data like the development of medicines for patients, COVID-19–affected cases and mortality rate for the particular country in a highly trusted and reliable manner.

4.2.6   5G

The 5G network is the 5th-generation mobile network which is the modern generation for cellular mobile communication technology. It is used to transfer data more efficiently than the older mobile networks, it provides high speed, large broadband and low latency. This technology provides amazing data competencies and infinite data transmission within the modern mobile operating systems [55, 56]. This will sustain with self-driving cars, smart devices, telemedicine, and the IoT. Internet of Vehicles (IoV) and smart devices become the key aspects of 5G networks [57]. AI, automation and augmented reality for troubleshooting in the era of 5G is the major factor for the manufacturing companies to convert their normal factories to smart factories to transform production and increase sales [58]. This novel technology also benefits the health sector by providing integrated mobility and advanced connectivity to the medical persons to monitor the patients everywhere every time. 5G technology allows the patients to use wearable devices to communicate their status of health time-to-time to the doctors. Doctors perform surgery by sitting at one end to another end by using this fastest bandwidth of 5G networks. In China, surgeons performed such a type of surgery by using an off-site surgical robot and operating in real-time [59]. 5G permitted robots for medical purposes that can supply medicines, monitor the temperature of sick persons and sanitize the rooms in the hospitals during the COVID-19 pandemic by minimizing the contact of health workers with this infection. Ava Robotics' created iRobot, especially for medical purposes that help the medical practitioner to deal with the persons that are home quarantined without the threat of being infected [60].

4.3   Challenges in Transforming Digital Technology

The Covid-19 pandemic has led to an unavoidable heave in the utilization of digital technologies due to the norms of maintaining social distance and global lockdown. The life of an individual's being has changed and has to

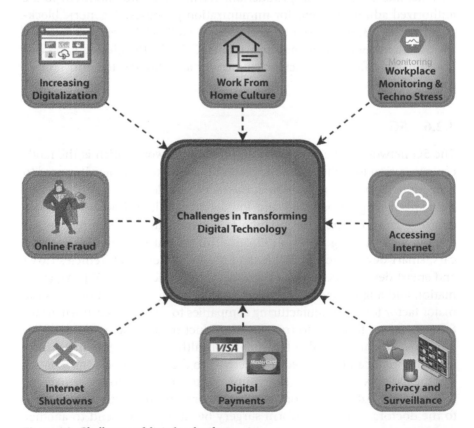

Figure 4.2 Challenges of digital technology.

adapt to the new methods of transformation for its survival. Following are the challenges that are being faced during a pandemic as shown in Figure 4.2.

4.3.1   Increasing Digitalization

Digitalization is like a heart in the body. For transforming the organizations into a digital platform, numerous digital technologies are being used like Blockchain, IoT, cloud, AI, 5G and much more. This requires huge investment in extending the bandwidth, technology used, network devices, smart gadgets, etc. as it is a need of an hour for survival than as an exemption. Employees have adopted the culture to work online, students are comfortable with e-learning by way of digital platforms like Zoom and Google Meet, business organizations have started conducting their meetings through video conferencing and Skype and customers have followed digital banking transactions, to go digital. Changes in these technological transformations are being used by various firms to move according to the present situation [61–63].

4.3.2   Work From Home Culture

Gig workers are independent online platform workers that work temporarily on-demand of companies to provide services to the clients or customers. When there was a total shutdown throughout the globe these gig workers were the great losers because their services were halted and their survival was very difficult [64]. During the emergence of smartphones from 2010 onwards, the journey of these workers commenced. But during and after the postpandemic their income level gradually declined as no fixed salaries were there for these workers. Swiggy, zomato, uber, ola were a few examples of these workers from the Indian scenario.

4.3.3   Workplace Monitoring and Techno Stress

Online culture demands the employees to be present at any time because of digital technology. This transformation of technology into digitalization will result in enhancing productivity, but it has also amplified the stress among the employees from all sectors [65]. This technology has proved stressful for students also, as they have to stay in touch with these appliances throughout the day which has affected their mental and physical health. Employees require proper training to adapt to this culture, should be accessible 24*7 and need to adopt multitasking techniques. All this digital culture has increased the work pressure in the lifestyle of every individual which ultimately leads to stress.

4.3.4 Online Fraud

Too much use of digital culture in every sector is witnessing huge online frauds and scams. Digital banking is the biggest example of online fraud. The main aim is to extract as much as possible by illegal methods. Individuals are facing a crisis during this pandemic phase, there is too much poverty all around because of the rise in the prices of commodities which have resulted in insecurity in the minds of humans [66]. The government is taking appropriate steps to reduce the excess of this theft and robbery on online platforms by way of digital technologies.

4.3.5 Accessing Internet

Access to the Internet has become very important during and after the post pandemic period. Individuals who were not associated with the Internet were facing complete abolition. Internet networks in developing countries are the need for survival in the present scenario, as everything is connected to this smart technology. Those not connected and have no access to the Internet because of various reasons such as technology, network issues, knowledge, skills, or any other reasons are left out [67, 68].

4.3.6 Internet Shutdowns

In the present scenario, people are dependent upon the Internet. Closure of these web services seems like putting life to an end [69]. The Internet services were interrupted in Jammu and Kashmir, a union territory in India, since August 5, 2019 and lasted till May 2020 was the major interruption in this democratic country [70]. Due to this closure, life of the people of UT has come to standstill. During a pandemic, Internet facilities were the primary ingredient for the survival of every citizen. The shutdown of this has become grimmer and has a huge impact ensuing from the pattern of ambiguity which can dishearten investors across the globe. Due to this most of the areas like healthcare, education, media, e-commerce, etc. faced unbearable consequences [71].

4.3.7 Digital Payments

Digital transactions played a vital role during and after the pandemic. Two different aspects are linked to digital money and helped to win this battle throughout this period and still the process is going on. Firstly, currency was alleged as the medium of spreading the infection and thus, digitalization came as a boon to society and was preferred by most individuals [72]. On one hand, consumers were making the purchases online and on the other were transferring the payments through plastic money or through Google apps which was made possible by the government also in several parts of India [73]. This ultimately resulted in a heap in using digital payments that will lead to work on the diffusion of this novel technology. Secondly, during the period of lockdown, most of the people were jobless and governments granted assistance through digital payment apps. These were a well-situated medium of transferring the funds from source to destination like in the earlier difficult situations [74]. Thus, both aspects were observed during the crisis and need to be further investigated.

4.3.8   Privacy and Surveillance

Privacy and surveillance have gained importance during lockdowns for managing mobile and digital apps. The infected patients were monitored and their contact tracing was recorded through these smart devices [75]. Privacy and surveillance was major concern. During the pandemic, the governments of various countries have launched different mobile applications for tracking symptomatic persons. The government of India launched AarogyaSetu app for tracking infected people in India [76], the Chinese government launched the Covid-19 tracker in China [77] and the US government launched contact tracking apps [78]. Thus, with these technological innovations, this worldwide pandemic could be fought and the issue of surveillance can be resolved.

4.4   Implications for Research

This section depicts the implications of the challenges that are discussed above.

- It is imperative to understand the implications of increased digitalization while deputing these digital technologies. These smart technologies are useful in every context of life in developing countries such as their incorporation in workflows, their efficiency in complex problems and many more.
- Many research aspects are taken into consideration for gig workers and employees working from home like the trust between them, to evaluate the performance, effective communication and teamwork.

- It is difficult for some of the employees to work virtually and understand new digital tools which create stress and tension in their minds. As their work gets overloaded. The latest and rigorous types of digital supervision are to be accepted with their determined implications.
- To understand the dimensions and effect of the digital divide great effort has been done. When all the activities throughout the globe were carried out digitally during the pandemic, it was difficult for those to survive without access to the Internet. Internet access is very important in today's present scenario.
- Measures for Internet shutdown need to be taken care of. As everything during and after the pandemic was heavily dependent upon the Internet, it posed various difficulties in the life of the well-being of an individual if services of the Internet were shut down.
- Gathering the information related to the online payment systems and their influence during the pandemic will provide assistance and support to the affected populations.
- Data has been collected by using various contact locating applications for surveillance that have become the major concern during this period. Problems related to determination and abolition of information, a compilation of related information, allocation of data among different apps and several other transactions were involved.

4.5   Conclusion

Digital transformation is like a heart in the body in the present times. Every technology has its own positive, as well as negative, aspects. During this pandemic, there was a drastic change in the process of digitalization that influenced not only the work culture but also the normal lifestyle of well-being. Thus, this chapter has provided an overview of some emerging technologies which have transformed the lives of individuals and made their future resilient and dependent upon these technologies. The study also highlights some of the challenges related to these technologies with their suitable implications.

References

1. Kholiya, P.S., Kapoor, A., Rana, M., Bhushan, M., Intelligent process automation: The future of digital transformation, in: *10th International Conference on System Modeling & Advancement in Research Trends (SMART)*, pp. 185–190, 2021.

2. Pandey, N. and Pal, A., Impact of digital surge during Covid-19 pandemic: A viewpoint on research and practice. *Int. J. Inf. Manage.*, 55, 102171, 2020.

3. Iivari, N., Sharma, S., Olkkonen, L.V., Digital transformation of everyday life – How COVID-19 pandemic transformed the basic education of the young generation and why information management research should care? *Int. J. Inf. Manage.*, 55, 102183, 2020.

4. Dong, and Yang, Business value of bid data analytics: A systems-theoretic approach and empirical test. *Inf. Manage.*, 57, 1, 103124–103132, 2019.

5. Wedel, M. and Kannan, P.K., Marketing analytics for data-rich environments. *J. Mark.*, 80, 6, 97–121, 2016.

6. Ng, I.C.L. and Wakenshaw, S.Y.L., The Internet-of-Things: Review and research directions. *Int. J. Res. Mark.*, 34, 1, 3–21, 2017.

7. Ng, K.K., Chen, C.H., Lee, C.K., Jiao, J.R., Yang, Z.X., A systematic literature review on intelligent automation: Aligning concepts from theory, practice, and future perspectives. *Adv. Eng. Inf.*, 47, 101246, 2021.

8. Bragazzi, N.L., Dai, H., Damiani, G., Behzadifar, M., Martini, M., Wu, J., How big data and artificial intelligence can help better manage the COVID-19 pandemic. *Int. J. Environ. Res. Public Health*, 17, 9, 3176, 2020.

9. Nalavade, A., Bai, A., Bhushan, M., Deep learning techniques and models for improving machine reading comprehension system. *IJAST*, 29, 04, 9692–9710, 2020.

10. Verma, K., Bhardwaj, S., Arya, R., Islam, M., Bhushan, M., Kumar, A., Samant, P., Latest tools for data mining and machine learning. *Int. J. Innov. Technol. Exploring Eng.*, 8, 9S, 18–23, 2019.

11. Vijai, C. and Nivetha, P., ABC technology - artificial intelligence, blockchain technology, cloud technology for banking sector. *Adv. Manag. (AIM)*, 13, 4, 19–24, 2020.

12. Loeckx, J., Blurring boundaries in education: Context and impact of MOOCs. *Int. Rev. Res. Open Distrib. Learn.*, 17, 3, 92–121, 2016.

13. Rong, G., Mendez, A., Bou Assi, E., Zhao, B., Sawan, M., Artificial intelligence in healthcare: Review and prediction case studies. *Engineering*, 6, 3, 291–301, 2020.

14. Reddy, S., Fox, J., Purohit, M.P., Artificial intelligence enabled healthcare delivery. *J. R. Soc. Med.*, 112, 1, 22–28, 2019.

15. Malik, A.P., Pathania, M., Rathaur, V.K., Overview of artificial intelligence in medicine. *J. Fam. Med. Prim. Care*, 8, 7, 2328–2331, 2019.

16. Liu, R., Rong, Y., Peng, Z., A review of medical artificial intelligence. *Glob. Health J.*, 4, 42–45, 2020.

17. Wang, Q. and Zhang, C., Can COVID-19 and environmental research in developing countries support these countries to meet the environmental challenges induced by the pandemic? *Environ. Sci. Pollut. Res.*, 28, 30, 41296–41316, 2021.

18. Liu, Y., Zhang, W., Pan, S., Li, Y., Chen, Y., Analyzing the robotic behavior in a smart city with deep enforcement and imitation learning using IoT. *Comput. Commun.*, 150, 346–356, 2020.

19. Haleem, A., Javaid, M., Vaishya, Effects of COVID 19 pandemic in daily life. *Curr. Med. Res. Pract.*, 10, 2, 78–79, 2020.

20. Bai, H.X., Hsieh, B., Xiong, Z., Halsey, K., Choi, J.W., Tran, T.M., Pan., I., Shi, L.B., Wang, D.C., Mei, J., Jiang, X.L., Performance of radiologists in differentiating COVID-19 from viral pneumonia on chest CT. *Radiology*, 292, 2, E46-E54, 2020.

21. Vaishya, R., Javaid, M., Khan, I.H., Haleem, A., Artificial intelligence (AI) applications for COVID-19 pandemic. *Diabetes Metab. Syndr.: Clin. Res. Rev.*, 14, 4, 337–339, 2020.

22. Sharma, S., Nanda, M., Goel, R., Jain, A., Bhushan, M., Kumar, A., Smart cities using Internet of Things: Recent trends and techniques. *Int. J. Innov. Technol. Exploring Eng. (IJITEE)*, 8, 9S, 24–28, 2019.

23. Khajenasiri, I., Estebsari, A., Verhelst, M., Gielen, G., A review on internet of things for intelligent energy control in buildings for smart city applications. *Energy Proc.*, 111, 770–9, 2020.

24. Sundareswaran, V. and M.S., Survey on smart agriculture using IoT. *Int. J. Innov. Res. Eng. Manage. (IJIREM)*, 5, 2, 62–66, 2018.

25. Tadejko, P., Application of Internet of Things in logistics-current challenges. *Ekon. Zarz dzanie*, 7, 54–64, 2015.

26. Mangla, M., Kumar, A., Mehta, V., Bhushan, M., Mohanty, S.N., *Real-Life Applications of the Internet of Things: Challenges, Applications, and Advances*, 1st Edition, p. 536, Apple Academic Press, New York, CRC Press, Taylor & Francis Group, 2022.

27. Mohammed, M.N., Syamsudin, H., Al-Zubaidi, S., Ramli, R., Yusuf, E., Novel COVID-19 detection and diagnosis system using IoT based smart helmet. *Int. J. Psychosoc. Rehabil.*, 24, 7, 2296–2303, 2020.

28. Vaishya, R., Javaid, M., Khan, I.H., Haleem, A., Diabetes & metabolic syndrome. Clinical research & reviews, in: *Artificial Intelligence (AI) applications for COVID-19 pandemic*, 2020.

29. Sfar, A.R., Natalizio, E., Challal, Y., Chtourou, Z., A roadmap for security challenges in the internet of things. *Digit. Commun. Netw.*, 4, 1, 118–37, 2018.

30. Minoli, D., Sohraby, K., Kouns, J., IoT security (IoTSec) considerations, requirements, and architectures. *14th IEEE Annual Consumer Communications & Networking Conference (CCNC)*, Las Vegas, NV, USA, pp. 8–11, 2017.

31. Arora, N., Banerjee, A.K., Narasu, M.L., The role of artificial intelligence in tackling COVID-19. *Future Virol.*, 15, 11, 717–724, 2020.

32. Rahman, S., Peeri, N.C., Shrestha, N., Zaki, R., Haque, U., Ab Hamid, S.H., Defending against the novel coronavirus (COVID-19) outbreak: How can the Internet of Things (IoT) help to save the world? *Health Policy Technol.*, 9, 136–138, 2020.

33. Singh, V.J., Bhushan, M., Kumar, V., Bansal, K.L., Optimization of segment size assuring application perceived QoS in healthcare. *Proceedings of the World Congress on Engineering*, vol. I, WCE, 2015.

34. Dorsey, E.R. and Topol, E.J., State of telehealth. *N. Engl. J. Med.*, 375, 2, 154–161, 2016.

35. Hau, Y.S., Kim, J.K., Hur, J., Chang, M.C., How about actively using telemedicine during the COVID-19 pandemic? *J. Med. Syst.*, 44, 1–2, 2020.

36. Greenhalgh, T., Wherton, J., Shaw, S., Morrison, C., Video consultations for COVID-19. *BMJ*, 368, m998, 020.

37. Greenhalgh, T., Koh, G.C.H., Car, J., COVID-19: A remote assessment in primary care. *BMJ*, 368, m1182, 2020.

38. Hong, Z., Li, N., Li, D., Li, J., Li, B., Xiong, W., Telemedicine during the COVID-19 pandemic: Experiences from Western China. *J. Med. Internet Res.*, 22, 5, e19577, 2020.

39. Zhou, X., Snoswell, C.L., Harding, L.E., Bambling, M., Edirippulige, S., Bai, X., The role of telehealth in reducing the mental health burden from COVID-19. *Telemed. J. E Health*, 26, 4, 377–379, 2020.

40. Torous, J., Jan, M.K., Rauseo, R.N., Firth, J., Digital mental health and COVID-19: Using technology today to accelerate the curve on access and quality tomorrow. *JMIR Ment. Health*, 7, 3, e18848, 2020.

41. Liu, S., Yang, L., Zhang, C., Xiang, Y., Liu, Z., Hu, S., Online mental health services in China during the COVID-19 outbreak. *Lancet Psychiatry*, 7, 4, e17–e18, 2020.

42. Kapoor, A., Guha, S., Das, M., Goswami, K.C., Yadav, R., Digital healthcare: The only solution for better healthcare during COVID-19 pandemic? *Indian Heart J.*, 72, 61–64, 2020.

43. Briscoe, G. and Marinos, A., Digital ecosystems in the clouds: Towards community cloud computing. *IEEE Int Conf Digit Ecosyst Technol*, pp. 103–108, 2009.

44. Ramachandra, G., Iftikhar, M., Khan, F.A., A comprehensive survey on security in cloud computing. *Proc. Comput. Sci.*, 110, 465–472, 2017.

45. Yadav, K., Role of cloud computing in education. *Int. J. Innov. Res. Comput. Commun. Eng.*, 2, 2, 3108–3112, 2014.

46. Patel, R. and Patel, M., Application of cloud computing in agricultural development of rural India. *Int. J. Comput. Sci. Inf. Technol.*, 4, 6, 922–926, 2013.

47. Hori, M., Kawashima, E., Yamazaki, T., Application of cloud computing to agriculture and prospects in other fields. *Fujitsu Sci. Technol. J.*, 46, 4, 446–454, 2010.

48. Underwood, S., Blockchain beyond bitcoin. *Commun. ACM*, 59, 11, 15–17, 2016.

49. Sharples, M. and Domingue, J., The blockchain and kudos: A distributed system for educational record, reputation and reward, in: *Adaptive and Adaptable Learning*, pp. 490–496, EC-TEL 2016, LNCS, Springer 9891, 2016.

50. Skiba, D.J., The potential of Blockchain in education and healthcare. *Nurs. Educ. Perspect.*, 38, 4, 220–221, 2017.

51. Rawat, D.B. and Ghafoor, K.Z., *Smart Cities Cybersecurity and Privacy*, Elsevier, Amsterdam, The Netherlands, 2018.

52. Sharma, S., Kumar, A., Bhushan, M., Goyal, N., Iyer, S.S., Is blockchain technology secure to work on?, ch. 5, in: *Blockchain and AI Technology in the Industrial Internet of Things*, S.K. Pani, S.L. Lau, X. Liu (Eds.), pp. 66–80, IGI Global, Hershey, PA, 2021.

53. Mettler, M., Blockchain technology in healthcare: The revolution starts here. *IEEE 18th International Conference on e-Health Networking, Applications and Services (Healthcom)*, Munich, Germany, pp. 1–3, 2016.

54. Rowan, N.J. and Laffey, J.G., Challenges and solutions for addressing critical shortage of supply chain for personal and protective equipment (PPE) arising from coronavirus disease (COVID19) pandemic–Case study from the republic of Ireland. *Sci. Total Environ.*, 725, 138532, 2020.

55. Bhalla, M.R. and Bhalla, A.V., Generations of mobile wireless technology: A survey. *Int. J. Comput. Appl.*, 5, 4, 26–32, 2010.

56. Mishra, A.R., *Fundamentals of network planning and optimization 2G/3G/4G: Evolution to 5G*, 2nd edition, John Wiley & Sons Ltd., Wiley, 2018.

57. Yang, Y. and Hua, K., Emerging technologies for 5G-enabled vehicular networks. *IEEE Access*, 7, 181117–181141, 2019.

58. Choi, T.M., Kumar, S., Yue, X., Chan, H.L., Disruptive technologies and operations management in the Industry 4.0 era and beyond. *Prod. Oper. Manag.*, 31, 1, 9–31, 2022.

59. Nowak, T.W., Sepczuk, M., Kotulski, Z., Niewolski, W., Artych, R., Bocianiak, K., Wary, J.P., Verticals in 5G MEC-use cases and security challenges. *IEEE Access*, 9, 87251–87298, 2021.

60. Ackerman, E., Telepresence robots are helping take pressure off hospital staff. *Robots Under COVID-19 Pandemic: A Comprehensive Survey*, 9, 1590–1615, 2021.

61. Nacheva, R. and Jansone, A., Multi-layered higher education e-learning framework. *Baltic J. Mod. Comput.*, 9, 3, 345–362, 2021.

62. Shaytura, S.V., Minitaeva, A.M., Ordov, K.V., Gospodinov, S.G., Chulkov, V.O., Review of distance learning solutions used during the covid-19 crisis, in: *2020 6th International Conference on Social Science and Higher Education (ICSSHE 2020)*, Atlantis Press, pp. 1–9, 2020.

63. Stone, D.L., Deadrick, D.L., Lukaszewski, K.M., Johnson, R., The influence of technology on the future of human resource management. *Hum. Resour. Manage. Rev.*, 25, 2, 216–231, 2015.
64. Sargeant, M., The gig economy and the future of work. *Ejournal Int. Comp. Labour Stud.*, 6, 2, 1–12, 2017.
65. Ayyagari, R., Grover, V., Purvis, R., Technostress: Technological antecedents and implications. *MIS Q.*, 35, 4, 831–858, 2011.
66. Revathi, P., Digital banking challenges and opportunities in India. *EPRA Int. J. Econ. Bus. Rev.*, 7, 12, 20–23, 2019.
67. Bonfadelli, H., The Internet and knowledge gaps: A theoretical and empirical investigation. *Eur. J. Commun.*, 17, 1, 65–84, 2020.
68. Scheerder, A., Deursen, A., Dijk, J., Determinants of internet skills, uses and outcomes. A systematic review of the second-and third-level digital divide. *Telemat. Inform.*, 34, 8, 1607–1624, 2017.
69. Dondhu, N. and Gustafsson, A., Effects of COVID-19 on business and research. *J. Bus. Res.*, 117, 284–289, 2020.
70. Qadir, S. and Jaggarwal, S., A study on reasons of internet shutdown in J&K. *J. Maharaja Sayajirao University of Baroda*, 55, 2, 172–179, 2021.
71. Khandelwal, A., Agrawal, A., Kumar, A., An outbreak of coronavirus (COVID-19) epidemic in India: Challenges and preventions. *J. Infect. Dis. Ther.*, 8, 421, 2, 2020.
72. Kicheva, T., Opportunities and challenges of remote work. *Izvestiya. Varna University Econ.*, 65, 2, 145–160, 2021.
73. Jain, A., Sarupria, A., Kothari, A., The impact of COVID-19 on E-wallet's payments in Indian economy. *Int. J. Creat. Res. Thoughts*, 8, 6, 2447–2454, 2020.
74. Pollach, I., Treiblmaier, H., Floh, A., Online fundraising for environmental nonprofit organizations. *Th Hawaii International Conference on System Sciences*.
75. Mbunge, E., Millham, R.C., Sibiya, M.N., Fashoto, S.G., Akinnuwesi, B., Simelane, S., Ndumiso, N., Framework for ethical and acceptable use of social distancing tools and smart devices during COVID-19 pandemic in Zimbabwe. *Sust. Oper. Comp.*, 2, 190–199, 2021.
76. Gupta, R., Bedi, M., Goyal, P., Wadhera, S., Verma, V., Analysis of COVID-19 tracking tool in India: Case study of Aarogya Setu mobile application. *Digit. Gov.: Res. Pract.*, 1, 4, 1–8, 2020.
77. AlTakarli, N.S., China's response to the COVID-19 outbreak: A model for epidemic preparedness and management. *Dubai Med. J.*, 3, 2, 44–49, 2020.
78. Urbaczewski, A. and Lee, Y.J., Information technology and the pandemic: A preliminary multinational analysis of the impact of mobile tracking technology on the COVID-19 contagion control. *Eur. J. Inf. Syst.*, 29, 4, 405–414, 2020.

Part II
PLANT PATHOLOGY

5

Plant Pathology Detection Using Deep Learning

Sangeeta V.[1], Appala S. Muttipati[2]* and Brahmaji Godi[3]

[1]Anil Neerukonda Institute of Technology and Sciences, Department of Computer Science and Engineering, Visakhapatnam, Andhra Pradesh, India
[2]iNurture Education Soultions Pvt. Ltd, Department of IT Academics, Bangalore, India
[3]Raghu Institute of Technology, Department of Computer Science and Engineering, Visakhapatnam, Andhra Pradesh, India

Abstract

As plant diseases are a major threat to developing countries like India, agriculture plays a significant or critical role in the economy. The prognosis of plant diseases at the very beginning reduces the risk to substance security. Biological examination of plants or visual inspection of plants by experts are expensive and have a lot of delays. This has paved a path to implement computer methodologies to detect diseases and suggest pesticides. Latest technologies like image processing and computer vision have developed many algorithms for early detection. Advances in computer vision are facilitated by deep learning (DL) for smart phone-based diagnosis systems. DL techniques use convolution neural networks with familiar residual neural network architecture, which is suitable to develop a model for early disease detection. The model was developed using a dataset of 7,000 healthy and unhealthy plant leaf images. The results were encouraging and showed 95% accuracy in comparison with existing methods.

Keywords: Plant leaf, leaf disease, convolution neural network, residual network, prediction

*Corresponding author: srinuvasu.mutti@gmail.com

Ashok Kumar, Megha Bhushan, José A. Galindo, Lalit Garg and Yu-Chen Hu (eds.) Machine Intelligence, Big Data Analytics, and IoT in Image Processing: Practical Applications, (103–122) © 2023 Scrivener Publishing LLC

5.1　Introduction

Countries like India, whose economies have a dependency on agriculture or crop yield, need to move towards using modern technologies in farming. These technologies have given the farmers the capability to produce enough and quality food at low cost and greater monetary outcomes. Even still food security leaves a threat to farmers, which is affected by many factors, including climate. Diseases and disablement of the plant hinder its life cycle like transpiration, photosynthesis, fertilization, pollination, germination, etc. Due to certain climatic conditions, the plants are affected by pathogens like fungi, bacteria, and viruses [1].

Plant diseases play a major role in food security and are even a threat to smallholder farmers as their livelihoods depend on healthy crops [2]. Early revelation of plant contagion and periodic surveillance of the plant has played a pivotal role in reducing the cost of farming and producing healthy food.

The available methods for plant disease recognition include simple naked-eye inspection by professionals, which allows for disease diagnosis and detection. This demands a large squad of professionals and also on-going plant monitoring, although both are quite costly when dealing with great farms. Meanwhile, farmers have limited necessary facilities or indeed even know how to contact specialists. As a result, consulting professionals is both costly and time-consuming. The proposed technique is effective for monitoring large fields of crops within those scenarios [3, 4]. Automatically identifying ailments by looking at the symptoms on plant leaves is cheaper and quicker.

The current methods for plant disease recognition are simple naked eye observations by professionals, which are used to find out and discover plant disease. This requires a significant team of specialists as well as on-going plant monitoring, both of which are quite expensive. Meanwhile, in certain nations, farmers lack adequate provision or even the knowledge of how to contact professionals. This consulting expert comes at a premium price and takes a long time to complete. In these circumstances, the recommended approach is useful for monitoring huge fields of crops. It is cheaper and less costly to identify pathogens electronically by inspecting plant leaves. Image-based automatic process control, surveillance, and robot guidance are also enabled by machine vision [5, 6].

Detecting plant diseases by eye is a time-consuming and incorrect operation that can only be done in specific settings. On the other side, automatic detection involves less work, consumes slightest effort, and enhances accuracy. Plant diseases include brown spots and yellow spots, early and late scorch, and fungal, viral, and bacterial infections. Image processing is used to analyse the change in color of the targeted limb and to quantify the affected area of illness [7, 8].

The approach of dividing or grouping visuals into various wedges is known as image segmentation. Portrait segmentation may be accomplished in a variety of ways, ranging from basic threshold to complex colour image segmentation approaches. These components are usually accompanying anything that mortals can readily discrete and see as independent things. Even though machines possess the capability to comprehend objects intelligently, several different ways for subdividing shots have been developed. The segmentation process is based on the image's numerous attributes. This might be colour data, picture borders, or a section of an image [9].

So, the authors are implementing upcoming technologies like image processing and computer vision. Many cost-effective procedures have been introduced to increase the yield and throughput. Various techniques have been introduced to determine plant diseases. Residual Neural Network (ResNet 152) is utilized for feature extraction and softmax for the categories of whether a plant is diseased or not.

The remaining content is organized into sections, where Section 5.2 deals with various plant leaf diseases, Section 5.3 contains background knowledge, Section 5.4 ResNet 512 V2 architecture, Section 5.5 introduces a proposed methodology that tells how to detect plant leaf disease, Section 5.6 deals with result analysis and the last section concludes the work.

5.2   Plant Leaf Disease

Plant infections and illnesses are vital parts of ecosystems, yet plant pathology research is heavily tilted toward agriculture and forestry. The majority of plant pathology research focuses on two categories of diseases: conventional and actively controlled illnesses and poorly understood diseases [1]. Plant disease emergence is influenced by a variety of atmospheres, including pathogen occurrences into new areas. In un-experienced situations, pathogen biology and ecology might be difficult to predict.

Existing information is essential for developing eradication methods, quarantine steps, and management plans, but it must be modified to new conditions in a short amount of time to be effective. While the academic researchers frequently rise to the technological challenges of reacting to a growing epidemic, what happens in the early phases of new disease increases is heavily impacted by traditional, social, economic conditions.

Important decisions, such as whether or not to pursue eradication operations and how to carry them out, must be taken fast and resolutely, and may not be wholly supported by scientific evidence. For better-studied diseases, research combining information from multiple fields has resulted in the development of disease-specific control techniques [1, 2].

These are some of the major reasons for plant diseases, which are subclassified as fungal, bacterial, and viral diseases [1, 3], which affect the plant growth at various stages of the plant development life cycle. The classification of plant diseases can be seen in Figure 5.1. These diseases are caused due to heavy rainfall, snow and sudden environmental changes occurred in the atmosphere, which affects plant life [10].

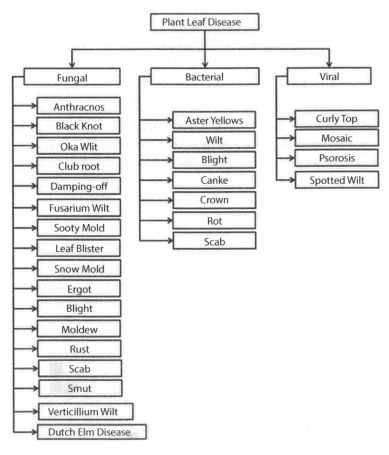

Figure 5.1 Classification of plant leaf diseases.

1) Early Blight: The fungus causes early blight on tomato and potato plants. Small brown spots with concentric rings that form a "bull's eye" pattern occur initially on the lower, older leaves. The disease spreads outward on the leaf surface as it matures, causing it to turn yellow, wither, and die. The stem, fruit, and upper portion of the plant will eventually get diseased. Crops can be seriously harmed.

2) Black Spot: Black spot is a fungus that affects roses and other flowers and fruits. While it does not kill plants, it weakens them and makes them more vulnerable to other ailments. Small black spots grow on foliage in cool, moist weather, and the foliage gradually turns yellow and falls off.

3) Rust: Rust, which appears as rusty spots on leaves and even stems, is another easy-to-identify fungal illness. Another easy-to-identify fungal infection is rust and leaf spots, which appear as rusty dots on leaves and even stems. Leaf spots are yellow or brown blemishes on the leaves that mimic burn marks. Leaf spots can be caused by pathogens, fungus, chemical damage, or insect feeding. Plant productivity and growth are frequently inhibited, and as a result, some plants become weak and deceased.

4) Leaf Spot: There are two forms of pathogen-induced leaf spot illnesses: bacteria and fungus. These diseases damage stone fruit trees as well as crops such as tomato, pepper, and lettuce. Both varieties of leaf spotting look and feel the same. Both forms of prevention and therapy typically employ the same techniques.

5) Verticillium Wilt: Verticillium wilt is a fungus that damages hundreds of trees, shrubs, and edible plants. Pathogens, which can persevere for years in the soil, enter the plant through the roots, blocking the vascular system and causing branches to wilt and undergrowth to turn yellow and fall off early. It can also cause growth to be inhibited.

6) Snow Mold: Snow Mold is a fungus that thrives in the cold, damp conditions that exist under the snow. It feeds on field grass. Symptoms appear once the snow melts and elements generate light tan spots of matted grass.

7) Fusarium Wilt: Fusarium wilt, which is source by a soil-borne fungus, affects improving the poison in various plants such as dianthus, beans, tomatoes, and peas. Wilted leaves and stunted plants, as well as root rot and darkened stem rot

are all indications of the illness. It is very active during the hot summer months.

8) Damping-off: Damping-off disease, which is caused by a variety of soil-borne fungi, is particularly troublesome in rainy, humid circumstances. It infects seedlings, causing them to wilt and die. It is most common in greenhouses, but it can also be found outside.

9) Mosaic: Mosaic viruses come in a variety of forms, but gardeners are most likely to come across two major viruses such as tomato mosaic virus and tobacco mosaic virus. Tomatoes, peppers, potatoes, apples, pears and cherries are infected by the earlier; cucumbers, tomatoes, beets, peppers, lettuce, petunias, and tobacco are infected deeply. Mosaic virus results in speckled yellow and green leaves that are twisted and deformed at times. Yellowing, stunted growth, deformed fruits, and concentrated output are all symptoms of some plants. Even in hot conditions, the mosaic virus is more common.

10) Anthracnose: On stems, leaves, or fruit, infected plants develop black, water-soaked sores. During moist, warm weather, the centres of these lesions frequently become covered with pink, gelatinous masses of spores. In just a few days, anthracnose may turn a magnificent harvest into rotten garbage.

11) Club Root: Plasmodiophora brassicae is a parasitic worm that infects plants by infecting their root hairs. Swollen, twisted, and distorted (clubbed) roots are vulnerable to breaking and rotting. Plants struggle to effectively absorb water and nutrients as a result. Most frequently it affects cabbage and cauliflower crops.

12) Downy Mildew: Many plants are affected, and the disease manifests itself as yellow to white patches on the upper surfaces of older leaves. These places are covered in white to greyish cotton-like fungus on the undersides. These "downy" masses are most noticeable after rain or heavy dew, and they vanish as soon as the sun returns. Even if the plant has enough water, leaves may get crisp and brown and fall off as the disease spreads. Mostly it affects cole crops and cauliflower crops.

13) Canker: An open wound infected with fungal or bacterial infections is frequently used to diagnose canker.

Some cankers are harmless, while others are deadly. Canker is mainly seen on woody landscaping plants. Sunken, bloated, cracked, dead patches on stems, limbs, or trunks are possible symptoms. Cankers can encircle and kill foliage. Cankers appear in stressed plants damaged by cold, insects, drought, nutritional imbalances, or root rot. The infections can also be spread by rodents.

5.3   Background Knowledge

Saharan *et al.* [1] used the concept of genes for creation of disease-resistant plants. He used the transformation techniques to transfer useful genes for creation of disease-resistant plants. He accomplished great success in the creation of transgenic crops. He proved that genetically engineered plants have better resilience to viral, fungal and bacterial pathogens. The farmers were more benefited with transgenic technology in plants. This has given great support to agro-food industries worldwide. Author required more inputs to get high throughput in functional genomics. Functional genomics facilitates to improve the disease management in plants.

Hassan *et al.* [12] have suggested and proved a deep convolutional neural network (CNN) to improve the plant production and in time or early identification of crop diseases because CNNs have marked their identity in achieving best results in the field of machine vision. Basic CNN models involve very high computation costs and require a larger number of parameters for processing. Authors proved that depth-separable convolution will take less number of parameters and in turn the computation cost. They used models, which were trained with an open dataset, consisting of 14 different plant species, and 38 different categorical disease classes and healthy plant leaves [11]. He evaluated the performance of his said model by taking into different parameters and time required per epoch. He achieved an accuracy rate of 99.56% and used different representations of the dataset. He observed that a segmented color image dataset has outperformed when compared with gray scale segmented dataset.

Mohanty *et al.* [13] used two different architectures, namely AlexNet and GoogleNet in the identification of plant diseases. The AlexNet architecture uses a design pattern similar to an existing LeNet-5 architecture. It makes use of more convolution layers followed by one or more fully connected layers. It optionally makes use of the normalization layer and a pooling layer after convolutional layer. These layers are generally associated with ReLu activation functioning. It made use of five convolution layers, in which the first two layers are called conv{1,2}, followed by a normalization and pooling layer. The network ends with a softmax layer. The softmax layer gets the feed from fc8 (fully connected 8 layers) with 38 outputs. Finally the softmax layer normalizes the input and gets its required output with 0.5 dropout ratio. According to GoogleNet architecture is wider and deeper architecture when compared to AlexNet architecture. It made use of 22 layers with less number of parameters when compared to AlexNet. GoogleNet architecture makes use of "network in network" architecture. The first module makes use of parallel convolutions for one, three, five layers along with a max-pooling layer. This enabled the architecture to capture different features, which were not possible earlier as the last step of the architecture uses a filter concatenation layer to concatenate the outputs of all these parallel layers.

Jun Liu and Xuewei Wang [14], given a review on comparison of traditional and current methods of plant disease and pests detection. The current methods made use of deep learning technology. The author has outlined the research on plant disease detection with deep learning, making use of three networks (classification network, detection network and segmentation network). In his literature, he summarized the advantages and disadvantages of each method. He made use of a two-stage detection network called as Faster R-CNN to get the feature map. Then, as a second stage, he calculated the anchor box confidence. Anchor box confidence is calculated using RPN. It makes use of ROI pooling and loss function. Based on the comparative study of the various methods he identified possible challenges in practical implementation. The two-stage detection network has improved the real time and practicability of the detection system. As a conclusion, the author has given various suggestions for facing the challenges of practical implementation.

Wei Wang et al. [15] have given a method for prevention of plant diseases efficiently in a complex environment. With the advancement in digitization and smart farming the plant disease detection has also digitized. It helps in enabling advanced decision making, analysing smartly and for proper planning. The author has made use of a mathematical model for improved accuracy and training efficiency. At the first step of implementation RPN is used to recognize the diseased portion of the leaves. Then, it makes use of Chan-Vese algorithm for image segmentation. As a final step the segmented leaves are inputted for transfer learning for training the model. The final model is tested with various diseases. The author has shown an accuracy of 83.57%, better than the traditional method. Even though this model has proved to be better at complex environments, the Chan-Vese algorithm's repetitive calculations made this model to run for a long time, which has paved a path for enhancement.

Shah *et al.* [16] have proposed a new model ResTS (Residual Teacher/Student) for improving the performance of detecting plant diseases, when compared to various existing methods. This model has given guarantee for food security as well. The author has enhanced the ResTS model when compared to Teacher/Student model, by using residual connections and carrying out batch normalization. The ResTS network consists of three components i.e., teacher, decoder, and student. It made use of VGG16 when the model is over fitting. The outputs of ResTS are precise and better to those of earlier methods.

5.4   Architecture of ResNet 512 V2

A CNN with a few hundred grand of convolutional layers is referred to as a ResNet [17]. Supplementary layers' productiveness was diminished by previous CNN architectures. ResNet has a considerable number of layers and is extremely fast. The foremost modification between ResNet V2 and the innovative (V1) is that V2 applies batch normalisation upon every weight layer prior to adding it. Figure 5.2 describes the Residual Network 152 version 2. ResNet excels at image identification and localization tasks, demonstrating the importance of a wide range of visual classification methods [18, 19].

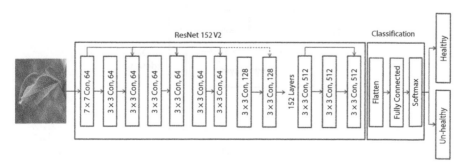

Figure 5.2 ResNet 152 V2 architecture.

5.4.1 Working of Residual Network

Kaiming He and Xiangyu Zhang invented ResNet in 2015 [20]. It is a deep residual learning neural network for image recognition [17, 20]. To address a complex problem, it usually stacks some extra layers in deep neural networks, which enhance accuracy and presentation. The goal of adding more layers is for them to learn more complex characteristics over time. There are two types of leftover connections. Multiple plant leaf visuals with 224*224 pixel resolutions are included in the data set:

i) The identity shorthand (x) can be used directly if the input and output dimensions are the same [30].

$$y = f(x, Wi + x) \tag{5.1}$$

It is used when the input and output dimensions of the residual block function are not the same.

ii) When the dimensions change, the residual connection must be recalculated. The shortcut is still used for identity mapping, but the larger dimension is used to fill in the missing zeros. The projection shortcut is used to match the dimension (done by 1*1 conv).

$$y = f(x, Wi + Wsx) \tag{5.2}$$

Equation (5.2) is because the input and output dimensions of the residual block function are not the same. Figure 5.3 depicts the residual learning process [13].

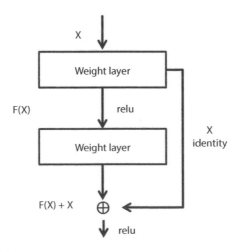

Figure 5.3 Residual learning process.

5.5   Methodology

The framework was classified into two phases, which is shown in Figure 5.4. First phase is image pre-processing. In pre-processing image resizing, data augmentation, data normalization and data splitting are done. Second phase is feature extraction and image classification, ResNet was used for characteristic extraction and for classification softmax classifiers are utilized [17].

5.5.1   Image Resizing

Since the images in the training dataset were of varying sizes, they have to be scaled before being utilised as model input. Images were scaled to 256 pixel squares. Rectangular photos were shrunk to 256 pixels on the shortest side, and then cropped out of the centre 256 X 256 squares.

After that, each pixel is centred by subtracting the mean pixel value. It is done per-channel, meaning that mean pixel values were calculated from the training dataset, one for each of the colour image's red, green, and blue channels [21].

5.5.2   Data Augmentation

Data augmentation methods generates many categories of an actual dataset that have been manipulated to make it larger. ML models can benefit from data augmentation tactics. An experiment revealed that, a deep learning model with shot augmentation results better when compared to a model without shot augmentation. Various measures like accuracy, training loss (also called as penalty for wrong prediction) and validation are used for image classification. Image augmentation is a technique for changing existing photos in order to provide more data for model training [18].

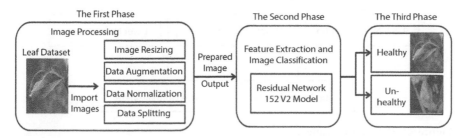

Figure 5.4 Proposed method for detecting healthy leaf or unhealthy leaf.

5.5.2.1   Types of Data Augmentation

Image rotation: The image's information remains constant independent of how it will be rotated.

Image shifting: Changing the photos allows us to customize the locale of the items in the image, giving the framework more variation. It might ultimately occur in a more general framework.

Image flipping: Gyration can be viewed as an add-on of flipping. It empowers us to flip the image both left and right, as well as up and down.

Image noise: Image noise is created by adding noise to the image. The model can learn to discriminate the indication from the fuss in the image using data augmentation. Thereby rendering the designed model to be greater resilience to image adjustments.

Image blurring: Because footage originates from a range of roots, their quality varies. Some images are excellent, while others are not. The original image may be altered in such instances, making the system more robust to the image quality used in the testing set.

5.5.3   Data Normalization

Normalization is a technique for ensuring that all pixel values have the same mean and standard deviation when data is being prepared. This allows the model to learn more quickly. In this chapter, various normalization methods have been discussed and shown in Figure 5.5. The weights of layer fc8 in AlexNet are categorised and are reinitialized, which can be used in DL models [18, 22].

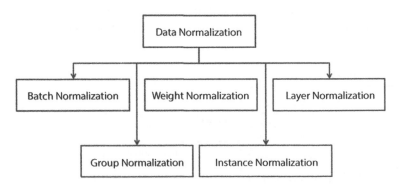

Figure 5.5 Data normalization.

In DL models, there are several ways to normalize that may be applied. The list is as follows:

1) Batch Normalization is one of the most often used normalization procedures for DL training. It permits quicker and stable training of deep neural networks by stabilizing the distributions of layer inputs throughout the training phase. Internal covariant shift is a key component of this method. Wherever internal covariate shift is suggested, that the amendment within the distribution of layer inputs is caused once the preceding layers are updated. The internal covariant shift must be condensed in order to improve the model's training. By adding network layers that limit the means and variances of the layer inputs, batch normalization reduces the internal covariate shift [16, 18].

2) Weight normalization is a technique for reparameterization of the load vectors in a deep neural community that goes with the aid of decoupling the duration and the directions of weight vectors. Weight normalization, according to the authors, is a technique for enhancing the optimization of a neural community model's weights [18].

3) Layer normalization is a technique for speeding up the education pace for diverse neural community models. Dissimilar batch normalization, this approach immediately calculates the normalization facts from the accumulated inputs to the hidden layer's neurons in real time. Layer normalization was created to overcome the disadvantages of batch normalizing, such as reliance on mini-batches.

4) As an alternative to batch normalisation, group normalisation may be described. This method divides the channels into companies and computes for normalisation (i.e. normalising the functions inside each company) within each business [23]. Dissimilar batch normalization, organization normalization is unaffected by batch sizes, and its correctness is consistent over a wide range of batch sizes.

5) Instance normalization, additionally referred to as evaluation normalization is nearly much like layer normalization. Dissimilar batch normalization, for example, normalization is implemented to an entire batch of photos as an alternative for a solitary one [24].

5.5.4   Data Splitting

In ML, data is the most important factor in resolving any problem. ML models without accurate data are akin to bodies without souls. Feeding enormous data to ML model is alone not sufficient but high quality data must be ensured to have accurate results [25].

The most difficulty faced by DL/ML specialists is dividing the data for training and testing. If the data set splitting is inappropriate in dividing it into training and testing sets, it can cause unexpected behavior from the model [26]. This may make the model to result in over-fitting or under-fitting of the given data, finally making it end up with biased results. The data set is generally divided into three sets i.e., train, test, and dev set (development set). The dev set is also called the holdout cross-validation set. Let us understand now what these sets are:

1. Train Set: This partition is used as an input to the model, where the model is required to learn from this input split.
2. Holdout Cross-validation set: The model is validated using this set to test whether it is trained appropriately or not. The model is said to be overfitted if it results in high variance. High variance may be because the difference of error from training set and holdout cross-validation set is very large.
3. Test Set: The model developed is now tested using this set. It gives a clear picture of how good the developed model is in predicting something, which is incorrect. It makes use of different performance measures for prediction.

One of the most contentious issues is the size of the train, development, and test sets. Though an 80:10:10 train, Cross validation, test set ratio is suitable for basic ML tasks, in today's big data environment, 20% of a dataset is a massive dataset. This data is easily used for training purposes, allowing the model to learn more complex attributes.

5.6   Result Analysis

The foremost step of this model is data collection, second step is data pre-processing, third is feature extraction, and finally disease detection from images.

5.6.1   Data Collection

Several plant leaf visuals with a resolution of 224*224 pixels are included in the data set. The collection includes leaf images from a variety of plants with various illnesses. The photos [27] are used to classify healthy and unhealthy processed leaves. For the crops indicated above, comprehensive leaf information is acquired from a variety of sources. Kaggle data set [29] is one of the sources. Photographs are obtained via the internet or simply taken using any camera-like gadget.

5.6.2   Feature Extractions

Various areas of research should be explored while picking feature characteristics depending on the numerous layers and constraints. In which it has a significant influence on system performance and growth. Due to the concern that each connection has its unique sort of architecture, they all have the equal goal of identifying new illnesses and determining their correctness and tangle. Individual object indicators will be combined with some of the model's object extractors. The dataset can be treated to a number of characteristic extracting, abatement, and selection procedures in order to decrease its intricacy by selecting the most important properties. Disclosure accuracy is occasionally affected by the equilibrium ratio of the dataset used for testing and outcome. Any asymmetrical dataset is balanced using hybrid balancing approaches before being given to classification [28].

5.6.3   Plant Leaf Disease Detection

The suggested practical examination methodology is run on data collection of 7,000 plant leaf visuals from the Kaggle collections [29]. It was discovered that the dataset comprised 7,000 digital images for training purposes, 1,000 digital images for testing purposes, and 3,000 digital images for validation purposes. The experiment was developed in the Windows operating system environment with Python version 3. By using known packages, namely, Numpy, Pandas, sklearn, TensorFlow, Keras, and matplotlib for plotting visualizations.

The Reset 152 v2 model (shown in Figure 5.6) was used to train the image dataset.

From the suggested experiment, 95% accuracy was obtained on the test, where the loss function model secured 0.186 and the accuracy function model secured 0.949 in 197 ms along with repetition steps. The leaf illness detection results of a plant using above mentioned techniques. Figures 5.7 and 5.8 show the supported graphs for accuracy epoch and loss function.

Figure 5.6 95% accuracy obtained with ResNet152V2.

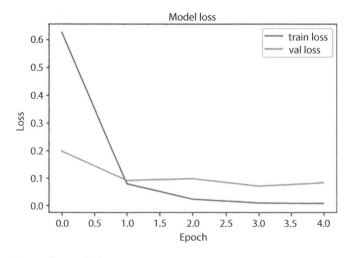

Figure 5.7 Curve for model loss accuracy.

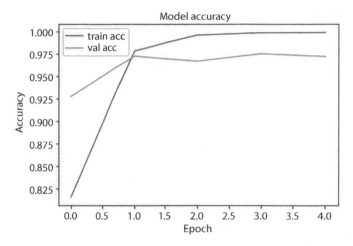

Figure 5.8 Curve for test accuracy.

5.7   Conclusion

There exist many methods for plant leaf disease detection and a lot of research is ongoing. But still there exists no proper solution, which is suitable to detect all types of diseases. In this work, a methodology of DL implemented ResNet architecture aimed at detecting most common diseases occurring on a plant leaf is a three-step procedure. At the first step, a plant leaf, which is required to be tested, its image is taken and it is processed. In the second step, ResNet is used for feature extraction and image classification in order to classify whether a leaf is healthy or unhealthy. The CNN strong feature extraction capability has given an accuracy of 95% compared to existing models with model loss function occurring with 0.1861. DL methods or ResNet outperforms the traditional methods, which have several steps in disease detection and pest identification. The required time to train the model was much less than that of other machine-learning approaches. This model can be extended for live capturing of plant images and sending them to labs through cloud and receiving the result immediately. This method has more equipment cost but the farmers do not have to wait for the solution. Even though a lot of research is going on to suggest the farmers for proper pesticides still there are problems that are yet to be solved.

References

1. Saharan, V. *et al.*, Viral, fungal and bacterial disease resistance in transgenic plants, in: *Advances in Plant Breeding Strategies: Agronomic, Abiotic and Biotic Stress Traits*, J. Al-Khayri, S. Jain, D. Johnson (Eds.), Springer, Cham, 2016, https://doi.org/10.1007/978-3-319-22518-0_17.
2. https://www.planetnatural.com/pest-problem-solver/plant-disease/.
3. Verma, K., Bhardwaj, S., Arya, R., Islam, M.S.U., Bhushan, M., Kumar, A., Samant, P., Latest tools for data mining and machine learning. *Int. J. Innov. Technol. Exploring Eng.*, 8, 9S, 18–23, 2019.
4. Ghaiwat Savita, N. and Parul, A., Detection and classification of plant leaf diseases using image processing techniques: A review. *Int. J. Recent Adv. Eng. Technol.*, 2, 3, 2347–812, 2014.
5. Khirade, S.D. and Patil, A.B., Plant disease detection using image processing. *2015 International Conference on Computing Communication Control and Automation*, IEEE, 2015.
6. Kulkarni Anand, H. and Ashwin Patil, R.K., Applying image processing technique to detect plant diseases. *Int. J. Mod. Eng. Res.*, 2, 5, 3661–4, 2012.
7. Sabah, B. and Navdeep, S., Remote area plant disease detection using image processing. *IOSR J. Electron. Commun. Eng.*, 2, 6, 31–4, 2012.
8. Bhanu, B. and Peng, J., Adaptive integrated image segmentation and object recognition. *IEEE Trans. Syst. Man Cybern. Part C*, 30, 427–41, 2000.
9. Singh, V. and Misra, A.K., Detection of plant leaf diseases using image segmentation and soft computing techniques. *Inf. Process. Agric.*, 4, 1, 41–49, 2017.
10. Jain, A., Sarsaiya, S., Wu, Q., Lu, Y., Shi, J., A review of plant leaf fungal diseases and its environment speciation. *Bioengineered.*, 10, 1, 409–424, 2019.
11. Akila, M. and Deepan, P., Detection and classification of plant leaf diseases by using deep learning algorithm. *Int. J. Eng. Res. Technol. (IJERT)*, 6, 7, 1–5, 2018.
12. Mahmudul Hassan, Sk, Maji, A.K., Jasiński, M., Leonowicz, Z., Jasińska, E., Identification of plant-leaf diseases using CNN and transfer-learning approach. *MDPI Electron.*, 10, 12, 1–19, 2021.
13. Mohanty, S.P., Hughes, D.P., Salathe, M., Using deep learning for image-based plant disease detection. *Front. Plant Sci.*, 7, 1419, 1–10, 2016.
14. Liu, J. and Wang, X., Plant diseases and pests detection based on deep learning: A review. *Plant Methods*, 17, 22, 1–18, 2021.
15. Guo, Y., Zhang, J., Yin, C., Hu, X., Zou, Y., Xue, Z., Wang, W., Plant disease identification based on deep learning algorithm in smart farming. *Hindawi Discrete Dyn. Nat. Soc.*, 2020, 2479172, 1–11, 2020.
16. Shah, D., Trivedi, V., Sheth, V., Shah, A., Chauhan, U., ResTS: Residual deep interpretable architecture for plant disease detection. *Inf. Process. Agric.*, 9, 2, 1–12, 2021.

17. Siva Krishna Reddy, A.V., Srinuvasu, M.A., Manibabu, K., Sai Krishna, Ch B V, Jhansi, D., Image based bird spices identification using deep learning. *Int. J. Creat. Res. Thoughts (IJCRT)*, 9, 7, 319–322, 2021.

18. Elshennawy, N.M. and Ibrahim, D.M., Deep-pneumonia framework using deep learning models based on chest X-ray images. *Diagnostics (Basel)*, 10, 9, 649, 2020 Aug 28.

19. Zhang, K., Sun, M., Han, T.X., Yuan, X., Guo, L., Liu, T., Residual networks of residual networks: Multilevel residual networks. *IEEE Trans. Circuits Syst. Video Technol.*, 28, 6, 1303–1314, June 2018.

20. He, K., Zhang, X., Ren, S., Sun, J., Deep residual learning for image recognition, in: *Proceedings of the IEEE Conference on Computer Vision and Pattern Recognition*, pp. 770–778, 2016.

21. Geetha, G., Samundeswari, S., Saranya, G., Meenakshi, K., Nithya, M., Plant leaf disease classification and detection system using machine learning. pp. 1–12, IOT Publishing, International Conference On Computational Physics in Emerging Technologies (ICCPET), Mangalore, India, 2020.

22. Chohan, M., Khan, A., Chohan, R., Katpar, S.H., Mahar, M.S., Plant disease detection using deep learning. *Int. J. Recent Technol. Eng.*, 9, 1, 908–914, 2020.

23. Saha, M. and Sasikala, E., Identification of plants leaf diseases using machine learning algorithms. *Int. J. Adv. Sci. Technol.*, 29, 9, 2900–2910, 2020.

24. Ramesh, S., Hebbar, R., Niveditha, M., Pooja, R., Prasad Bhat, N., Shashank, N., Vinod, P.V., Plant disease detection using machine learning. *International Conference on Design Innovations for 3Cs Compute Communicate Control*, pp. 41–45, 2018.

25. Godi, B., Viswanadham, S., Muttipati, A.S., Samantray, O.P., Gadiraju, S.R., E-healthcare monitoring system using IoT with machine learning approaches. *2020 International Conference on Computer Science, Engineering and Applications (ICCSEA)*, pp. 1–5, 2020.

26. Samantray, O.P. and Narayan Tripathy, S., A knowledge-domain analyser for malware classification. *International Conference on Computer Science, Engineering and Applications (ICCSEA)*, pp. 1–7, 2020.

27. Pawar, S., Bhushan, M., Wagh, M., The plant leaf disease diagnosis and spectral data analysis using machine learning – A review. *Int. J. Adv. Sci. Technol.*, 29, 9s, 3343–3359, May 2020.

28. Nalavade, A., Bai, A., Bhushan, M., Deep learning techniques and models for improving machine reading comprehension system. *IJAST*, 29, 04, 9692–9710, Oct. 2020.

29. www.kaggle.com/tomato-leaf-disease-detection.

30. Godi, B., Muttipati, A.S., Rao, M.P., Satyanarayana, G.V., Padmaja, G.M., Gadiraju, S.R., ResNet model to forecast plant leaf disease. *2022 International Conference on Computing, Communication and Power Technology (IC3P)*, pp. 38–43, 2022.

Smart Irrigation and Cultivation Recommendation System for Precision Agriculture Driven by IoT

N. Marline Joys Kumari[1], N. Thirupathi Rao[2] and Debnath Bhattacharyya[3]*

[1]Department of Computer Science and Engineering, Anil Neerukonda Institute of Technology and Sciences, Sanghivalasa, Visakhapatnam, AP, India
[2]Department of Computer Science and Engineering, Vignan's Institute of Information Technology, Visakhapatnam, AP, India
[3]Department of Computer Science and Engineering, Koneru Lakshmaiah Education Foundation, Vaddeswaram, Guntur, Andhra Pradesh, India

Abstract

Agriculture is a distinct sector of the economy from the rest of the country's economy. India is ranked second in the world for food production. Irrigation, fertilizer and crop rotation are the three most critical aspects of farming to consider. Farmers are unable to produce as much food as they could in the modern world due to advances in technology. The primary objective should be to maximize the usage and collaboration of new technology in agriculture. The relevant vector analysis (RVA) algorithm used for accurate farming of farming cultivation recommender and smart irrigation system initial module is called recommendations. It is divided into two sections: agricultural factors and vector models of importance. Precision agriculture enables farmers to manage irrigation in such a way that they consume less water while producing more food. By utilizing this system's many sensors and forecasts, farmers can determine how much water plants require to survive. This is how the forecasting process works, the soil moisture content, temperature, and humidity of the soil are utilized to predict what would happen. The experiments demonstrate that smart irrigation, which produces a higher agricultural yield while using less water than conventional irrigation, is the most efficient technique to increase crop yields while conserving water.

Corresponding author: debnathb@gmail.com

Ashok Kumar, Megha Bhushan, José A. Galindo, Lalit Garg and Yu-Chen Hu (eds.) Machine Intelligence, Big Data Analytics, and IoT in Image Processing: Practical Applications, (123–150) © 2023 Scrivener Publishing LLC

Keywords: Soil moisture content, agricultural yield, Internet of Things, machine learning, smart irrigation

6.1   Introduction

According to the United Nations, an estimated 83 million individuals are added to the world's population each and every year [1]. The world's population is predicted to reach 9.8 billion people by 2050. Researchers have long dismissed overpopulation worries, yet the fast increase in human population is posing significant and urgent challenges to the world's economies, agricultural systems, and social structures, to name a few. Their challenge is to increase the quantity of food they can produce on a limited area of land while simultaneously coping with adverse weather conditions, climate change, environmental degradation, and market volatility [2, 3]. Farming must adapt to emerging technology, such as the Internet of Things (IoT), artificial intelligence (AI), machine learning (ML), and robots among other things in order to remain competitive. These new technologies, such as IoT, not only help us better understand and manage growth variables on farms, such as irrigation, water and soil conservation, and fertilizer use minimization, but also change the way in which this study is about thinking about agriculture in general, according to the World Bank.

According to research by the Confederation of Indian Industry (CII) [4], India's agricultural technological momentum is increasing. In the agriculture sector, more than 500 Aggrotech start-ups are using technology, such as AI, ML, and other advanced techniques to increase efficiency, productivity, yield, agricultural finance, and other critical agricultural services. A rising number of growers are embracing new technologies made available by IoT in order to satisfy the demands of a growing population and improve the efficiency of their IoT operations. Agriculture will be able to reduce waste and boost output as a result of new "smart farming" applications that make use of IoT technology. Irrigation is very vital in agriculture. Before the 19th century [5], inhabitants in ancient Egypt employed a variety of methods to irrigate their plants. The Egyptians were the first people to install sprinklers in their houses, and they did it thousands of years ago.

There are many nozzles on this sort of sprinkler head, and the tube is in the form of a bow, which travels back and forth. Due to technological advancements, basic operations like irrigation became less repetitive and time-consuming in the 1950s, resulting in the creation of an autonomous irrigation system. Sprinkler systems were originally utilized on tiny lawns in the 1950s, when they were first introduced. Because this autonomous watering system does not need you to keep an eye on a particular area of land, it is quite convenient. The smart irrigation system is the next big thing in the irrigation industry. It makes it possible to reduce water use by taking crops into consideration. The most prevalent method in which water is used is for irrigation. Intelligent Irrigation is a method of artificially watering the soil in order for crops to flourish. The use of automatic systems makes it simpler to manage and monitor the flow of water to crops, resulting in less water being used by the plants. According to experts, smart irrigation systems and controllers are preferable to ordinary irrigation controllers when it comes to water conservation.

The findings of a large amount of controlled study have shown that significant amounts of water may be conserved, ranging from 40% to 70%. Smart irrigation controllers, according to research conducted by the Irrigation Association (IA) and the International Center for Water Technology at California State University [6], Fresno, may save up to 20% more water than regular irrigation controllers, which consume less water. IoT is responsible for the operation of the smart irrigation system IoT. When data is transported over a network without needing to be exchanged between people or computers, this is referred to as IoT. An irrigation system based on this approach would be able to interface with computer equipment such as sensors and also transmit real-time data to the cloud through the internet, allowing it to function properly and efficiently. Because of this, the individual who is utilizing this system [7] will not have to go around on land that is utilized for agricultural or irrigation purposes. However, plant irrigation is sometimes a lengthy and time-consuming task that needs the participation of a large number of people in order to be completed in an acceptable period. Previously, all of the jobs were completed by a single individual. In order to reduce the number of employees and the amount of time spent watering plants, more and more technologically advanced methods will be used. Control over these systems is quite restricted, yet a large amount of money is squandered as a result.

Water is one of the resources that humans depend on a daily basis. Using a large amount of water to water the plants is one method of doing so. Because the volume of water provided exceeds the amount required by the plant, this technique results in a significant amount of money being lost [8]. Water is currently seen as a nonrenewable, limitless resource that may be exploited in massive quantities without depleting its supply. In Malaysia, on the other hand, water is a premium, and so charges. Consequently, it is possible that it will become an expensive resource in the next several years as a consequence of these developments. Water is becoming more expensive, and the cost of labor is becoming more expensive as well. Therefore, if you do not take use of these resources, the same product will cost you more money. There is a significant likelihood that technological advancements will aid in cost reduction and resource conservation [9, 10]. Another issue arises when agricultural property is located a considerable distance away from a person's residence. The water pumps were occasionally turned on and off many times a day, resulting in a large number of visits to turn on and off the pumps each day. In order to address this issue, the primary purpose of this project is to reduce agricultural water use while also establishing a system that is totally based on IoT with all of the devices being linked together.

It is possible to find and alter information with a single swipe of a finger. Consider the following items: (i) project cost savings, (ii) labor cost savings, and (iii) energy consumption reductions are all possible. It all boils down to how much you use and how long it lasts you (iv). There are two primary objectives: I am developing an irrigation system that conserves water, and I am also developing an application system that might aid in the development of an irrigation system that conserves water.

Agriculture accounts for a significant portion of Bangladesh's economy, and it is the primary source of income for the majority of the country's population. Bangladesh, on the other hand, must support and promote the use of modern agricultural technologies. Ahead with breakneck pace, cloud control technology is taking over and strengthening the reputation of every firm in which it is implemented. In recent years, cloud computing and IoT technologies have grown increasingly prevalent in agriculture. Farmers are integrating these two technologies in order to enhance yields while also lowering expenses. In other words, precision farming is a slang term for intelligent farming. Precision agriculture makes use of a wide range of sensors to assist farmers in keeping an eye on their crops and adapting to changes in the environment, among other things. Sensors, such as location sensors, optical sensors, electrical sensors, mechanical sensors, dielectric soil moisture sensors, airflow sensors, agricultural weather stations, humidity sensors, proximity sensors, and pH sensors, are all often used in agricultural applications. Farming is one of the industries that consumes a significant amount of water. Water waste is a significant issue in agriculture. When a large amount of water is sprayed on the crops. When compared to conventional irrigation systems, a smart irrigation system regulates the amount of water that is delivered to the fields and crops depending on their specific needs and requirements. Users of smart irrigation systems may keep a watch on their irrigation, start and stop the process with a single button press, and get updates regarding moisture levels, weather conditions, and the health of their crops. Smart irrigation systems are becoming more popular. A complex irrigation system is thus required in order to satisfy these several objectives.

6.1.1   Background of the Problem

6.1.1.1   Need of Water Management

Water management is critical in countries with limited water resources. This also has an effect on agriculture, as a significant amount of water is used for that purpose. Global warming's [11] potential implications necessitate the development of water adaptation techniques to assure the availability of water for food production and consumption. As a result, studies targeted at reducing water consumption during the irrigation process have expanded in recent years. Commercial sensors for agriculture irrigation systems are prohibitively expensive, making this type of equipment unaffordable to small farmers. Manufacturers, on the other hand, are currently developing low-cost sensors that may be connected to nodes to enable the implementation of economical irrigation management and agriculture monitoring [12] systems due to recent advancements in IoT and wireless sensor networks (WSN) technologies that can be used to construct these systems. In this study, a survey outlining the current state of the art in the field of smart irrigation systems is presented. This chapter establishes the irrigation system parameters that are monitored in terms of water quantity and quality, soil properties, and weather conditions. This chapter summarizes the most frequently used nodes and wireless technologies. Finally, the work will explore the implementation problems and best practices for sensor-based irrigation systems.

6.1.1.2   Importance of Precision Agriculture

Precision agriculture techniques enabled by the Internet of Things provide farmers with valuable tools for maximizing each farming operation. These technology-driven efforts aim to boost agricultural yields and income while reducing the amount of traditional agricultural inputs required (water, fertilizer, insecticides, and herbicides). Tractors with Global Positioning System (GPS) [13] systems, for example, let farmers to sow crops more efficiently and optimize travel between and within their fields, saving time and fuel. Weather, soil, pest, and hydration data collected by sensors on farm equipment may be sent to a centralized smart farm platform for analysis and predictive agricultural decisions. Fields may be levelled using lasers controlled by the Internet of Things, allowing for more effective water application and less liquid waste to flow into nearby streams and rivers. Artificial intelligence (AI) might be used to forecast pest behavior, which could aid pest management planning. A pest control that is effective results in less crop and environmental harm. To distinguish the weed from the crop, a combination of remotely sensed data, effective image classification systems, climatic data, and other relevant data points can be used. This will limit the application of weedicide to locations where it is required. Crop health [14] and insect assaults may both be tracked via remote satellites.

When it is time to spray fields with crop protection agents, automated drones can take the place of crop dusters, saving money and eliminating the risk of piloted aircraft flying too close to the ground. Crop dusting drones may also collect continuous photographs and video of fields while flying, allowing farmers to check plant health without dispatching scouts. Precision agriculture farming technologies are a benefit to farmers in terms of resource efficiency and effectiveness, and they have a lot of promise to make agriculture more sustainable and increase food availability. Throughout the previous few decades, agricultural implements have undergone numerous technical revolutions, becoming more industrialized and reliant on cutting-edge technology. The introduction of smart agriculture gadgets enables farmers to keep optimal control over the growth of crops and livestock. IoT and its associated devices have begun to pervade every aspect of human lives, from healthcare, automation, automotive, and logistics, to smart cities and industrialization. IoT ushers in a new era of precision agriculture sample collection.

6.1.1.3   Internet of Things

Precision agriculture is a broad word that encompasses all services that rely on digital systems and technological advancements to meet the needs of modern farmers [15] for production optimization, waste reduction, and environmental quality maintenance. Farmers can use IoT sensors implanted in the crop to assist them in properly allocating herbicides and fertilizers, as well as the following support:

- Optimization of harvest times
- Crop sustainability
- Monitoring of greenhouse temperatures, light levels, and relative humidity
- Measuring the soil's quality and humidity level

Numerous smartphone applications have been identified to include IoT concepts, data aggregation, and speed of processing, which may bring the data up to date. Information can be sent to small farmers regarding watering, sowing, fertilizing, and weeding. These apps collect data from sensors, most notably remote sensors and weather stations. It enables in-depth data analysis and generates useful recommendations.

After the invention and use of IoT technology, seeding is no longer a guessing game. A programmable intelligent gadget can determine the optimal location for a seed to be planted and cultivated. When the harvest is mature, the crops are collected by intelligent tractors with greater efficiency and care. At the moment, the amount of energy required for crop cultivation by fixing tractor damage ranges between 80% and 90%. Work has been accomplished the following by utilizing the GPS-guided steering system and route planning based on the input data:

- Cost savings in fuel
- Accuracy of operations is increased

Small-scale farmers may benefit in a variety of ways from applications built for them, the identification of plant illnesses and their transmission to experts for correction. The amount of nutrients required by fertilizers is determined by the color of the leaves and the condition of the soil. Additionally, the soil's pH value and other parameters can be determined. The plants' water requirements were determined based on observations of their leaves. Crop harvesting readiness using ultraviolet and white light-based pictures can assist in preventing maturity.

6.1.1.4   Application of IoT in Machine Learning and Deep Learning

ML-based agriculture applications are highly efficient and simple to install. The ML [16] process is divided into three stages: data collection, model creation, and generalization. When human talent is weak, machine learning algorithms are routinely used to handle complex problems. In agriculture, machine learning can be used to evaluate soil parameters like organic carbon and moisture content, anticipate crop yields, identify disease and weeds in crops, and identify species. Deep learning improves standard machine learning by increasing model complexity and modifying input with various functions that enable data representation in a hierarchical form, via multiple layers of abstraction, depending on the network architecture utilized. Feature learning, or more accurately, the autonomous extraction of features from raw data, is a significant advantage of deep learning. A fundamental component of the deep learning model, which uses the homogeneous qualities of an agricultural field to detect remote, heavily obstructed

[17], and new objects, is the ability to recognize unknown objects such as anomalies rather than a collection of current items. Blockchain technology has fast become a key component of a wide range of precision agriculture applications.

Researchers have proposed building blockchain-based IoT systems for precision agriculture in response to the requirement for intelligent peer-to-peer systems capable of verifying, securing, monitoring, and analyzing agricultural data, as shown in Figure 6.1. Blockchain technology is critical for transforming traditional methods of storing, sorting, and disseminating agricultural data into a more dependable, immutable, transparent, and decentralized digital format. When IoT and blockchain [18] are utilized in precision farming, a network of smart farms is developed. This linkage gives the system more autonomy and flexibility.

IoT, Data Science, Machine Learning, Deep Learning, and Blockchain are all technologies that deal with data and are particularly useful for understanding and providing actionable insights. As a result, these advanced technologies are used in a variety of agricultural practices, such as determining the best crop for a specific location [19], identifying factors that can devastate crops, such as weeds, insects, and crop diseases, to gain insights into crop growth and aid in decision-making, and determining the best crop for a specific location. Land management, soil preparation, water monitoring, weed identification, pesticide recommendation, disease diagnosis, and cost calculation are the seven key phases in agriculture. Land management is the process of keeping track of physical elements including weather and geological features. This is significant since climatic conditions affect agricultural productivity all across the world. Rainfall is an important part of the earth's climate system, and its unpredictability has a big impact on agricultural production, water management, and biological processes. As a result, a technology that can predict rainfall in advance is required to make crop management easier.

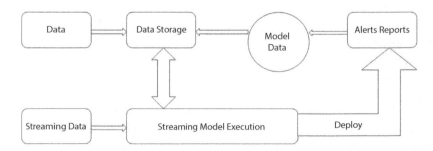

Figure 6.1 IoT analytics life cycle.

6.2   Related Works

Farmers confront a difficult problem when it comes to recognizing agricultural insects, as pest infestations damage a significant amount of the crop and degrade its quality.

This effort was merely an implementation of a model using the dataset. The web interface must be designed in such a way that even the most inexperienced users can utilize it effectively. All inputs must be entered manually for the model to forecast the crop. Utilizing web scraping, the suggested work assists in obtaining temperature and humidity measurements. As a result, no manual entry of values is required. Temperature and humidity data are automatically retrieved and fed into the optimal model, which consists of 10 algorithms with hyper parameter tweaking. The projected outcomes, coupled with metadata, are displayed in the online interface, allowing the user to more effectively comprehend the results. Growing degree days (GDD) can be calculated using the base temperature of a given crop.

Compared to previous methodologies, these new mathematical formulas can quickly and accurately calculate the base temperature. The GDD base temperature for any crop at any developmental stage can be calculated using these calculations

In this study, machine learning approaches, such as decision trees, K closest neighbors, linear regression models, neural networks, naive Bayes, and support vector machines, were utilized to suggest a crop to the user. In comparison, it exposed people to more algorithms. The production value was forecasted using a linear regression model in conjunction with meteorological data like as rainfall, temperature, and humidity. As shown in Figure 6.2, none of these algorithms scored higher than 90%.

K-means and convolutional neural network (CNN) models can be used to detect weeds planted alongside soybeans. K-means clustering was used to identify the image's features, and a CNN was used to classify the weeds and soybeans. It also means that fine-tuning the CNN model can boost accuracy. In agricultural areas, the CNN model is a useful tool for weed detection. The pictures and augmentations are clustered first using K-means, and then the CNN model is utilized to precisely identify the weed. A Resnet152V2 pretrained model is used in the suggested method, which includes important layers, such as a skip layer and an identity layer. These layers' primary objective is to ensure that the output image matches the input. This increases precision and ensures accurate predictions. The suggested model not only assists the spectator in anticipating the image but

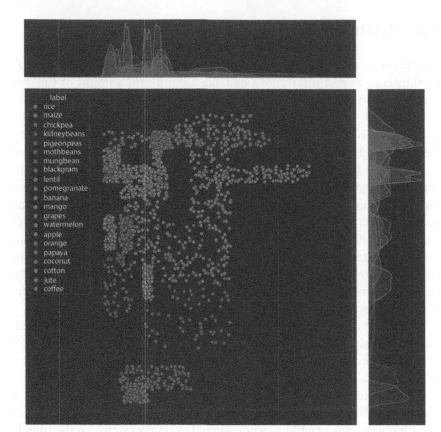

Figure 6.2 Ratio of rainfall versus humidity across various crops.

also includes information on the herbicides that can be employed, which is useful information. To detect weeds, existing deep learning techniques are applied.

The various machine learning and deep learning strategies for weed recognition are discussed in this chapter. It mainly focuses on pretrained models. As a result, pretrained models offer a number of benefits and can be used to classify images. It also goes through how to work with datasets and how to optimize them for model building. As a result, a variety of public datasets is available on various platforms. Image scaling, data augmentation, and image segmentation are some of the strategies that can lead to correct classifications, according to the study, and pretrained models have a stronger tendency to improve their accuracy. Because this research shows how to apply deep learning techniques, the suggested model comprises image resizing and data augmentation procedures before building the deep learning model for weed prediction.

6.3   Challenges of IoT in Smart Irrigation

Farmers that use IoT technology may increase their harvests, increase their profits, and improve their environmental stewardship all at the same time. Precision agriculture, which makes use of IoT technologies, may help farmers raise their yield while also lowering their operating costs. IoT in agriculture comprises wirelessly linked specialized equipment, software, and other kinds of information technology, as well as other types of information technology. According to BI Intelligence, the number of IoT devices in agriculture is estimated to reach 75 million by 2020, growing at a rate of 20% [20] each year. By 2025, it is projected that "smart agriculture" would be widely implemented around the world. When it reaches $15.3 billion, it will represent a huge rise from the $5 billion it reached in 2016. Smart farming facilitated by IoT helps growers and farmers to decrease waste and increase production in a variety of ways, including reducing the amount of fertilizer needed, the number of farm vehicles on the road, and the amount of water and electricity utilized. Smart agricultural systems based on IoT deploy sensors to monitor fields and automate irrigation chores such as watering and fertilizing (light, humidity, temperature, soil moisture, crop health, and so on). Landowners may monitor the progress of their businesses from anywhere in the globe. As a last resort, they may undertake data-driven operations either manually or automatically, depending on the situation. A farmer, for example, may use sensors to water the land if the moisture level in the soil drops below a certain threshold. Farming that is efficient outperforms traditional farming and saves money is known as smart farming or precision farming.

IoT may have a substantial impact on farming in a variety of other ways as well, but these are the most crucial. This kind of farm management consists of a number of essential components. Sensors, controllers, robots, and self-driving automobiles are just a few of the technologies that are available. Among the additional technologies that might be employed are variable rate technology, motion sensors, button cameras, and wearable gadgets, just to name a few examples. The health of a firm, the productivity of its employees, and the effectiveness of its equipment may all benefit from the monitoring provided by the use of this data. A deeper knowledge of industrial production may pave the way for more imaginative approaches to the marketing of consumer products in the future.

Global population is shown in Figure 6.3, and there is not enough land or resources to provide enough food to satisfy everyone's needs. Natural resources, such as fresh water and agricultural areas suitable for food production, are in limited supply, exacerbating the already dire situation. Individuals' working conditions in agriculture are also a source of worry. In addition, the number of people working in agriculture has decreased in the majority of countries throughout the world. As the number of people engaged in agriculture has declined, physical labor has become less important to maintain the industry. As a result, agriculture is becoming more reliant on high-speed Internet connections.

Drones are used in agriculture to harvest crops. Agricultural drones undertake a wide range of tasks, including crop health assessment, irrigation, crop monitoring, spraying and planting crops, as well as soil and field analysis. Drones are also used in forestry and forestry-related operations. Putting up geo-fences and keeping an eye on the cattle are important tasks. In certain cases, keeping track of precise location, health, and well-being may be accomplished by simply installing an IoT app on Individual smartphone. Along with assisting in the prevention of disease transmission, this information also has the added benefit of saving money by lowering labor costs. In greenhouses, there is an increase in intelligence. In a smart greenhouse driven by IoT, human interaction is no longer required to maximize efficiency IoT.

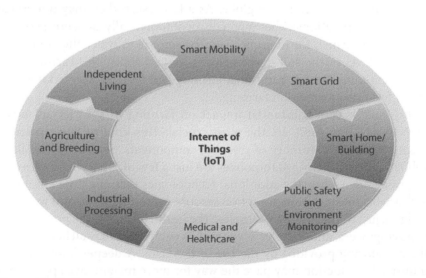

Figure 6.3 Challenges related to IoT in smart irrigation.

Predictive analytics is being used in smart farming. Farmers can make better decisions about how to produce, store, market, and protect their crops if they are able to anticipate crop yields in advance of time. Artificial networks are used to anticipate agricultural productivity based on data collected from sensors installed on fields. These data include, among other things, information on the soil, the temperature and pressure, the amount of rainfall, and the humidity in the air. They may be able to get trustworthy soil data by using a dashboard or a mobile app created specifically for farmers. Increasing numbers of farmers are starting to see how IoT might help them be more efficient and save money in the field.

6.4   Farmers' Challenges in the Current Situation

As India's population continues to rise, the agricultural sector is under increasing pressure. A lack of modern farming technology in India has resulted in a high rate of disease losses for farmers. Consumers and the environment are both harmed by excessive usage of insecticides and pesticides. The aim is to tackle problems by providing farmers with timely information and detecting diseases in their early stages. Chemical pesticides pose a serious threat to both farmers and the general public. Effects on the natural world soil, water, turf, and other vegetation can all be harmed by pesticides. Pesticides can harm birds, fish, beneficial insects, and nontarget plants in addition to destroying insects and weeds. Pesticides classified as insecticides have the highest acute toxicity levels. Employers and employees need to know that investing in preventative health measures and health promotion programs is a worthwhile endeavour that benefits both parties and contributes to long-term economic growth. As a result of the limited amount of direct and/or inferential information. Pesticides appear to be fraught with uncertainty when it comes to long-term exposure. Thus, there is a strong need for developing and disseminating community-based pesticide education programs based on knowledge, aptitude and behaviors. Thus after experimentation, study found that reducing pesticide use using innovative technology solutions may lead to a total abandonment of chemical compounds in many circumstances, such as urban green spaces. In light of the dangers that chemical pesticides pose to human health and the environment, it is obvious that a new approach to farming is required. Chemical pesticides must be drastically reduced in this new concept for it to have positive effects on human health, environmental quality, and the economy. To create a framework for farmers that can serve as a disease detection system, bringing experts to the farmers. Plant recognition computer vision (CV) algorithms can be used for a variety of purposes, including the following:

a) The creation of a computational approach for disease prediction.
b) Implementation of freshly designed software that is user-friendly.

Deep learning methods would be used to train models that would detect diseases in plant leaves, IoT sensors and devices would be used to provide up-to-date information to farmers without the need for an always connected device, and cloud computing would be used to update the models using new information submitted by farmers and verified by experts.

a) Providing farmers in impoverished countries with reliable internet connectivity can be difficult.
b) Cellular coverage on farms is often limited.
c) They are vulnerable to Internet outages caused by weather.

6.5   Data Collection in Precision Agriculture

The crop growth characteristics observed in the field, such as soil temperature, soil humidity, soil wetness, and soil pH, are gathered as a dataset and used as training samples. Sprinkler irrigation was used in three of the subareas, while drip irrigation was used in the other two. The data was gathered on a cotton-growing field in Southern Andhra Pradesh, with different irrigation treatment subplots. Furthermore, statistical downscaling of precipitation was examined for anticipating precise measurements and giving decision-making help for irrigation resource optimization. Many soil properties are measured by the sensors. The measured observations are sent to the server via the sensors that are directly linked to the Arduino UNO. Water is by far the most important concern in this case.

6.5.1   Algorithm

A feature is a characteristic of an object of interest that distinguishes it from other types of objects in an image. Color is a widely used characteristic in the sorting and analysis of agricultural products and foods. Color-defined images have a number of advantages, including their robustness, efficiency, and simplicity of processing. Generally, the color attribute is most commonly used for discriminating objects.

Depending on the type of objects to be recognized, object detection in an image can range from simple to sophisticated.

Begin
    Collection of field and web datasets as a starting point;
Begin/Training
    SLR train;
    PSM prediction;
    SVM should be trained;
    PSM prediction;
    Training has come to an end.
Begin/Prediction
    Determine the PSMD (Predicted Soil Moisture Difference).
    P0, P1, P2, P3, P4, P5, P6, P7, P8, P9, P10, P
    PSMD for day 1, 2, and Pn, prediction difference 3…n;
    Set the Centroid value to Pi, and the Threshold T to Tmax.
        (i=0 to n; i++) for (i=0 to n; i++)
        K SLR should be used.
    GSM and MSE predictions.
    In the case of (Tmax>GSM),
    Irrigation that is smart ()
    Finally,
Finally,
End/Prediction
End

Objects with clear features are easy to recognize, while those with blurred features are difficult to detect. The major objective of the data collecting phase of the Precision Agriculture system is to collect data. Information can be thought of as data about items of interest within a particular domain. It could be soil, crop, or production data. This data is analyzed to find areas where methods should be changed for a better outcome and to avoid superfluous inputs. Numerous research has detailed how CV may be used to extract relevant information from agricultural photographs. The study employs a detection approach to determine the presence of items in photographs for the purpose of counting. Numerous research in agriculture over the last few years have successfully utilized CV. While CV has demonstrated superior performance in areas, such as fruit quality evaluation, disease detection, and flower processing, very few studies have employed CV to estimate area or production. CV is a critical component of numerous automated harvesting systems for agricultural goods such as roses and citrus. These works utilize image segmentation to extract information about the object of interest from photographs. Segmentation focuses on extracting foreground objects from background objects using a variety of criteria. Detection of disease using deep learning models has increased exponentially since two large (and open to use) datasets were published: Plant Village. While Digi Pathos is a relatively new dataset, it is used to train and evaluate deep learning models in a number of the articles described below. The used images downloaded from the Internet to train a Caffe-based deep CNN [19]. Using image augmentation, the dataset was increased in size. Caffe Net architecture was employed. An early study that took advantage of the plant village dataset constructed a deep learning model utilizing more than 54,000 photos. On a controlled dataset, they were 99.9% accurate. 31.4% accuracy was achieved when the model was evaluated on an entirely new batch of photos. Their deep learning models performed better when the training dataset was collected under more varied conditions such as different backgrounds, lighting, and so on. A CNN trained with transfer learning correctly identified 93% of the cassava in a training dataset of the crop. A CNN model called InceptionV3 was utilized as a starting point for transfer learning, and the dataset they used was their own.

A method developed uses deep CNNs to identify apple leaf diseases. Their own dataset had 13689 images, which they used to compare the performance of a deep CNN based on AlexNet with the performance of a deep CNN based on AlexNet.

In Figure 6.4, according to the precision agriculture cycle in IoT, transfer learning can be used to detect illnesses in the grain of millet. With the help of the VGG16 architecture and feature extraction, they were able to reach 95% accuracy in the transfer learning process. Deep learning was used to create a mobile app that can detect infections in cassava. It was trained on the COCO dataset using the Mobile Net architecture. When tested in the actual world, their model's accuracy suffered a 32% loss in performance, despite its good results under controlled conditions. For plant categorization, tested the applicability of transfer learning using four models and four datasets, respectively. They came to the conclusion that the models and datasets employed had no bearing on the performance.

The literature suggests that classification results can be improved even if promising findings have been obtained from various investigations. Using a neural network for categorization results in more accurate findings. Therefore, the current study is focusing on neural network categorization.

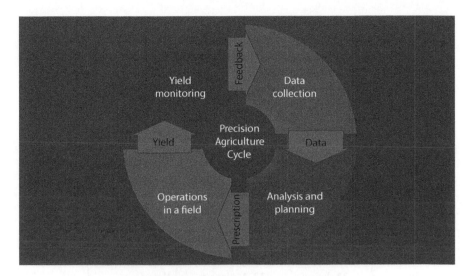

Figure 6.4 Precision agriculture cycle in IoT.

There are several processes involved in the categorization process for CV utilizing artificial neural networks. To estimate the current Production of flowers ready for harvest, these extracted objects must be divided into two classes: first, "Flower," and second, "Bud," in order to estimate the advance production for the following few days. A dataset of flowers and buds is created from the results of the object detection method.

The next phase in classification is to build a classifier that is more accurate than the previous one.

- Extraction of objects: A training image with objects has a consistent black background that contrasts with the objects. The training image's objects are segregated using the threshold segmentation method.
- Determining the regions of objects: With blob analysis, a pixel list containing the respective x-y coordinates for these regions is obtained in a n x 2 array, where n is the number of pixels in the object region, the first column contains the relevant x locations, and the second column has the corresponding y locations.
- Color values of pixels in an object region extracted: Once the pixels of object regions in a color image are detected, the RGB color values of these pixels are extracted independently for each item.

- Features: The features generate the vector source image from several objects to categories and each object contains thousands of pixels. The idea of simply using the color values of all pixels in all objects in the training neural network is deemed impractical in terms of memory and processing time.

6.5.1.1   Environmental Consideration on Stage Production of Crop

Artificial Neural Network (ANN), and a soft computing tool, has been found to be an effective classifier in a variety of research reviewed in the

Table 6.1 Environmental considerations impacting different stages of production.

Stages of cultivation	Temperature of the air	Temperature of the soil	Soil moisture	Humidity
Sprouting	26-33 degrees Celsius (Best). At least 18 degrees Celsius (Minimum)	23-28 degree Celsius (Optimum)	Based on the water absorbed	-
Growth	30-33 degree Celsius (Best) and poor when < 20 degree Celsius	23-29 degree Celsius (Optimum) and poor when < 21 degree Celsius	Adequate moisture is essential	Better
Ripening	Preferred cold nights or Optimum < 15 degree Celsius	Low temperature is best	Minimum moisture provokes	Preferred dry climate
Flowering	Preferred warm nights or with 18 degrees Celsius	Utmost in warm soil	Best in moist soil and stop on drought condition	Better
Over Ripening	Provokes at hot season	Favored by high temperature	Favored by water availability during dry season	-

literature study. As a result, an artificial neural network is used in this study to help with the categorization task, where the environmental consideration and their impact on different stages of the crop production was observed as shown in Table 6.1.

The basic goal of the suggested module, farming cultivation recommender and smart irrigation system, is to use the Relevant Vector Analysis (RVA) approach to examine favorable environmental conditions such as temperature and humidity to plan irrigation modes, either planned or automatic. As a result, farmers may determine the amount of water required for cultivation to maintain healthy plants while using the least amount of water possible.

6.5.2 Implementation Measures

6.5.2.1 Analysis of Relevant Vectors

The new RVA recommendation algorithm was used for all data analytics for the ARV recommendation System. The technique is a mixture of two well-known precision agriculture assessment methods: Root Mean Squared Error (RMSE) and Nash–Sutcliffe Efficiency (NSE). The unique algorithm predicts significant factors based on observed sensor values and delivers high yield farming recommendations, with the accuracy of observed sensor data determined using RMSE and NSE analysis.

The RMSE is a statistic that compares the observed and anticipated values. The residuals are the RMSE's deviations. Assume that the data samples used for estimation are calculated incorrectly. The RMSE is a crucial accuracy statistic used in many of the best models for forecasting a certain dataset. The amount of water in a container is a fraction between 0 and 1.

6.5.2.2 Mean Square Error

The predictor's quality is determined by the Mean Square Error. N denotes the predicted value, yp denotes the observed vector, and yp denotes the observed vector.

6.5.2.3 Potential of IoT in Precision Agriculture

By 2050, the world's current population of 8.3 billion people is expected to reach 9.7 billion. Each year, approximately 95 million individuals are added to the global population. This rise in population must be matched by an increase in food production. Agricultural output would have to increase by at least one-third in comparison to existing yields in order to meet food demand. Due to the inability of farm land acreage to keep up with population increase, farmers have begun to move to precision agriculture.

Precision agriculture is fundamentally about deploying intelligence to fields in order to provide farmers with visibility over crop yields and cost considerations as they produce crops. Smart gadgets IoT objects outfitted with sensors collect intelligence in real time and communicate it to data processing centres. Farmers are now adopting newly developed techniques to manage crops on micro-samples of their fields to assess soil composition, regulate fertilizer and pesticide kinds, and achieve accurate yield estimations.

- Fertilizers and the timing methods of application.
- Patterns of irrigation and precipitation.
- Temperature and relative humidity of the air.
- Composition and management of soil.
- Processes for crop care.
- The seeds that were used.

Precision farming would not be possible without IoT-enabled technologies, which accelerate farmers' awareness of the primary causes of crop yield changes, factors affecting measured yields, and production cost drivers. Measuring and collecting data from fields, crops, and the surrounding environment are critical capabilities afforded by mature IoT solutions.

The bulk of farms are situated in rural areas with limited access to digital networks. Sensor and device failure might be caused by frequent weather changes. These difficulties necessitate the adoption of powerful, simple, and easy-to-maintain IoT technologies. Sensors, control systems, robotics, autonomous vehicles, automated hardware, variable rate technologies, motion detectors, button cameras, and wearable gadgets are all part of this farm management strategy. This information may be used to track the company's overall health, as well as employee productivity and equipment efficiency. Product distribution strategy benefits from the ability to anticipate manufacturing output.

- **Drones in agriculture:** Drones are being deployed in agriculture to improve a range of agricultural activities, including crop health assessment, irrigation, crop monitoring, crop spraying, planting, and soil and field analysis.
- **Geofencing and livestock tracking:** Farm owners can utilize wireless IoT applications to collect information about

their cattle's location, well-being, and health. This information aids in the prevention of disease transmission while also cutting labor expenditures.

- **Greenhouses with Intelligence:** Smart greenhouse constructed with IoT monitors and adjusts the climate automatically, removing the need for manual intervention.
- **Smart farming with predictive analytics:** Crop prediction is significant because it supports the farmer in making future decisions about crop production, storage, marketing methods, and risk management. Artificial networks are used to estimate crop output rates using data obtained from field sensors. Parameters such as soil, temperature, pressure, rainfall, and humidity are included in this data. Farmers can utilize the dashboard or a customized mobile application to gather correct soil data.

Farmers have started to realize that IoT can help them boost agricultural productivity at a reduced cost.

6.5.3   Architecture of the Proposed Model

Plant irrigation system designers take into account a variety of aspects while designing irrigation systems for their plants, including soil type and conditions such as weather and temperature. Using a network of sensors, data on soil moisture and temperature may be collected. Ontology is used in order to establish which plant, climate, and soil type is optimal for a certain plant, climate type, and soil type. The ontological result accounts for 50% of the final decision on whether to water or not to water plants in the long run.

The other half of the workload is handled by the ML system. It is easy to see how well the ML model was planned and developed for the irrigation system. Several layers of proposed IoT architecture are interconnected: the application, the processing, the transportation, and the perception tiers. Unlike traditional IoT design, which has three levels, this design has just two layers (application layer, network layer, and perception layer). In the physical layer, which is also referred to as the perception layer, data collecting sensors may be installed. Things like humidity, soil moisture, and weather are all caught up by the air and carried away with it.

Wireless, 2G, 3G, and local area network (LAN) networks are used to transmit data to the processing layer of the network stack. The transport layer is in charge of ensuring that data is delivered to the correct location.

The processing layer is responsible for saving, processing, and interpreting a significant amount of data from the transport layer. There are databases, cloud computing, and even edge computing used in this process. The application layer is responsible for assisting users who are utilizing an application. Sensors, a Global System Mobile (GSM) module, an edge server, and an IoT server are just a few of the components that make up the system. Each of the layers has a certain function. Cloud computing and cognitive computing are two of the four layers of computing.

In Figure 6.2, you can see how the sensors acquire the data displayed there. Measurements of soil moisture, air humidity, and temperature are taken as part of this procedure, as well. The perception layer includes sensors, actuators, and the microcontroller, among other things. The last three layers bring the structure to a close. Together, they give advice on watering strategies, crop monitoring, and other useful suggestions to the farmers they work with. Data is acquired, categorized, and kept in a data center so that it may be used for future research.

Clearly visible in the photograph, a comprehensive design review of the physical components was carried out before they were assembled. Every component is easily available and reasonably priced in the marketplace. As a result, it will be easy to find the device that is now in use in the physical world. It has sensors for humidity, light, and wetness in addition to the usual humidity and light sensors. Every 30 seconds, data is sent to a data center using a GSM module known as the SIM808 (Global System for Mobile Communications). Users of android app will be able to observe the results of decision-making process, and they will be able to use the system's actuators to release or shut off water from the valve as a result.

In Figure 6.5, the proposed model and working of the ML smart decision system, which is running on an IoT server, ensures that plants are adequately hydrated. In this environment, the kind of crops that may be grown and how they grow are all affected by the soil and climatic conditions. By incorporating ontology into the smart system, it improves these characteristics. With the use of these technologies, the proposed model is able to completely automate the system and eliminate human error. Later on, the algorithm will have a look at the semantic knowledge base that the smart irrigation system has. The following section is located here. Subscribe data management (SDM) is intended to be used in conjunction with real-world data. The categorization and assessment of ideas are based on logical levels when they are included in semantic data. Informed decisions on what to do next may be made by looking at what happens in the outputs of propositional logic systems that have been presented in the ontology.

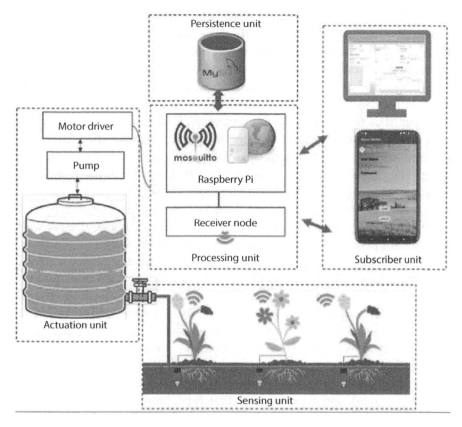

Figure 6.5 Proposed model and its working.

To predict how much water a crop requires based on its sort, climate and soil type, This study applies concepts from the proposed ontology. Ontology may be used to create decisions based on various kinds of information [21–24] since they are organized data that can be used to make judgements. Sensors and SPARQL are the two parts of running a watering system that you should be aware of (a query language for RDF). The quantity of water required to water a certain plant may be influenced by the temperature and kind of soil present in the environment. When the valve is turned on, water is sent to the crops. It is possible that it will be completely disabled depending on the field's layout. Sensor data is collected at a variety of levels across a large area using a variety of sensors. Measurements of surface and deep soil moisture are taken at the yard level, while measurements of surface and deep soil moisture are also taken at the quadrant level. The RDF data delivered to the control agent includes ontology information on crops, climates, and soils, among other things. This is the format in which you must furnish the agent with the information. Additionally, it learns about the water needs of the plant in a certain soil texture.

The system's ontology is rather extensive and sophisticated since there are a huge number of aspects to consider while deciding whether or not to water a plant in the first place. The ontology at a high degree of abstraction. Pakistan is divided into four separate climate zones. As with every geographical location, each has its own specific characteristics: hills, deserts and lowland regions abound, as do coastal areas and other bodies of water. The varied humidity levels seen in different parts of the world may be responsible for a large variation in irrigation water needs.

Temperature and humidity are both key factors in determining how much water the plants need, but the type of soil is also crucial. Clay soils are great for producing wheat because they drain well and retain moisture. The look and function of soil may be classified into four or five separate categories, depending on the kind of soil. Various crops, such as sugarcane, rice, and wheat, need different quantities of water to grow well. For example, wheat and cotton need far more water than the other crops.

Figure 6.6 depicts the process through which algorithm decision-making support is put in place. It is critical to understand how much water each kind of crop needs in order to develop well. The actual irrigation supply required for the field is determined by analyzing data from pasture and agricultural land, soil type, and weather type, among other factors. It is

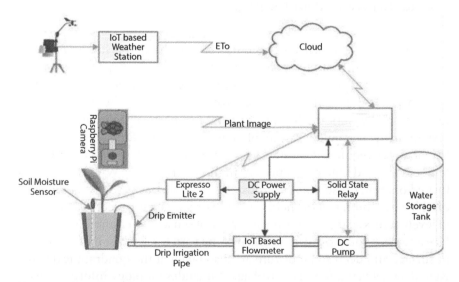

Figure 6.6 Data flow of smart irrigation.

feasible for a farmer to get watering recommendations via the use of an android application. Farmers will be able to activate a valve in the field by pressing the "active" button on their smartphone while in the field.

6.6   Conclusion

Farmers can be connected to the recommender system via a desktop program or mobile phone in the future to exchange information. The current study is an approach to test the application of IoT approaches in the field of precision agriculture using CV technologies. Precision agriculture is a relatively new paradigm in agricultural techniques that aims to reduce costs associated with the production of agro products while maintaining or improving output. Precision agriculture is a large domain that involves numerous steps of processing, beginning with data collection of field variations in needs or production, and then assessing information to see whether any changes are necessary.

References

1. Solanki, M.S., Precision irrigation in agriculture using an IoT based smart water management platform: A review chapter. *Int. J. Innov. Res. Comput. Sci. Technol.*, 10, 219–224, 2021.
2. Doshi, M. and Varghese, A., *Smart agriculture using renewable energy and AI-powered IoT. AI, edge and IoT-based smart agriculture*, pp. 205–225, 160, Pandit Deendayal Petroleum University, Gandhinagar, India, 2022.
3. Bavkar, S., Patil, N., Birje, Y., IoT enabled smart irrigation system using arduino. *SSRN Electronic Journal*, 2020.
4. Pezol, N.S., Adnan, R., Tajjudin, M., Design of an IoT (IoT) based smart irrigation and fertilization system using fuzzy logic for chili plant. *2020 IEEE International Conference on Automatic Control and Intelligent Systems (I2CACIS)*, 2020.
5. Saha, H.N., Roy, R., Chakraborty, M., Sarkar, C., Crop management system using IoT, in: *AI, Edge and IoT-Based Smart Agriculture*, pp. 125–141, 2022.
6. Bhattacharyya, D., Long term prediction of rainfall in Andhra Pradesh with deep learning. *J. Med. Pharm. Allied Sci.*, 10, 4, 3132–3137, 2021.
7. Aggarwal, S. and Kumar, A., A smart irrigation system to automate irrigation process using IoT and artificial neural network. *2019 2nd International Conference on Signal Processing and Communication (ICSPC)*, 2019.
8. M, M.B., S, R., S, R., P, S., R, S., Evergreen based agriculture irrigation system using IoT. *SIJ Trans. Comput. Sci. Eng. Appl. (CSEA)*, 05, 02, 11–14, 2017.

9. Balakrishna, K., Mohammed, F., Ullas, C.R., Hema, C.M., Sonakshi, S.K., Application of IoT and machine learning in crop protection against animal intrusion. *Glob. Transit. Proc.*, 2, 2, 169–174, 2021, https://doi.org/10.1016/j.gltp.2021.08.061.

10. Katta, S., Ramatenki, S., Sammeta, H., Smart irrigation and crop security in agriculture using IoT, in: *AI, Edge and IoT-Based Smart Agriculture*, pp. 143–155, 2022.

11. Marline Joys Kumari, N., Bhattacharyya, D., Thirupathi Rao, N., Improving the diagnostic accuracy using amplification and sequencing of the SARS-CoV-2 genome, in: *Digital Innovation for Healthcare in COVID-19 Pandemic: Strategies and Solutions*, pp. 331–350, 2022.

12. Raut, S. and Chitre, V., Soil monitoring and testing using IoT for fertility level and crop prediction. *SSRN Electronic Journal*, 202.

13. Hashem, I., Yaqoob, I., Anuar, N., Mokhtar, S., Gani, A., Ullah Khan, S., The rise of big data on cloud computing: Review and open research issues. *Inf. Syst.*, 47, 98–115, 2015.

14. Katta, S., Ramatenki, S., Sammeta, H., Smart irrigation and crop security in agriculture using IoT, in: *AI, Edge and IoT-Based Smart Agriculture*, pp. 143–155, 2022.

15. Aggarwal, S. and Kumar, A., A smart irrigation system to automate irrigation process using IoT and artificial neural network. *2019 2nd International Conference on Signal Processing and Communication (ICSPC)*, 2019.

16. Agnihotri, N., Santra, S., Sen, S., Crop and animal farming IoT (CAF-IoT), in: *Machine Learning and IoT*, pp. 327–340, 2018.

17. Ahmed, A., Omari, S.A., Awal, R., Fares, A., Chouikha, M., *A distributed system for supporting smart irrigation using IoT technology*, Wiley Online Library, 3, 7, 478–493, 2020.

18. R, R., S, K., L, V., IoT based smart irrigation system using image processing. *Int. J. Electr. Electron. Eng.*, 4, 3, 5–8, 2017.

19. Karthikamani, R. and Rajaguru, H., IoT based smart irrigation system using raspberry Pi, in: *2021 Smart Technologies, Communication and Robotics (STCR)*, 2021.

20. Difallah, W., Benahmed, K., Draoui, B., Bounaama, F., Design of a solar powered smart irrigation system (SPSIS) using WSN as an IoT device. *Proceedings of the 2018 International Conference on Software Engineering and Information Management - ICSIM2018*, 2018.

21. Bhushan, M., Goel, S., Kumar, A., Improving quality of software product line by analysing inconsistencies in feature models using an ontological rule-based approach. *Expert Syst.*, 35, 3, e12256, 2018.

22. Megha, Negi, A., Kaur, K., Method to resolve software product line errors, in: *International Conference on Information, Communication and Computing Technology*, Springer, pp. 258–268, 2017.
23. Bhushan, M., Goel, S., Kumar, A., Negi, A., Managing software product line using an ontological rule-based framework, in: *2017 International Conference on Infocom Technologies and Unmanned Systems (Trends and Future Directions) (ICTUS)*, IEEE, pp. 376–382, 2017.
24. Bhushan, M. and Goel, S., Improving software product line using an ontological approach. *Sādhanā*, 41, 12, 1381–1391, 2016.

Machine Learning-Based Hybrid Model for Wheat Yield Prediction

Haneet Kour[1]*, Vaishali Pandith[1], Jatinder Manhas[2] and Vinod Sharma[1]

[1]Department of Computer Science & IT, University of Jammu, Jammu, India
[2]Department of Computer Science & IT, Bhaderwah Campus,
University of Jammu, Jammu, India

Abstract

Machine learning (ML) approaches are significant in resolving real-world issues. These methods have been used in the agricultural industry to help farmers anticipate crop yields, detect crop illnesses, and perform other tasks. In this chapter, a hybrid model for wheat crop yield prediction is presented, which combines feature selection with ML technique. The proposed hybrid model has two stages: the first uses a feature selection strategy to find the best features for wheat crops, and the second uses ML to estimate crop yield based on those best features. In this study, 12 different hybrid models have been implemented by integrating two feature selection techniques, each with six ML techniques in order to determine the optimum hybrid technique for wheat yield prediction. The applied feature selection techniques were genetic algorithm (GA) and ReliefF algorithm, while ML approaches used were K-nearest neighbor (KNN), Naïve Bayes, artificial neural network, logistic regression, support vector machine, and linear discriminant analysis. Five performance metrics, accuracy, recall, precision, f-score, and kappa coefficient, were used to assess the performance of these hybrid models. Among the 12 implemented hybrid approaches, GA-KNN model predicted the highest performance with accuracy of 98.3%.

Keywords: Crop yield, feature selection, machine learning, soil, wheat

*Corresponding author: haneetkour9@gmail.com

Ashok Kumar, Megha Bhushan, José A. Galindo, Lalit Garg and Yu-Chen Hu (eds.) Machine Intelligence, Big Data Analytics, and IoT in Image Processing: Practical Applications, (151–176) © 2023 Scrivener Publishing LLC

7.1   Introduction

Agriculture is an enormously essential part of any country's economy. Various challenges such as population growth, exhaustion of natural resources, and environmental changes; place a premium on a country's agricultural structure for enhancing the amount and quality of crops grown. Thus agriculture is incorporating digital technology to increase crop output while minimizing environmental impact. It also makes use of sensors to collect the data from the soil, crop, and environment in order to better understand how to boost crop productivity. Machine Learning (ML) technologies have proven to be valuable tools in the agriculture industry for a variety of tasks including weeds identification, crop recognition, soil management, and so on. These methods offer farmers and agriculturalists cost-effective measures to a variety of issues being faced by them. Apart from this, researchers also implemented ML techniques to aid crop yield prediction [1].

Crop yield prediction is a significant task in precision agriculture where it includes yield map and estimation along with crop supply-demand matching. It can assist the farmers to make decisions in advance about when and which crops to cultivate based on the prevailing market's requirements. It is tremendously helpful for food production all around the world because it allows for fast import and export decisions. Several factors including the atmosphere, rainfall, soil type, fertilizers, temperature and genotypic traits of the crop, can impact crop yield. As a result, a fundamental understanding of the correlation between these parameters and yield is required. Various yield predictor models are available to forecast accurate crop production, but their efficiency is still an issue. Under these circumstances, ML based models [2] play an important role in yield prediction as these models learn knowledge from the datasets by identifying the patterns [3, 4]. Various researchers have used ML approaches to forecast agricultural productivity. These approaches have been shown to accurately predict crop yields [5, 6]. Traditional ML algorithms have failed to achieve the desired results when handling large and complicated data. Researchers have proposed hybrid ways to solve these challenges, which have proven to be more efficient than traditional methods since they improve efficiency while limiting their drawbacks [7].

The current research study presents a hybrid model through concatenating attribute selection with classification technique for wheat crop yield prediction. Attribute Selection identifies the optimal features for the wheat crop, and Classification entails applying ML techniques to those selected optimal features to forecast crop yield as low, medium or high. In order to determine the optimal hybrid technique for wheat yield prediction, 12 variations of the hybrid model were implemented by combining two attribute selection techniques each with six ML techniques. The undertaken feature selection techniques were Genetic Algorithm (GA) and ReliefF algorithms (RF) because these approaches are effective with multiple output classes and numerical and nominal input features. ML approaches used were K-nearest neighbor (KNN), support vector machine (SVM), Naïve Bayes (NB), artificial neural network (ANN), logistic regression (LR), and linear discriminant analysis (LDA). The experiments were implemented using a benchmark dataset of wheat crops.

The current chapter is presented as: Section 7.2 covers prior research linked to the current study, Section 7.3 discusses materials and techniques utilized in the experimental setup, Section 7.4 shows and analyzes the experimental results, and Section 7.5 displays the conclusion of the chapter.

7.2   Related Work

A lot of research has been carried out in the agricultural domain for the recent past implementing ML techniques to predict production of different crops. The primary feature of the ML technique is that the output (yield prediction) is seen as an implicit function of the input parameters (genotypes, soil parameters, and external environmental condition), that can be nonlinear and intricate. Study done by various researchers for crop yield prediction using ML techniques during past few years, is presented as under:

Authors have implemented a NB approach for the classification of agricultural land soil [8]. They also evaluated the performance of NB with the Bayesian network and J48. The experimental results predicted the NB approach to be superior in their study. In Veenadhari et al. [9], C4.5 algorithm was applied for the yield prediction from climatic parameters in some districts of Madhya Pradesh. Researchers performed comparative analysis between KNN and NB [10]; and NB, J48 (C4.5) and JRip [11] for prediction of soil fertility rate on the soil data. Using data from 110 samples, the same work was conducted to assess the performance of JRip, NB, and J48 for soil type prediction [12].

A study was carried out on rice crop yield prediction in 27 districts of Maharashtra using SVM [13]. The performance of SVM was evaluated with other ML techniques like NB, multiperceptron, and Bayes net. A crop yield prediction model was proposed to estimate the plant growth and diseases on the basis of various parameters such as crop name, land area, soil type, soil pH, pest details, weather, water level, and seed type; by applying data mining techniques [14]. A yield prediction model was developed by applying Counter-Propagation Artificial Neural Networks (CP-ANNs), XY-fused Networks (XY-Fs) and Supervised Kohonen Networks (SKNs), on soil parameters and satellite imagery crop growth characteristics [15]. This model was evaluated on winter wheat, and the overall average accuracy for SKN was 81.65%, 78.3% for CP-ANN and 80.92% for XY-F, demonstrating the best overall results for the SKN model.

In Rajak *et al.* [16], a hybrid model was proposed based on SVM and ANN with a majority voting approach in order to recommend a crop for site specific parameters. ML based models were implemented for crop yield prediction using k-means clustering with LDA [17]; fuzzy-c means clustering with neural network [18] and Principal Component Analysis with Radial Basis Neural Network [19] for wheat crop; Multilayer Perceptron was applied for crop yield prediction in Gujarat state [20]. Descriptive analytics on Soil, Rainfall and Yield datasets were performed to reveal the most effective technique among KNN, SVM, and Least Squared SVM (LS-SVM) for sugarcane yield estimation [21]. Results indicated that the accuracy of the LS-SVM model is higher and has a minimum error rate. Crane-Droesch [22] implemented a semi-parametric neural network for corn yield prediction from weather parameters.

C5.0, Random forest and KNN were implemented for prediction of soil fertility with high accuracy and efficiency [23]. A deep neural network was implemented to predict corn hybrids yield from genetic markers and environment data [24]. Moreover, SVM and random forest techniques were applied on agricultural data for soil classification and crop yield prediction system in Bondre and Mahagaonkar [25]. A two-tiered ML model was presented based on neural networks and k-means clustering for prediction of nitrogen, potassium, and phosphorus deficiencies of paddy crop [26]. K-medoid clustering with PCA was applied for prediction of crop yield in some districts of Chhattisgarh [27]. A neural network with new activation functions named DharaSig, DharaSigm, and SHBSig was proposed, for wheat yield prediction based on weather parameters [28].

According to the literature review, ML approaches could be used to assess agricultural productivity. These algorithms have proven to be valuable when working with simple datasets. Conventional ML algorithms, on the other hand, have failed to attain good performance in case of increasingly complicated data. Researchers have proposed hybrid systems as a possible solution to these issues. Thus, various research questions arise after going through literature study, which are given below:

RQ1. Which ML technique is the best for wheat yield prediction?

RQ2. Which feature selection approach is suitable to find out the optimum attributes towards wheat yield prediction?

RQ3. Do hybrid methods enhance the performance of traditional classifiers for wheat yield prediction?

The current research focuses on above research questions. In order to address these research questions, the authors developed a hybrid method for predicting wheat yield, by applying ML techniques integrated with feature selection.

7.3   Materials and Methods

In the current work, hybrid architecture has been introduced for analyzing wheat crop yield based on ML and feature selection. Experiments were implemented using the "Matlab" tool. The proposed hybrid model concatenated attribute selection with prediction technique for wheat crop yield prediction. Attribute Selection identifies the optimal features for the wheat crop, and Prediction entails applying ML techniques to those selected optimal features to forecast crop yield as "low, medium or high." The applied feature selection techniques were GA and RF; while ML approaches used were KNN, NB, ANN, LR, SVM and LDA. In this study, 12 hybrid variations were implemented to ascertain the optimal hybrid methodology for wheat yield prediction by combining two feature selection strategies with six ML techniques.

7.3.1   Methodology for the Current Work

The methodology adopted to perform the proposed work is depicted in Figure 7.1. It consists of various steps: data collection, data pre-processing, data partition, model training, testing and model evaluation. All these steps are explained in detail below.

7.3.1.1   Data Collection for Wheat Crop

For the current research study, data was gathered from https://soil-health.dac.gov.in/PublicReports/NSVW [29] as part of the Model Village Programme 2019 to 2020 under the supervision of Soil Chemist. This collected data for wheat crop was transformed into .csv format. This dataset

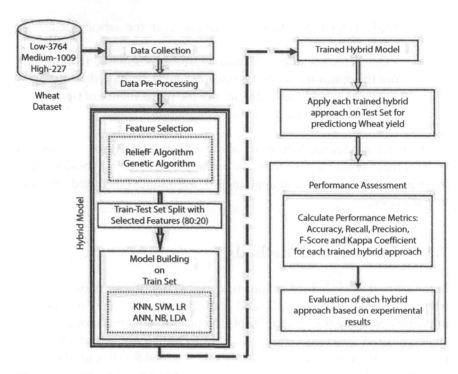

Figure 7.1 Methodology for the proposed work.

has 5000 records, eleven input parameters and one output parameter (with three classes for yield estimation - Low, Medium or High). There are 3764, 1009 and 227 instances occurring in low, medium or high classifications respectively. Table 7.1 and Table 7.2 represent the details of input parameters and the instances of the dataset respectively.

7.3.1.2   Data Pre-Processing

Following data collection, data pre-processing was done. Statistical methods were applied for imputation and outlier detection. Min-max approach was used to normalize each feature in the range [0, 1], as shown in the equation below.

$$Z' = \frac{z - mn(I)}{(I) - mn(I)} \tag{7.1}$$

Table 7.1 Input parameters in the wheat crop dataset.

S. no.	Attribute no.	Input attributes	Description	Unit
1	F_1	pH	Hydrogen potential – pH value of soil	--
2	F_2	EC	Electrical Conductivity- amount of salt in soil	dsm^{-1}
3	F_3	OC	Organic Carbon- amount of organic and inorganic component in soil	%
4	F_4	N	Nitrogen	kgha^{-1}
5	F_5	P	Phosphorus	kgha^{-1}
6	F_6	K	Potassium	kgha^{-1}
7	F_7	S	Sulphur	kgha^{-1}
8	F_8	Zn	Zinc	kgha^{-1}
9	F_9	Fe	Iron	mgkg^{-1}
10	F_{10}	Cu	Copper	mgkg^{-1}
11	F_{11}	Mn	Manganese	mgkg^{-1}

where
Z = the actual value of the input feature I
Z' = normalized value of the respective feature I
$mn(I)$ = the smallest value of the input feature I
$mx(I)$ = the largest value of the input feature I

7.3.1.3   Implementation of the Proposed Hybrid Model

The proposed hybrid framework has been implemented after data pre-processing. This hybrid approach consists of two steps: the first step involves application of attribute selection to identify optimal features for the wheat crop, and the second one involves predicting wheat yield as low, medium or high by applying ML technique on those identified optimal features. The most important parameters in the dataset were selected by applying the attribute selection approach. In the case of a dataset having

Table 7.2 Few instances of wheat crop dataset.

S. no.	pH	EC	OC	N	P	K	S	Zn	Fe	Cu	Mn	Class
1	7.5	0.56	0.37	257.15	10.68	196.42	16.6	1.36	6.86	0.42	3.36	Low
2	7.66	1.11	1.86	902.1	76.16	715.68	5.43	4.82	3.24	1.02	4.54	High
3	6.6	0.187	0.48	417	23.27	249.82	15.94	0.37	4.21	0.19	2.25	Medium
4	7.1	0.132	0.678	471.21	27.03	214.39	27.15	1.54	6.38	0.44	9.12	Medium
5	6.6	0.216	0.735	323.17	23.27	160.47	12.49	0.57	5.82	0.36	2.25	Medium
6	6.84	0.91	1.69	676	229.56	330.4	11.76	2.12	4.24	0.68	1.7	High
7	7	0.63	0.2	139	17.2	156.9	11.7	9.18	9.48	0.28	3.18	Low
8	7	0.278	0.82	573.37	101.43	264.91	0	0.65	2.23	0.16	1.46	High
9	7.3	0.56	0.29	201.55	14.31	246.2	17.8	0.72	7.22	0.28	2.52	Low
10	7.15	0.76	0.67	318.05	69.35	49.02	9.11	2.36	2.96	0.82	2.52	Medium

enormous parameters with limited instances, all of the parameters are not important for prediction tasks thus affecting the model's efficiency due to increased space and time complexity. Hence it may lead to overfitting in the trained model. In these cases, feature selection is found to be noteworthy as it decreases the computing cost of the learning method. For the current experimental study, two feature selection techniques including GA and RF have been taken. After feature selection, ML approaches including KNN, NB, ANN, LR, SVM and LDA were used to train the model.

In this study, 12 hybrid approaches have been implemented by integrating two feature selection techniques each with six ML techniques for revealing the most appropriate hybrid technique for yield prediction of wheat crop. Each hybrid approach was evaluated for its performance based on different metrics such as accuracy, recall, precision, f-score, and kappa coefficient.

7.3.2   Techniques Used for Feature Selection

In this chapter, two feature selection techniques including GA and RF have been taken. The details of these techniques are presented below:

7.3.2.1   ReliefF Algorithm

This attribute selection approach figures out the weights of the input attributes in case of multiclass problems. This method works by penalizing the predictors having dissimilar values for the neighbors in the same class, while rewarding those that produce different values for separate classes. It works by estimating the quality of features and how effectively they can discriminate between nearby samples. The weights of all predictors are initially set to zero by this approach. These weight values get updated on the basis of difference between the attribute values of target and its closed-by instances [30]. The general steps for RF feature selection approach is presented in Algorithm 1.

Algorithm 1. ReliefF algorithm

Input: Feature Array I for each sample in the dataset, the value of K, and the Output Class O

Output: Weight Array $W_t[$] for all the features in the dataset

Steps:

1. $W_t[I] = 0.0$; Initialize the weight of all attributes to zeros
2. For J = 1 to N

    a. Select a random sample S_j from the dataset
    b. Explore the K-nearest neighbors with the same class of S_j: Nearest Hit (H)
    c. Explore the K-nearest neighbors with the class different from S_j: Nearest Miss (M)
    d. Update the weight of each feature: $W_t[I] = W_t[I]$ - difference$(I, S_j, H)/N$ + difference$(I, S_j, M)/N$

3. End For

The current research work applies the RF by setting all the features as predictors with Class attribute as response. The rank and the weight predicted for each feature by the RF are depicted in Table 7.3. Figure 7.2 presents the weight of each input attribute as predicted by the Relief algorithm. Table 7.3 and Figure 7.2 depict that feature 5 "phosphorus" contributes the most to wheat yield forecast, while feature 10 "copper" contributes the least, thus phosphorus nutrient has a great impact on the yield of wheat crop whereas copper does not affect wheat production. Among eleven features

Table 7.3 ReliefF algorithm determines the rank and weight for each feature.

S. no.	Feature	Weight	Rank
1	pH (Hydrogen potential)	0.03157	4
2	EC (Electrical Conductivity)	0.02480	8
3	OC (Organic Compound)	0.02354	9
4	N (Nitrogen)	0.02538	7
5	P (Phosphorus)	0.08404	1
6	K (Potassium)	0.02689	6
7	S (Sulphur)	0.01673	10
8	Zn (Zinc)	0.03121	5
9	Fe (Iron)	0.03828	2
10	Cu (Copper)	0.01124	11
11	Mn (Manganese)	0.03467	3

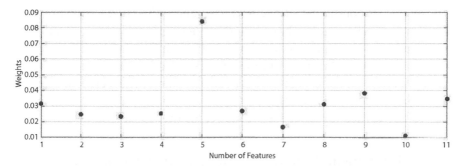

Figure 7.2 Weight value for each feature as predicted by ReliefF algorithm.

in the wheat crop dataset, the top five most contributing features namely "P, Fe, Mn, pH, Zn" were chosen for training the presented hybrid model.

7.3.2.2   Genetic Algorithm

It is a nature-inspired optimization technique used for feature selection. It is based on the idea of natural evolution, which means organisms' genes continue to evolve through generations in order to better adapt to their surroundings. It uses Darwin's notion of survival of the fittest as part of a search algorithm to find the optimum answer among all the options. It keeps track of the population of potential solutions for a certain problem, known as individual. It uses selection, crossover, and mutation, among other genetic operators, to create a better solution over numerous generations. Each solution is assessed using a fitness function which is crucial in optimizing the solution [31]. The general steps involved in GAs are presented in Algorithm 2.

Algorithm 2. Genetic algorithm

---

1. Initialize node population
2. Estimate node population by employing *Fitness Function*
3. *Select* parent nodes for reproduction
4. Carry out *crossover* and *mutation*
5. Estimate population
6. Reiterate 3 to 5 until stopping criteria is reached

In the current study, GA involves identification of optimal subset of features in the data where each individual represents the subset of features to be selected in order to augment the competence of the classifier.

All the features of the data set refer to the genes. The binary array presents a feature set where genes have binary values reflecting the inclusion (1) or exclusion (0) of a particular feature in the learning model. The value of various parameters taken as: population = 30, number of generations = 200 with mutation rate of 0.01, crossover rate = 0.8 and fitness criteria = "error loss." GA has selected feature subset "1 2 3 5 8 10" i.e., "pH, EC, OC, P, Zn, Cu" to be the most optimal feature subset for the wheat crop.

7.3.3 Implementation of Machine Learning Techniques for Wheat Yield Prediction

In the current study, most commonly and important ML algorithms have been applied for prediction of wheat crop yield. These algorithms are: KNN [32], NB [33], ANN [34], LR [35], SVM [36], and LDA [37]. In the present study, each ML model was implemented as a classifier and has been trained on a Train Set with optimal input parameters (as selected by feature selection techniques taken in this study). The output label has three classes.

7.3.3.1 K-Nearest Neighbor

KNN presents supervised learning techniques implemented for the tasks of classification and regression. The assumption behind this approach is that each instance in the dataset refers to points in D-dimensional space. The k data instances in close proximity are used for evaluating the output label of a new data instance. The closed-by k data points are found for tracing out the similar patterns through the application of distance measure including Euclidean, Cityblock, Manhattan, etc. These metrics evaluate the distance between a given instance and each sample in the training set, with the majority of class label for the k-nearest neighbors representing the output class for the given instance. In this algorithm, k refers to a positive integer number, and the best value for k can usually be obtained using heuristic techniques.

Assume M be the instances having D input attributes in the training set; and L be a new sample with the attribute array $[l_1, l_2, l_3,....l_d]$, where t_v points to the vth attribute. The euclidean metric between M and L is computed as:

$$v \quad Euclideani(M, L) = \sqrt{\sum_{v=1}^{D} (m_v - i_v)^2} \qquad (7.2)$$

For the current work, 10-fold cross validation and Euclidean distance metric were used to create the *KNN* based hybrid system. The number of nearest neighbors (k) that is optimal was discovered to be 7.

7.3.3.2   Artificial Neural Network

The neural network is a mathematical model of a simplified brain-like system made up of a parallel distributed network of artificial neurons that can do simultaneous computations for pattern recognition. The artificial neuron mimics the action of biological neurons. ANN consists of the input layer, the hidden layers and the output layer where the input layer takes the input parameters from the dataset, processes them and transfers them to the hidden neurons. In the hidden layer, each neuron processes and then propagates the input signal obtained from the layer above it and the strength of the propagated signal depends on the weight of the neuron, activation function and bias value. The final output is obtained at the output layer. Each input data is related to certain weight values. Neural network is initialized with certain random values for weight and then final weights are evaluated by applying any learning algorithm. Also, the output of each neuron is limited to some range using the activation function. Bias term refers to the constant added to the weighted input before the activation function to be applied. It only affects the output values; not interfering with the real input data and does not depend on previous layers' outputs.

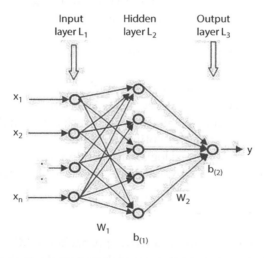

Figure 7.3 General architecture of ANN.

Generally ANN is composed of an input layer, one or two hidden layers with few neurons, and an output layer. Figure 7.3 represents a general framework for ANN architecture. The first layer (L1) represents the input layer; the second layer (L2) refers to the hidden layer and the third layer (L3) is called the output layer; each consisting of neurons connected to each other.

In the above network, the input layer has n neurons where n represents the number of features in the input vector $X=[x_1, x_2, ..., x_n]$; W_i refers to the weight of connections between neurons between layers and $b_{(i)}$ represents the bias value of neurons within the layer. The output $\mathbf{y}$ is represented as:

$$y = f(\textstyle\sum_1^2 x_i \cdot w_i + b_i) \tag{7.3}$$

where $f(.)$ represents activation function.

In this work, *ANN* model consisted of one input layer, one hidden layer and one output layer. The hidden layer was made up of 8 neurons, and the learning rate was set to 0.01. The activation function "linear" was applied for the input layer and the hidden layer whereas "sigmoid" was used for the output layer. The training of ANN model was done using the "Scale Conjugate Gradient" learning algorithm for wheat crop yield prediction. This model was trained for 200 epochs.

7.3.3.3　Logistic Regression

LR represents an ML model under supervised learning, which is applied to predict the target value of a categorical dependent variable from a group of independent variables. It refers to a predictive analysis technique based on the probability notion. In case of target value having more than two categories with order, it is referred to ordinal or multiclass logistic regression. For ordinal logistic regression, the output label (class) with N classes each represented as integer in the range 1, 2, ..., N. The "softmax" function is used to calculate the score for all categorical variables, and the final output class is chosen using a majority voting technique.

For the current work, ordinal logistic regression was applied using function "mnrfit" through 10-fold cross validation as the output label has three classes.

7.3.3.4　Naïve Bayes

NB is one of the effective and inductive approaches of ML. This algorithm reflects a probabilistic learning approach based on Bayes theorem, with the principle of conditional independence assumption that each pair of features to be classified is independent and each feature has the same weight. Thus, each parameter contributes independently to predict the final result. Mathematically Bayes theorem can be stated as:

$$P\left(\frac{T}{S}\right) = \frac{P\left(\frac{S}{T}\right) \cdot P(T)}{P(S)} \qquad (7.4)$$

where T represents the output variable, $P(T)$ refers to the prior probability, $P(S/T)$ refers to the conditional probability or likelihood ratio, $P(T/S)$ refers to the posteriori probability of S, and S represents the input feature vector with n features and S is presented as: $S = (s_1, s_2, s_3, ..., s_n)$.

Thus there is intrigue in the numerator part, having regard to the fact that the denominator part does not rely upon the class label and features values are provided, with the goal that the denominator is adequately constant. The numerator part follows the concept of the joint probability model. NB algorithm classifies data in two steps: Training Step: This step estimates the parameter of a probability distribution using the training data, assuming the predictors are conditionally independent given the class; and Testing Step: For new instance, this step computes the posterior probability of that sample for each class; and then classifies the test data with highest posterior probability. In this work, NB was introduced with "Gaussian" function and "empirical" prior probability.

7.3.3.5 Support Vector Machine

It is a supervised approach applied to classification and regression applications, which is based on structural risk minimization. It traces out for an optimal N-dimensional hyper plane that divides the data into two classes. This optimal hyperplane attempts to maximize the margin between closest observations for independent classes, creating boundaries for the positive and the negative class. It can be considered as a linear discriminative approach trying to optimize the inter-class margin during learning the model. This hyperplane can be represented by a linear equation as:

$$f(a) = w^T a + b \qquad (7.5)$$

where a refers to data points, w is a N-dimensional coefficient vector that defines an orthogonal vector to the hyperplane, and b points to the offset from the origin.

In the case of linear SVM, $f(a)\cdot \geq 0$ for the nearest point belongs to one class, and $f(a)\cdot \leq 0$ for the nearest point belongs to another class.

The standard SVM approach for classification problem is represented as:

$$maximize\left[\sum_{i=1}^{N} x_i - \frac{1}{2}\sum_{i,j}^{N} x_i x_j w_i w_j \left(S_i, T_i\right)\right] \tag{7.6}$$

$$subject\ to: \left[\sum_{i=1}^{N} x_i w_j = 0, 0 \leq \forall x_i \leq C\right] \tag{7.7}$$

where S presents the training vector, T is the label associated with the training vectors, and x presents the parameters vector of the classifier hyperplane. $K(S, T)$ refers to the kernel function equal to $[\phi(S_i), \phi(T_i)]$ where $\phi(S)$ is the mapping of input vectors onto the kernel space to compute the distance between the training vector S_i and T_i, and C is the penalty parameter to control the misclassification rate.

SVM was applied with error-correcting output codes (ECOC) model having "onevsone" coding since the output attribute is multiclass. Ten-fold cross validation and linear kernel function was applied to implement the SVM approach.

7.3.3.6   Linear Discriminant Analysis

LDA represents a straightforward and efficient classification technique which is based upon two assumptions: each input attribute follows Gaussian distribution and each attribute has the same variance. It computes the mean and variance of every attribute for each class in the dataset. It can be used for both binary and multi-class problems. This method makes predictions for a new data instance by estimating its probability of falling into every class. The final output label is chosen from the class with the highest occurrence.

In this experimental study, LDA was applied with 10-fold cross validation, "empirical" for prior probability and "logit" for score transformation.

7.4   Experimental Result and Analysis

This study provides a hybrid model for wheat yield prediction based on optimum attribute selection and classification. Based on the results of the experiments, it has been concluded that the proposed hybrid technique could be employed for the wheat yield estimation. Agriculturists and farmers require a system that can forecast precise crop yields. The performance of the provided hybrid techniques was therefore validated using five evaluation parameters: accuracy, recall, precision, f-score, and kappa coefficient. All these metrics provide a clear view of the real classification, as well as the rate of misclassification, and address the issue of overfitting in imbalanced datasets. These metrics can be computed as presented below:

If

tp_i = number of true positive for the class C_i
tn_i = number true negative for the class C_i
fp_i = number of false positive for the class C_i
fn_i = number of false negative for the class C_i
N = total instances
P_i = overall predicted values for the class C_i
T_i = total number of truth values for the class C_i

Then

$$\text{Average Classification Accuracy} = \frac{\sum_{i=1}^{L} \frac{tp_i + tn_i}{tp_i + tn_i + fp_i + fn_i}}{L} \quad (7.8)$$

$$\text{Average Precision} = \frac{\sum_{i=1}^{L} \frac{tp_i}{tp_i + fp_i}}{L} \quad (7.9)$$

$$\text{Average Recall} = \frac{\sum_{i=1}^{L} \frac{tn_i}{tn_i + fp_i}}{L} \quad (7.10)$$

$$F-score = 2 * \frac{Precision * Recall}{Precision + Recall} \quad (7.11)$$

$$\text{Kappa Coefficient} = \frac{N\sum_{i=1}^{L} tp_i - \sum_{i=1}^{L}(T_i P_i)}{N^2 - \sum_{i=1}^{L}(T_i P_i)} \qquad (7.12)$$

Table 7.4 reflects the outcomes of the experimental studies for conventional ML techniques (taken in this study) for wheat yield prediction. All of the ML approaches that were considered can be used to forecast yield. However these techniques suffer from the issues of high false positive and high false negative rate. Out of these six ML techniques, LR and LDA achieved good results with accuracy of 95.5%. The former predicted recall of 81.18% with precision of 82.53%, f-score of 0.8185 and kappa value of 0.885; whereas the latter obtained recall of 84.98% with precision of 84.32%, f-score of 0.8465 and kappa value of 0.884. NB predicted the lowest accuracy of 87.6% with f-score and kappa value of 0.7202 and 0.713, respectively. As a result, some methodology for improving the performance of current ML techniques is required. The authors of the current research have recommended a hybrid approach in this regard. By combining two feature selection strategies with six ML algorithms; this strategy has 12 distinct implementations.

Table 7.5 presents the experimental results for each hybrid model for wheat yield prediction. Out of 12 implemented models, best results were predicted by GA-KNN with accuracy of 98.3%, recall of 94.32% with precision of 93.41%, f-score of 0.9386 with kappa coefficient of 0.957. These results were better than those achieved by conventional LR and LDA methods.

Table 7.4 Experimental results for traditional ML techniques for wheat yield prediction.

Traditional ML technique	Accuracy	Recall	Precision	F-score	Kappa coefficient
KNN	94.1%	75.21%	81.57%	0.7826	0.852
SVM	92.4%	73.39%	71.87%	0.7262	0.807
LR	95.5%	81.18%	82.53%	0.8185	0.885
ANN	94.7%	74.53%	83.15%	0.7860	0.861
LDA	95.5%	84.98%	84.32%	0.8465	0.884
NB	87.6	73.28%	70.8%	0.7202	0.713

Table 7.5 Experimental results for each hybrid model for wheat yield prediction.

Model no.	Feature selector	ML technique	Hybrid model	Accuracy	Recall	Precision	F-score	Kappa
M_1	ReliefF Algorithm	KNN	RF-KNN	98%	93.39%	94.14%	0.9376	0.949
M_2		SVM	RF-SVM	95.3%	75.30%	82.40%	0.7869	0.88
M_3		LR	RF-LR	94.9%	71 48%	78.46%	0.7481	0.869
M_4		ANN	RF-ANN	97.1%	91 99%	89.21%	0.9058	0.928
M_5		LDA	RF-LDA	91.4%	72 18%	82.06%	0.7681	0.763
M_6		NB	RF-NB	85.9%	65 64%	68.61%	0.6709	0.662
M_7	Genetic Algorithm	KNN	GA-KNN	98.3%	94.32%	93.41%	0.9386	0.957
M_8		SVM	GA-SVM	96.2%	80.78%	84.87%	0.8277	0.903
M_9		LR	GA-LR	96%	79.09%	82.81%	0.8091	0.897
M_{10}		ANN	GA-ANN	97.2%	93.62%	89.11%	0.9131	0.921
M_{11}		LDA	GA-LDA	96.6%	84.56%	85.89%	0.8522	0.913
M_{12}		NB	GA-NB	89.3%	75.48%	79.43%	0.7741	0.75

Almost all the hybrid models obtained better results than their respective conventional methods. RF based hybrid approaches have enhanced the performance of standard KNN, SVM, and ANN for wheat yield prediction; but it has decreased the performance of remaining ML techniques. In some cases, hybrid models like RF-SVM, RF-LDA and RF-LR achieved accuracy greater than 90% but these approaches predicted recall of 75.30%, 72.18% and 71.48% respectively. Thus there are more cases of false negatives in RF based hybrid models. This may be due to the reason that all the features selected by RF approach in the feature subset will not be most contributing towards yield prediction of wheat. Hybrid approach based on GA achieved better results for all conventional ML techniques. It has improved the accuracy of all ML techniques taken in this study for yield prediction. The accuracies of KNN, SVM, LR, ANN, LDA and NB have enhanced from 94.1%, 92.4%, 95.5%, 94.7%, 95.5% and 87.6% to 98.3%, 96.2%, 96%, 97.2%, 96.6% and 89.3% respectively. GA based NB approach also encountered the false negative and false positive cases; but these cases are very lesser than that in traditional NB. F-score and Kappa value achieved by GA based hybrid method are greater than their respective traditional ML approaches. Thus GA based hybrid approaches have been demonstrated to enhance the efficiency of all ML techniques.

Figures 7.4 to 7.9 present the comparative analysis of all the traditional ML approaches with their respective hybrid models for the yield prediction of the crop under study. In the case of KNN, all the hybrid approaches enhanced its accuracy, f-score and kappa value; thus improving its performance (as shown in Figure 7.4). Figure 7.5 and Figure 7.7 depict that all the undertaken feature selection techniques have improved the efficiency of SVM and ANN approaches respectively. In the case of the traditional LR approach, GA has been found to have a significant impact on its performance (as shown in Figure 7.6). GA also improved efficiency of LDA (see Figure 7.8), and NB (see Figure 7.9) for wheat yield prediction.

According to the findings of the experimental studies, the presented hybrid approach outperformed the traditional ML techniques in terms of all performance metrics. In comparison to traditional approaches, the presented hybrid approach projects a higher classification rate. Compared to classic ML algorithms, this hybrid strategy produced less misclassification rate. As a result, the hybrid model outperformed traditional ML approaches. This is because traditional ML algorithms employed all of the features in the dataset to estimate wheat yield. Hence, the data has become higher dimensional with irrelevant attributes that causes larger computational cost leading to overfitting in the trained model. In the presented hybrid model, only significant attributes were considered, leading to

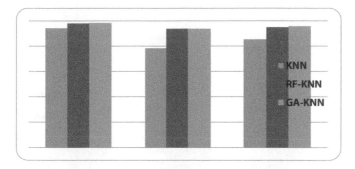

Figure 7.4 Comparative analysis of traditional KNN vs hybrid models.

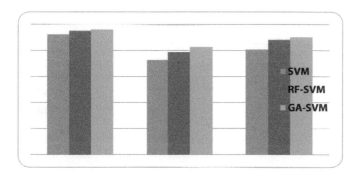

Figure 7.5 Comparative analysis of traditional SVM vs hybrid models.

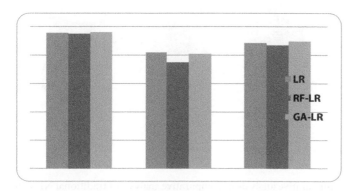

Figure 7.6 Comparative analysis of traditional LR vs hybrid models.

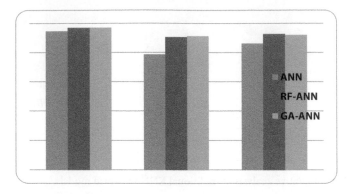

Figure 7.7 Comparative analysis of traditional ANN vs hybrid models.

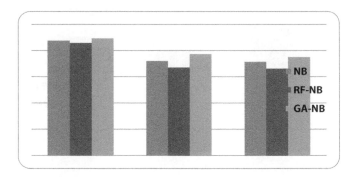

Figure 7.8 Comparative analysis of traditional LDA vs hybrid models.

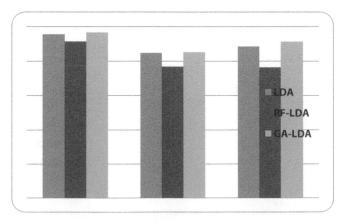

Figure 7.9 Comparative analysis of comparative analysis of traditional NB vs hybrid models.

reduced dataset dimensions and an improvement in the performance of trained hybrid models.

The answers to all of the research questions (given in the section 7.2) have been discovered after doing all of the experiments and assessing their results. For RQ1, LR and LDA classifiers have obtained the best performance for wheat yield prediction with accuracy of 95.5%. For RQ2, the most optimum features in the wheat dataset were found by GA for yield prediction. In the case of RQ3, hybrid technique has improved the performance of all traditional classifiers enabling them to obtain a maximum accuracy of 98.3%.

7.5   Conclusion

As per the outcomes of the current study, ML algorithms can be utilized to predict wheat crop yield. Techniques for feature selection have been discovered to have a considerable impact on each classifier's performance because it selects the most optimal attributes in the dataset. These selected optimal attributes are used to build the model. Thus, when compared to existing ML algorithms used in this work, the proposed hybrid architecture performed better. Farmers will be able to forecast production based on soil factors using these effective algorithms. These models will be updated in the future to include fertilizer recommendations, assisting soil analysts and farmers in making informed decisions. Only soil factors are taken into account in this study for predicting yields. The future work aims to use genetic markers and climatic parameters affecting crop production in respective regions to predict actual production well in advance. It also aims to analyze the nutrient status and environmental conditions for particular region to maximize crop production.

Acknowledgment

The authors thank Mr. Ashwani Kotwal who assisted the authors in performing the research. He works as Soil Chemist in the Department of Agriculture, Talab Tillo, Jammu.

References

1. Role of machine learning in modern age agriculture, https://technostacks. com/blog/machine-learning-in-agriculture.
2. Verma, K., Bhardwaj, S., Arya, R., Islam, M.S., Bhushan, M., Kumar, A., Samant, P., Latest tools for data mining and machine learning. *Int. J. Innov. Technol. Exploring Eng.*, 8, 9S, 2019.
3. Klompenburg, T.V., Kassahun, A., Catal, C., Crop yield prediction using machine learning: A systematic literature review. *Comput. Electron. Agric.*, 177, 2020, https://doi.org/10.1016/j.compag.2020.105709.
4. Mucherino, Papajorgi, P., Pardalos, P.M., A survey of data mining techniques applied to agriculture. *Oper. Res.*, 9, 2, 121–140, 2009, https://doi.org/10:1007/s12351-009-0054-6.
5. Ardabili, S., Mosavi, A., Annamaria, R., Advances in machine learning modeling reviewing hybrid and ensemble methods, in: *Lecture Notes in Networks and Systems*, vol. 101, pp. 215–227, 2020, https://doi.org/10.1007/978-3-030-36841-8_21.
6. Pawar, S., Bhushan, M., Wagh, M., The plant leaf disease diagnosis and spectral data analysis using machine learning – a review. *Int. J. Adv. Sci. Technol.*, 29, 9s, 3343–3359, 2020.
7. Suri, R.S., Dubey, V., Kapoor, N.R., Kumar, A., Bhushan, M., Optimizing the compressive strength of concrete with altered compositions using hybrid PSO-ANN. *4th International Conference on Information Systems and Management Science*, University of Malta, 2021.
8. Bhargavi, P. and Jyothi, S., Applying naïve bayes data mining technique for classification of agricultural land soils. *Int. J. Inf. Secur.*, 9, 8, 2010.
9. Veenadhari, S., Misra, B., Singh, C., Machine learning approach for forecasting crop yield based on climatic parameters. *International Conference on Computer Communication and Informatics*, Coimbatore, 2014, https://doi.org/10.1109/ICCCI.2014.6921718.
10. Paul, M., Vishwakarma, S.K., Verma, A., Analysis of soil behavior and prediction of crop yield using data mining approach. *International Conference on Computational Intelligence and Communication Networks*, 2015, https://doi.org/10.1109/CICN.2015.156.
11. Ganesh, S.H., Data mining technique to predict the accuracy of the soil fertility. *Int. J. Comput. Sci. Mob. Comput.*, 4, 7, 330–333, 2015.
12. Rajeshwari, V. and Arunesh, P.K., Analyzing soil data using data mining classification techniques. *Indian J. Sci. Technol.*, 9, 19, 1–4, 2016, https://doi.org/10.17485/ijst/2016/v9i19/93873.
13. Gandhi, N., Armstrong, L.J., Petkar, O., Tripathy, A.K., Rice crop yield prediction in India using support vector machines. *13th International Joint Conference on Computer Science and Software Engineering*, 2016, https://doi.org/978-1-5090-2033-1/16.

14. Sujatha, R. and Isakki, P., A study on crop yield forecasting using classification techniques. *International Conference on Computing Technologies and Intelligent Data Engineering*, 2016, https://doi.org/10.4236/ijis.2013.3014.

15. Pantazi, X.E., Moshou, D., Alexandridis, T., Whetton, R.L., Mouazen, A.M., Yield prediction using machine learning and advanced sensing techniques. *Comput. Electron. Agric.*, 121, 57–65, 2016, https://doi.org/10.1016/j.compag.2015.11.018.

16. Rajak, R.K., Pawar, A., Pendke, M., Shinde, P., Rathod, S., Devare, A., Crop recommendation system to maximize crop yield using machine learning technique. *Int. Res. J. Eng. Technol.*, 4, 12, 950–953, 2017.

17. Geetika, and Bajaj, R., Crop yield prediction using data mining: An efficient data modeling approach. *Int. J. Eng. Technol.*, 7, 2, 128–131, 2018, https://doi.org/10.14419/ijet.v7i2.27.13157.

18. Verma, A., Jatain, A., Bajaj, S., Crop yield prediction of wheat using fuzzy c-means clustering and neural network. *Int. J. Appl. Eng. Res.*, 13, 11, 9816–9821, 2018.

19. Adnan, M., Rehman, A., Latif, M.A., Ahmad, N., Nazir, M., Akhter, N., Mapping wheat crop phenology and the yield using machine learning (ML). *Int. J. Adv. Comput. Sci. Appl.*, 9, 8, 301–306, 2018, https://doi.org/10.14569/IJACSA.2018.090838.

20. Bojani, S. and Bhatt, N., Applications of data mining techniques for wheat crop yield forecasting for districts of Gujarat state. *Int. J. Sci. Res. Publ.*, 8, 7, 302–306, 2018, https://doi.org/10.29322/IJSRP.8.7.2018.p7948.

21. Kumar, A., Kumar, N., Vats, V., Efficient crop yield prediction using machine learning algorithms. *Int. Res. J. Eng. Technol.*, 5, 6, 3151–3159, 2018.

22. Crane-Droesch, A., Machine learning methods for crop yield prediction and climate change impact assessment in agriculture. *Environ. Res. Lett.*, 13, 11, 2018.

23. Jayalakshmi, R. and Devi, M.S., Relevance of machine learning algorithms on soil fertility prediction using R. *Int. J. Comput. Intell. Inf.*, 8, 4, 193–199, 2019.

24. Khaki, S. and Wang, L., Crop yield prediction using deep neural networks. *Front. Plant Sci.*, 10, 10, 2019, https://doi.org/10.3389/fpls.2019.00621.

25. Bondre, D.A. and Mahagaonkar, S., Prediction of crop yield and fertilizer recommendation using machine learning algorithms. *Int. J. Eng. Appl. Sci. Technol.*, 4, 5, 371–376, 2019, https://doi.org/10.33564/ijeast.2019.v04i05.055.

26. Shidnal, S., Latte, M.V., Kapoor, A., Crop yield prediction: Two-tiered machine learning model approach. *Int. J. Inf. Technol.*, 13, 1983–1991, 2019, https://doi.org/10.1007/s41870-019-00375-x.

27. Khan, H. and Ghosh, S.M., Machine learning approach for crop yield prediction emphasis on k-medoid clustering and pre-processing. *International Conference on Intelligent Computing and Smart Communication,*

Algorithms for Intelligent Systems, Springer, Singapore, 2020, https://doi.org/10.1007/978-981-15-0633-8_27.

28. Bhojani, S.H. and Bhatt, N., Wheat Crop yield prediction using new activation functions in neural network. *Neural Comput. Appl.*, 32, 13941–13951, 2020, https://doi.org/10.1007/s00521-020-04797-8.

29. https://soilhealth.dac.gov.in/PublicReports/NSVW.

30. Robnik-Sikonja, M. and Kononenko, I., Theoretical and empirical analysis of reliefF and rreliefF. *Mach. Learn.*, 53, 23–69, 2003.

31. Koehler, G.J., New directions in genetic algorithm theory. *Ann. Oper. Res.*, 75, 49–68, 1997, https://doi.org/10.1023/A:1018928017332.

32. Brownlee, J., K-nearest neighbors for machine learning, 2016, https://machinelearningmastery.com.

33. Shubham, J., Naïve bayes theorem, 2018, https://becominghuman.ai/naïve-bayes-theorem-d8854a41ea08.html.

34. Chauhan, N.S., Introduction to Artificial Neural Networks (ANN), 2019, https://towardsdatascience.com.

35. Nagesh, S., Real world implementation of logistic regression, 2019, http://towardsdatascience.com.

36. Support vector machine, https://towardsdatascience.com/support-vector-machine-introduction-to-machine-learning-algorithms-934a444fca47.

37. Introduction to linear discriminant analysis in supervised learning, https://www.analyticssteps.com/blogs/introduction-linear-discriminant-analysis-supervised-learning.

8

A Status Quo of Machine Learning Algorithms in Smart Agricultural Systems Employing IoT-Based WSN: Trends, Challenges and Futuristic Competences

Abhishek Bhola[1]*, Suraj Srivastava[2], Ajit Noonia[3], Bhisham Sharma[3] and Sushil Kumar Narang[3]

[1]Department of Computer Science and Engineering, Koneru Lakshmaiah Education Foundation, Vaddeswaram, Andhra Pradesh, India
[2]Chitkara University Institute of Engineering and Technology, Chitkara University, Punjab, India
[3]Chitkara University School of Engineering and Technology, Chitkara University, Himachal Pradesh, India

Abstract

Wireless sensor network (WSN) is widely utilized in real-time practices due to its low cost, large geographical coverage, and easy deployability. The job of WSNs is to monitor a field of interest, collect data, and send it back to the base station for post-processing analysis. This field comprises various challenges, such as selection of network routing strategies due to its dynamic nature, quality of service, decreasing throughput, security, etc. In the recent past, various machine learning approaches are successfully utilized in the field of WSN to overcome the above issues. In smart agriculture, this network is used for monitoring field temperature, measuring soil quality, irrigation systems, crop production, and so on. In this work, the study and discussions about the current trends, various challenges, and their possible machine learning-based solutions for smart agriculture using WSN are presented with their futuristic scope.

Keywords: Wireless sensor network, machine learning, smart agriculture, Internet of Things

*Corresponding author: abhishek_bhola@kluniversity.in

Ashok Kumar, Megha Bhushan, José A. Galindo, Lalit Garg and Yu-Chen Hu (eds.) Machine Intelligence, Big Data Analytics, and IoT in Image Processing: Practical Applications, (177–196) © 2023 Scrivener Publishing LLC

8.1   Introduction

As the world's population grows, increasing crop production becomes a challenging task in the 21st century [1]. Agriculture is not only important to the global economy but is also the only way for the entire world to survive. In India, the agriculture sector employs 53% of the population, while 61.5% of the population relies on agriculture for their survival. Conventional farming techniques and processes are outperformed. Traditional ways of planting the same crop year after year deplete the soil's nutrients and there are no effective means for detecting the illness in its early stages, resulting in lower agricultural production [2].

Agricultural problems are on the rise, affecting farm economics and the environment. Some issues presently faced by the agriculture sector are due to rising population, market demands, and limited land, water, and energy [3]. Smart agricultural technology is designed to assist in the search for a solution. The rapid growth of wireless communication networks, as well as the availability of a wide range of new remote, proximal, and touch sensors, are creating new opportunities in the agriculture sector. Smart farming's purpose is to replace traditional farming techniques with modern information and communication technologies-based solutions [4].

These technologies facilitate the gathering and transmission of real-time information in an agricultural context at a minimal cost. These data, once collected, processed, and analyzed, can help determine the state of the agricultural environment (e.g., soil, crop, water, and climate). All of these factors are included in the paradigm of "smart farming," to boost agricultural goods production efficiency and improve quality on a long-term basis. Smart agriculture is achieved with a variety of modern technologies, including cloud computing, image processing, machine learning (ML), Big Data, wireless sensor network (WSNs), unmanned aerial vehicles (UAVs), and unmanned ground vehicles (UGVs) [5].

In agriculture, WSNs help in providing real-time monitoring and intelligent decision making to increase yields while lowering costs [6]. A WSN node used in smart agriculture can continuously monitor and control physical data such as temperature, humidity, water, soil, crop, energy, and so on. The primary purpose of these sensor nodes is to collect data and transfer it to the key WSN gateway node, where it is analyzed and computed [7–9].

WSN faces a lot of challenges in the agriculture domain including sensor node size and placement, energy consumption [10], node coverage and connectivity, routing issues, sensor data aggregation, event monitoring and target detection [11]. In the last few years, researchers have frequently used ML approaches in aggregation with wireless sensor networks. Farm management systems are growing into true artificial intelligent systems by applying ML to sensor data, offering optimal insights for decisions and actions. ML improves agricultural productivity by automating operations and maximizing the use of natural resources in smart agriculture.

This chapter is organized as follows: Section 8.2 discusses various types of wireless sensor used in smart agriculture to measure and monitor various parameters. Section 8.3 provides a literature survey of machine learning in WSN. In the subsequent Section 8.4, machine learning–based solutions using the WSN in smart agriculture are described with their applications. Section 8.5 presents futuristic applications in agriculture domain and finally, Section 8.6 provides the conclusion and future work.

8.2   Types of Wireless Sensor for Smart Agriculture

The section discusses various types of sensors with their applications which are utilized in the field of smart agriculture to monitor and measure various parameters. These sensors help in increasing the overall capabilities of the wireless sensor platform. For a better overview, this work mainly categorized these sensors into three types i.e., soil, environmental, and plant-related. Table 8.1 presents different sensors, along with a range of measuring parameters related to the soil, environment and plants in smart agriculture.

Environmental sensors help in monitoring humidity, temperature, and wind speed to be used in conjunction with soil and plant sensors for a variety of agricultural issues. Hybrid environmental sensors lead to intelligent and better decision-making. Plant Sensors work by attaching to the plants in smart agricultural applications which are mentioned in Table 8.1. Plant's sensors help in controlling fertilizer use, control quality of the crop, pesticides, cattle motion monitoring and control.

The next section discusses application of ML and WSN in the smart agriculture domain.

8.3   Application of Machine Learning Algorithms for Smart Decision Making in Smart Agriculture

WSN comprises of number of sensor devices that are geographically distributed that aim to gather environmental data through exposed sensor

Table 8.1 Various sensors employed in smart agriculture.

Types	Sensor names	Measuring parameters
Soil Sensors	Pogo portable, Hydra Probe II s and EC-250	Temperature, Moisture, Rain and water flow, Conductivity, Salinity; Water level
	ECH_2O EC-5	Moisture
	VH-400	Moisture, water-level
	THERM200	Temperature
	Tipping Bucket Rain Gage	Rain and Water Flow
	AquaTrak 5000	Water level
	WET-2	Temperature, Conductivity, Salinity
Environmental Sensors	WXT520 compact, CM-100 compact and Met Station One (MSO)	Direction of wind, Humidity, High temperature, Pressure of Atmosphere, Speed of Wind, Rain
	All-In-One (AIO) Weather	High temperature, humidity, high temperature, speed of wind, direction of wind
	XFAM-115KPASR	High temperature, Humidity, Pressure of Atmosphere
	RM Young (model 5103)	Direction of wind speed of wind
	Met -One-Series 380 Rain and RG13 and RG13H	Rain
	LI-200 and CS300-L Pyranometer	Solar radiation
Plant Sensors	LW100 and 237-L, leaf wetness, SenseH2™ hydrogen	Moisture, Temperature, Wetness
	TPS-2 portable, Cl-340 hand-held and PTM-48A Photosynthesis	Moisture, Temperature, Wetness, CO_2, Photosynthesis

nodes for future analysis. Wireless Sensor Networks are explicitly programmed which makes their working untrustworthy and ineffective in the dynamic environment. When WSN are exposed to work in a dynamic environment such as agriculture, it can affect network routing strategies, node localization, delay, area coverage, fault detection, etc. the promising problems. Therefore, ML methods are applied in this field as these techniques are based on the concept of self-learning from experiences and do

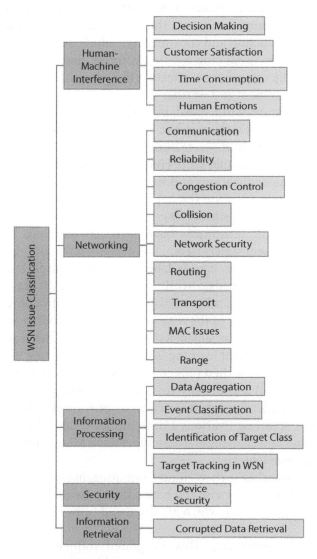

Figure 8.1 Classification of WSN issues.

not need explicit programming. ML makes computing procedures more efficient, trusty and low-priced. On other hand, ML advances the performance of the WSN and also limits human interventions. In this section, various WSN issues are firstly classified as shown in Figure 8.1.

The prediction and monitoring of crop yields, weather conditions and detection of plant diseases make a promising task for farmers in increasing crop production. There are various factors associated with the ground which affects crop production: the depth of the soil, structure and composition of the soil organic matter content, and presence of nitrogen, phosphorus and potassium fertilizers. The factors associated with irrigation include water production and consumption, which depends upon the type of irrigation system used, crop rotation patterns, and pest and weed control methods. The other factors affecting the productivity of the crop also depend on environment and weather data such as air temperature, precipitation, solar radiation, wind-speed, pest presence, weed control, and use of biological diversity. By collecting and analyzing the above-mentioned data using wireless sensors and ML techniques, helps in predicting the overall productivity of the particular crop. Human Interface can be reduced by the application of ML applications and helps to resolve various issues such as decision making in irrigation (by applying regression) and crop selection (using clustering and classifications).

Machine Learning techniques have been classified into "supervised," "unsupervised," "semisupervised," and "reinforcement" learning. In "supervised learning," the data is labeled (input & output), and it finds the relationship between data while system training on the other hand in unsupervised learning, where unlabeled data (i.e., no output is associated) is utilised. But most of the real-world problems consist of both types of data: labelled as well as unlabeled and this makes the learning semisupervised. Reinforcement learning is learned by interacting with the environment and actions are then ranked based on rewards (when appropriate action is performed) and penalty (in case of the wrong action) concepts. An attempt is made through this article to provide solutions to various WSN challenges through various ML techniques. Table 8.2 presents various WSN issues and challenges where supervised ML methods are utilized to overcome some issues.

The new approaches such as cloud computing, big data and the Internet of Things (IoT), are on the rise and have a huge potential for smart agricultural applications [45–48], which are discussed below:

Cloud Computing: This integrated framework provides huge computing power, storage and resources reusability.

Table 8.2 Machine learning solutions for various WSN issues.

Solution types	References	Issues	Method applied
Supervised Machine Learning	[12]	Node localization	Regression
	[13]	Coverage problem	Random Forest
	[14, 15]	Anomaly and fault detection	Decision tree, Deep Learning
	[16]	Routing	Artificial Neural Network
	[17]	MAC	Random Forest
	[18]	Data aggregation	Decision tree, Regression, ANN
	[19]	Synchronization	Bayesian network
	[20]	Congestion control	Artificial Neural Network
	[21]	Target tracking	Bayesian
	[22]	Event detection	Bayesian
	[23]	Energy harvesting	Regression
	[24]	Mobile Sink Path Selection	Decision tree
	[25]	Connectivity Problem	Regression
	[26]	Detecting faulty sensor nodes	Artificial Neural Network
	[27]	Data Quality Estimation	Deep Learning
Unsupervised Machine Learning	[28]	Routing	Singular Value Decomposition
	[29]	Data Aggregation	Principal Component Analysis
	[30]	Mobile Sink Path Selection	K-mean Clustering
	[31]	Synchronization	Hierarchical Clustering

(Continued)

Table 8.2 Machine learning solutions for various WSN issues. (*Continued*)

Solution types	References	Issues	Method applied
	[32]	Energy harvesting	Hierarchical Clustering
	[33, 34]	Localization	Principal Component Analysis, Fuzzy c-mean Clustering
	[35]	Connectivity	Fuzzy c-mean Clustering
	[36]	Fault Detection	Principal Component Analysis
	[37]	Target Tracking	Principal Component Analysis
Semi-supervised Machine Learning	[38]	Localization	Semi-supervised hidden Markov Models
	[39]	Fault detection	Semi-supervised local kernel density estimation
Reinforcement Machine Learning	[40]	Localization	Artificial Bee Colony Algorithm
	[41]	Coverage	Decomposition based multiobjective evolutionary algorithm
	[42]	Routing	Ant colony optimization
	[43]	Target racking	Genetic algorithm
	[44]	Mobile sink	Particle swarm optimization

Some important applications in smart agriculture are Spatial diversity and ecosystems for different seasons; Health monitoring and crop speculation; Irrigation system for the large agricultural area; to produce seasonal vegetables and flowers in tropical agriculture.

Internet of Things (IoT): This concept defines "objects" that can identify, and communicate with surrounding objects. IoT provides a flexible control method and potential solution for smart Agriculture. Some IoT applications for smart agriculture include RFID tags for Inexpensive agricultural buying; Remote monitoring of livestock movement; Remote reporting on farms for pest count; Automatic pesticide spray; Remote water flow control and Leakage detection [12].

Big Data Analytics: These are used to obtain relevant insights from vast and heterogeneous data. Big-data analytics helps in finding hidden relationships, customer preferences, anonymous patterns, styles of business, crime detection and disasters, etc. Few applications of big data analytics in smart agriculture are: plant growth and disease management; developing advanced agricultural information; smart agricultural control machinery systems; predicting crop yield and climate data; good policy determination in terms of data statistics.

There are several Wireless Communication approaches and technologies utilized in smart agriculture, such as ZigBee, Lora, Wi-Fi, 4G/5G, WiMAX, and SigFox. Various parameters associated with these technologies include fuel consumption, range, data transfer rate, cost, complexity etc. Electricity usage is measured as a constraint in the design of WSNs applications in smart agriculture. ZigBee wireless protocol is designed for communication, operation, and low energy consumption. SigFox, LoRa, Wi-Fi,4G/5G, and WiMAX operate with low energy on long-range distance [49].

8.4 ML and WSN-Based Techniques for Smart Agriculture

Table 8.3 provides a case study of algorithms and methods used to predict crop yields, irrigation management, and the detection of plant diseases, using ML and WSN technology.

Table 8.3 Analysis of different ML and WSN techniques for smart agriculture.

ML algorithms	WSN	Approach	Applications in smart agriculture
Kernel-based-Extreme Learning Machine [50]	ZigBee	Compares lower back (bp), SVM and Elman	Predicts temperature and moistness
Fuzzy logic [51]	GSM and GPS	Fuzzy rules	Predicts crop growth using soil *ph*
ARIMA [52]	LoRa	ARIMA version is compared with ANN and RF	Predicts crop health disorders
Data mining techniques, linear regression, and priori algorithm [53]	WSN	linear regressive and Associative rules for data mining	Predicts the correct temperature, humidity, moisture and crop yield
Naives Bayes Kernel approach [54]	ZigBee	Various Sensor data are utilised	Predicts diseases, pests' detection and quality of the crop
Bayesian statistics [55]	WiFi and MQTT Sensor	Optimization algorithm, based on sensors data	Prediction of crop quality
Machine vision and statistical analysis [56]	WiFi, ZigBee	ANOVAs	Find the team growth, and increasing their productivity
Support Vector Machine Linear regression, and neural networks [57]	ZigBee	Data mining techniques	Predicts optimal conditions for the cultivation of Roses in a greenhouse

(Continued)

Table 8.3 Analysis of different ML and WSN techniques for smart agriculture. (*Continued*)

ML algorithms	WSN	Approach	Applications in smart agriculture
Feedforward Long short-term memory and gated recurrent unit [58]	ZigBee and GSM	Training of the ML model	Predicts and determine the appropriate crops for sowing
KNN, RF, Logistic regression, and linear regression [59]	ZigBee	KNN to be very effective in the case of each of the methods of ML	Predicts disease outbreaks
Neural networks and Multi-Layered Perceptron [60]	WSN (pH, soil moisture, salinity, electromagnetic sensors)	MLP, and a neural network model for decision making	Suitability assessment of the land
Regression vector, different nonlinear model structures emerge, ANN model [61]	WSN/IoT	The neural network model for training	Forecasting max. and min. T On the field level
Gradient Boosting Regression Trees, Multiple Linear Regression [62]	WSN/IoT	The soil temperature was determined with the help of the ENR, and MLR	soil moisture prediction
Ensemble modelling [63]	MQTT Sensors	ML of the use of the collected data in real-time to make the right decision.	Smart Irrigation, conserving water, labor and plant nutrients

(*Continued*)

Table 8.3 Analysis of different ML and WSN techniques for smart agriculture. (*Continued*)

ML algorithms	WSN	Approach	Applications in smart agriculture
ML traditional linear regression and two nonparametric methods [64]	WSN	ML is suitable for use as a traditional, linear regression, and two nonparametric approaches GBRT, and BTC	Precisely forecasting irrigation decisions
Logistic Regression models [65]	WSN	The regression models have lower accuracy, including information about the neighbours	frost Prediction
supervised machine learning techniques [66]	IoT sensors dataset	Decision trees, KNN and neural networks	Environmental factors (weather, humidity, wind)
Support vector regression and k-means clustering [67]	ZigBee	SVR & k-means approach	Soil moisture prediction

8.5   Future Scope in Smart Agriculture

Smart agriculture is a powerful technology to meet the talent standards growing needs from limited resources. Apart from the number of benefits, many challenges in agriculture are faced by farmers, researchers, and scientists, which are listed below:

> Economic issues: while any smart agricultural solution is being developed, cost analysis should be done. As it is known

from ML and WSN-based technologies, initial expenses are unreasonable due to the variety of sensors, gates, and channel configurations. Following this change is also a significant part of cost breaking. Here, the scientist and researcher's goal is to consistently increase the crop productivity while minimizing the need for the cost [67].

Technical Issues: Agricultural sector, most of the computer systems are revealed to the external region. This forms it easier for nerves to be destroyed for a variety of reasons such as rain, wind, snow, etc. Briefly, sometimes rest periods can be triggered by a power failure. Therefore, a backup of everything needs to be kept to meet the requirement. Failure of one point has never led to success. There are other important technical issues in rural areas around the world especially in countries in growth and advancement process, such as India with lightning speed and trustworthy Internet connection does not exist [68]. As a result of wireless communication diversified issues such as multipath multiplication, occurring as a result of structural trees, etc. The system cannot work reliably. Therefore, there is requirement to develop resilient and dependable technology.

Data analytics issues: Today's world is making many businesses and organization more profitable through data. The same is true for smart agriculture also. Different promising issues faced by researchers and scientists in data analytics are listed below:

a) Data integration: It means combining data from distinct sources of dissimilar formats within a single concept. Fine details are obtained with good details that are obtained through proper integration. However, it is a promising work to compile information from a variety of origins. Correcting the formation of slow and unstructured data is also a major issue that needs to be addressed [69]. New data mining approaches and methodology have been investigated in the form of technologies concerning computers using algorithm integration methods, distributed and packaged. Performance analysis can be carried out using computation complexity, energy consumption, and communication requirements. Integration of heterogeneous data efficiently is a key challenge in smart agriculture [70].

b) Mining information: It refers to the prediction of efficient and significant resolutions for large data that can be replicated to new data objects. Data extraction and data testing are promising challenges due to large data platforms and clouds.

c) Visibility: It is the main component of technical data when managing IoT systems. This is a very difficult task when working on the Internet of Things due to its unique and chaotic structure, which leads to inaccurate results [71].

8.6   Conclusion

This chapter presented the ML and WSN methods that are used to predict crop yields, irrigation management, and the detection of plant diseases. Various researchers have investigated the prediction of yield by machine learning and WSN but differ in application and approaches. Studies have also shown that multidimensional models did not always give the optimal yield forecasting results. Several models have been tested to find the most effective model. The results suggest that no sole decision can be taken about which model is the best, some ML-based WSN models are definitely more frequently implemented than others. Much research has tested which model provides the greatest prediction using a range of learning tools. It is noted that ANN, logistic, and linear regression, SVM, and classification are the most popular algorithms used with WSN technology. This chapter presented all the research development on predicting crop yields. In future studies, focus on the development of an ML and WSN-based crop prediction model can be considered.

References

1. Navarro, E., Costa, N., Pereira, A., A systematic review of IoT solutions for smart farming. *Sensors*, 20, 15, 4231, 2020.
2. Kour, V.P. and Arora, S., Recent developments of the Internet of Things in agriculture: A Survey. *IEEE Access*, 8, 129924–57, 2020.

3. Mekonnen, Y., Namuduri, S., Burton, L., Sarwat, A., Bhansali, S., Review—Machine learning techniques in wireless sensor network based precision agriculture. *J. Electrochem. Soc.*, 167, 3, 2019.

4. Moysiadis, V., Sarigiannidis, P., Vitsas, V., Khelifi, A., Smart farming in Europe. *Comput. Sci. Rev.*, 1, 39, 2021.

5. Balafoutis, A.T., Evert, F.K.V., Fountas, S., Smart farming technology trends: Economic and environmental effects, labor impact, and adoption readiness. *Agronomy*, 10, 5, 2020.

6. Zervopoulos, A., Tsipis, A., Alvanou, A.G., Bezas, K., Papamichail, A., Vergis, S. *et al.*, Wireless sensor network synchronization for precision agriculture applications. *Agriculture*, 10, 3, 1–20, 2020.

7. Chand, T. and Sharma, B., HRCCTP: A hybrid reliable and congestion control transport protocol for wireless sensor networks, in: *2015 IEEE Sensors*, pp. 1–4, IEEE, Busan, Korea (South), 2015.

8. Kharb, K., Sharma, B., Trilok, C.A., Reliable and congestion control protocols for wireless sensor networks. *Int. J. Eng. Technol. Innov.*, 6, 1, 68, 2016.

9 Bajaj, K., Sharma, B., Singh, R., *Integration of WSN with IoT applications: A vision, architecture, and future challenges*, Springer, Switzerland AG, 2020.

10. Rajasekaran, T. and Anandamurugan, S., Challenges and applications of wireless sensor networks in smart farming—A survey, in: *Advances in Big Data and Cloud Computing*, J.D. Peter, A.H. Alavi, B. Javadi (Eds.), pp. 353–61, Springer., Singapore, 2019, Advances in Intelligent Systems and Computing.

11. Sharma, H., Haque, A., Blaabjerg, F., Machine learning in wireless sensor networks for smart cities: A survey. *Electronics*, 10, 9, 2021.

12. Banihashemian, S.S., Adibnia, F., Sarram, M.A., A New range-free and storage-efficient localization algorithm using neural networks in wireless sensor networks. *Wirel. Pers. Commun.*, 98, 1, 1547–1568, 2018.

13. Shu, J., Liu, S., Liu, L., Zhan, L., Hu, G., Research on link quality estimation mechanism for wireless sensor networks based on support vector machine. *Chin. J. Electron.*, 1, 26, 377–84, 2019.

14. Xie, M., Hu, J., Han, S., Chen, H.-H., Scalable hypergrid k-NN-based online anomaly detection in wireless sensor networks. *IEEE Trans. Parallel Distrib. Syst.*, 24, 8, 1661–70, 2013 Aug.

15. Ma, T., Wang, F., Cheng, J., Yu, Y., Chen, X., A hybrid spectral clustering and deep neural network ensemble algorithm for intrusion detection in sensor networks. *Sensors*, 16, 10, 1701, 2016.

16. Mehmood, A., Lv, Z., Lloret, J., Umar, M.M., ELDC: An artificial neural network based energy-efficient and robust routing scheme for pollution monitoring in WSNs. *IEEE Trans. Emerging Top. Comput.*, 8, 1, 106–14, 2020.

17. Alotaibi, B. and Elleithy, K., A new MAC address spoofing detection technique based on random forests. *Sensors*, 16, 3, 281, 2016.

18. Edwards-Murphy, F., Magno, M., Whelan, P.M., O'Halloran, J., Popovici, E.M., b+WSN: Smart beehive with preliminary decision tree analysis for

agriculture and honey bee health monitoring. *Comput. Electron. Agric.*, 1, 124, 211–9, 2016.

19. Pérez-Solano, J.J. and Felici-Castell, S., Improving time synchronization in wireless sensor networks using bayesian inference. *J. Netw. Comput. Appl.*, 82, 47–55, 2017.

20. Rezaee, A.A. and Pasandideh, F., A fuzzy congestion control protocol based on active queue management in wireless sensor networks with medical applications. *Wirel. Pers. Commun.*, 98, 1, 815–42, 2018.

21. Zhou, B., Chen, Q., Li, T.J., Xiao, P., Online variational bayesian filtering-based mobile target tracking in wireless sensor networks. *Sensors*, 14, 11, 21281–315, 2014.

22. Avci, B., Trajcevski, G., Tamassia, R., Scheuermann, P., Zhou, F., Efficient detection of motion-trend predicates in wireless sensor networks. *Comput. Commun.*, 101, 26–43, 2017.

23. Sharma, A. and Kakkar, A., Forecasting daily global solar irradiance generation using machine learning. *Renewable Sustainable Energy Rev.*, 82, 2254–69, 2018.

24. Kim, S. and Kim, D.-Y., Efficient data-forwarding method in delay-tolerant P2P networking for IoT services. *Peer Peer Netw. Appl.*, 11, 6, 1176–85, 2018.

25. End-to-end data delivery reliability model for estimating and optimizing the link quality of industrial WSNs [Internet]. *IEEE Transactions on Automation Science and Engineering*, 15, 3, 1127–1137, 2021, Available from: https://ieeexplore.ieee.org/document/8022974/.

26. Chanak, P. and Banerjee, I., Fuzzy rule-based faulty node classification and management scheme for large scale wireless sensor networks. *Expert Syst. Appl.*, 45, 307–21, 2016.

27. Wang, Y., Yang, A., Chen, X., Wang, P., Wang, Y., Yang, H., A deep learning approach for blind drift calibration of sensor networks. *IEEE Sens. J.*, 17, 13, 4158–71, 2017.

28. Guo, P., Cao, J., Liu, X., Lossless in-network processing in WSNs for domain-specific monitoring applications. *IEEE Trans. Ind. Inf.*, 13, 5, 2130–9, 2017.

29. Morell, A., Correa, A., Barceló, M., Vicario, J.L., Data aggregation and principal component analysis in WSNs. *IEEE Trans. Wireless Commun.*, 15, 6, 3908–19, 2016.

30. Ray, A. and De, D., Energy efficient clustering protocol based on K-means (EECPK-means)-midpoint algorithm for enhanced network lifetime in wireless sensor network. *IET Wireless Sens. Syst.*, 6, 6, 181–91, 2016.

31. Neamatollahi, P., Abrishami, S., Naghibzadeh, M., Yaghmaee, Moghaddam, M.H., Younis, O., Hierarchical clustering-task scheduling policy in cluster-based wireless sensor networks. *IEEE Trans. Ind. Inf.*, 14, 5, 1876–86, 2018.

32. Awan, S.W. and Saleem, S., Hierarchical clustering algorithms for heterogeneous energy harvesting wireless sensor networks, in: *2016 International Symposium on Wireless Communication Systems (ISWCS)*, pp. 270–4, 2016.

33. Li, X., Ding, S., Li, Y., Outlier suppression via non-convex robust PCA for efficient localization in wireless sensor networks. *IEEE Sens. J.*, 17, 21, 7053–63, 2017.

34. Zhu, F. and Wei, J., Localization algorithm for large-scale wireless sensor networks based on FCMTSR-support vector machine. *Int. J. Distrib. Sens. Netw.*, 12, 10, 1550147716674010, 2016.

35. Qin, J., Fu, W., Gao, H., Zheng, W.X., Distributed k -means algorithm and fuzzy c -means algorithm for sensor networks based on multiagent consensus theory. *IEEE Trans. Cybern.*, 47, 3, 772–83, 2017.

36. Islam, M.R., Uddin, J., Kim, J.M., Acoustic emission sensor network based fault diagnosis of induction motors using a gabor filter and multiclass support vector machines. *Ad Hoc Sens. Wirel. Netw.*, 34, 273–287, 2016.

37. Oikonomou, P., Botsialas, A., Olziersky, A., Kazas, I., Stratakos, I., Katsikas, S. *et al.*, A wireless sensing system for monitoring the workplace environment of an industrial installation. *Sens. Actuators B: Chem.*, 224, 266–74, 2016.

38. Kumar, S., Tiwari, S.N., Hegde, R.M., Sensor node tracking using semi-supervised Hidden Markov Models. *Ad Hoc Networks*, 33, 55–70, 2015.

39. Zhao, M. and Chow, T.W.S., Wireless sensor network fault detection via semi-supervised local kernel density estimation, in: *2015 IEEE International Conference on Industrial Technology (ICIT)*, pp. 1495–500, 2015.

40. Hashim, H.A., Ayinde, B.O., Abido, M.A., Optimal placement of relay nodes in wireless sensor network using artificial bee colony algorithm. *J. Netw. Comput. Appl.*, 64, 239–48, 2016.

41. Xu, Y., Ding, O., Qu, R., Li, K., Hybrid multi-objective evolutionary algorithms based on decomposition for wireless sensor network coverage optimization. *Appl. Soft Comput.*, 68, 268–82, 2018.

42. An improved routing algorithm based on ant colony optimization in wireless sensor networks [Internet]. *IEEE Communications Letters*, 21, 6, 1317–1320, 2021, Available from: https://ieeexplore.ieee.org/abstract/document/7862755/.

43. Elhoseny, M., Tharwat, A., Farouk, A., Hassanien, A.E., K-coverage model based on genetic algorithm to extend WSN lifetime. *IEEE Sens. Lett.*, 1, 4, 1–4, 2017.

44. Wang, J., Cao, Y., Li, B., Kim, H., Lee, S., Particle swarm optimization based clustering algorithm with mobile sink for WSNs. *Future Gener. Comput. Syst.*, 76, 452–7, 2017.

45. Sharma, S., Nanda, M., Goel, R., Jain, A., Bhushan, M., Kumar, A., Smart cities using Internet of Things: *Recent Trends and Techniques. Int. J. Innov. Technol. Exploring Eng. (IJITEE)*, 8, 9S, 24–28, 2019.a, V., Bhushan, M., Mohanty, S.N., *Real-life applications of the Internet of Things: Challenges, applications, and advances*, CRC Press, Taylor & Francis Group, Florida, USA January 2022, 9781774638477, Available at: https://www.appleacademicpress.com/real-life-applications-of-internet-of-things-challenges-applications-and-advances/9781774638477.

46. Mangla., M., Kumar, A., Mehta, V., Bhushan, M., Mohanty, S.N., *Real-life appli-cations of the Internet of Things: Challenges, applications, and advances*, CRC Press, Taylor & Francis Group, Florida, USA, January 2022, 9781774638477, Available at: https://www.appleacademicpress.com/real-life-applications-of-internetof-things-challenges-applications-and-advances/9781774638477.

47. Goel, R., Jain, A., Verma, K., Bhushan, M., Kumar, A., Mushrooming trends and technologies to aid visually impaired people. *International Conference on Emerging Trends in Information Technology and Engineering (ic-ETITE'20)*, February 24-25, 2020, IEEE, Vellore Institute of Technology, Vellore, India.

48. Kumar, A., Sharma, A., Kumar, R., SERVmegh: Framework for green cloud. *Concurr. Comput.: Pract. Exp.*, 29, 1–20, 2016, https://doi.org/10.1002/cpe.3903.

49. Ojha, T., Misra, S., Raghuwanshi, N., Wireless sensor networks for agricul-ture: The state-of-the-art in practice and future challenges. *Comput. Electron. Agric.*, 118, 66–84, 2015.

50. Liu, Q., Jin, D., Shen, J., Fu, Z., Linge, N., A WSN-based prediction model of microclimate in a greenhouse using extreme learning approaches, in: *2016 18th International Conference on Advanced Communication Technology (ICACT)*, pp. 730–5, 2016.

51. Channe, H., Kothari, S., Kadam, D., Multidisciplinary model for smart agriculture using Internet-of-Things (IoT). Sensors, cloud-computing, mobile-computing & big-data analysis. *Int. J. Computer Technology & Applications (IJCTA)*, 6, 9, 2015.

52. dos Santos, U.J.L., Pessin, G., da Costa, C.A., da Rosa Righi, R., AgriPrediction: A proactive internet of things model to anticipate problems and improve production in agricultural crops. *Comput. Electron. Agric.*, 161, 202–13, 2019.

53. Muangprathub, J., Boonnam, N., Kajornkasirat, S., Lekbangpong, N., Wanichsombat, A., Nillaor, P., IoT and agriculture data analysis for smart farm. *Comput. Electron. Agric.*, 156, 467–74, 2019.

54. Wani, H. and Ashtankar, N., An appropriate model predicting pest/dis-eases of crops using machine learning algorithms, in: *2017 4th International Conference on Advanced Computing and Communication Systems (ICACCS)*, pp. 1–4, 2017.

55. Machine learning based cloud integrated farming. *Proceedings of the 2017 International Conference on Machine Learning and Soft Computing* [Internet], 2017, Available from: https://dl.acm.org/doi/abs/10.1145/3036290.3036297.

56. Liao, M.-S., Chen, S.-F., Chou, C.-Y., Chen, H.-Y., Yeh, S.-H., Chang, Y.-C. *et al.*, On precisely relating the growth of Phalaenopsis leaves to greenhouse environmental factors by using an IoT-based monitoring system. *Comput. Electron. Agric.*, 136, 125–39, 2017.

57. Rodríguez, S., Gualotuña, T., Grilo, C., A system for the monitoring and pre-dicting of data in precision agriculture in a rose greenhouse based on wire-less sensor networks. *Proc. Comput. Sci.*, 121, 306–13, 2017.

58. Aruul Mozhi Varman, S., Baskaran, A.R., Aravindh, S., Prabhu, E., Deep learning and IoT for smart agriculture using WSN, in: *2017 IEEE International Conference on Computational Intelligence and Computing Research (ICCIC)*, pp. 1–6, 2017.

59. Materne, N. and Inoue, M., IoT monitoring system for early detection of agricultural pests and diseases, in: *2018 12th South East Asian Technical University Consortium (SEATUC)*, pp. 1–5, 2018.

60. Vincent, D.R., Deepa, N., Elavarasan, D., Srinivasan, K., Chauhdary, S.H., Iwendi, C., Sensors driven AI-based agriculture recommendation model for assessing land suitability. *Sensors*, 19, 17, 3667, 2019.

61. Aliev, K., Jawaid, M.M., Narejo, S., Pasero, E., Pulatov, A., Internet of plants application for smart agriculture. *Int. J. Adv. Comput. Sci. Appl. (IJACSA)*, 9, 4, 421–429, 2018, Available from: https://thesai.org/Publications/ViewPaper?Volume=9&Issue=4&Code=IJACSA&SerialNo=58.

62. Singh, G., Sharma, D., Goap, A., Sehgal, S., Shukla, A.K., Kumar, S., Machine learning based soil moisture prediction for Internet of Things based smart irrigation system, in: *2019 5th International Conference on Signal Processing, Computing and Control (ISPCC)*, pp. 175–80, 2019.

63. Goap, A., Sharma, D., Shukla, A.K., Rama Krishna, C., An IoT based smart irrigation management system using machine learning and open source technologies. *Comput. Electron. Agric.*, 155, 41–9, 2018.

64. Goldstein, A., Fink, L., Meitin, A., Bohadana, S., Lutenberg, O., Ravid, G., Applying machine learning on sensor data for irrigation recommendations: Revealing the agronomist's tacit knowledge. *Precis. Agric.*, 19, 3, 421–44, 2018.

65. Diedrichs, A.L., Bromberg, F., Dujovne, D., Brun-Laguna, K., Watteyne, T., Prediction of frost events using machine learning and IoT sensing devices. *IEEE Internet Things J.*, 5, 6, 4589–97, 2018.

66. Balducci, F., Impedovo, D., Pirlo, G., Machine learning applications on agricultural datasets for smart farm enhancement. *Machines*, 6, 3, 38, 2018.

67. Marjani, M., Nasaruddin, F., Gani, A., Karim, A., Hashem, I.A.T., Siddiqa, A. et al., Big IoT data analytics: Architecture, opportunities, and open research challenges. *IEEE Access*, 5, 5247–61, 2017.

68. Precision agriculture: Top 15 challenges and issues [Internet], 2021, Available from: https://teks.co.in/site/blog/precision-agriculture-top-15-challenges-and-issues/.

69. Savaglio, C., Gerace, P., Di Fatta, G., Fortino, G., Data Mining at the IoT edge, in: *2019 28th International Conference on Computer Communication and Networks (ICCCN)*, pp. 1–6, 2019.

70. Verma, K., Bhardwaj, S., Arya, R., Ul Islam, M.S., Bhushan, M., Kumar, A., Samant, P., Latest tools for data mining and machine learning. *Int. J. Innov. Technol. Exploring Eng.*, 8, 9S, 18–23, 2019.

71. Mukhopadhyay, A., Maulik, U., Bandyopadhyay, S., Coello, C.A.C., A survey of multiobjective evolutionary algorithms for data mining: Part I. *IEEE Trans. Evol. Comput.*, 18, 1, 4–19, 2014.

Part III
SMART CITY AND VILLAGES

Impact of Data Pre-Processing in Information Retrieval for Data Analytics

Huma Naz[1]*, Sachin Ahuja[2], Rahul Nijhawan[1] and Neelu Jyothi Ahuja[1]

[1]School of Computer Science, University of Petroleum and Energy Studies, Dehradun, India
[2]Chitkara University Institute of Engineering and Technology, Chitkara University, Punjab, India

Abstract

In recent years, data-driven decision making has emerged as the main focus of research due to the extensive use and availability of data-driven approaches. The accuracy of such research studies is completely dependent on the quality of data available for the research. To enhance the performance of the model, diverse "data pre-processing" techniques are adopted by the researchers. This chapter attempts to provide an insight into the application of data pre-processing techniques and their effects on information retrieval. That further takes into consideration a few chosen problems involving huge amounts of data. This chapter covers the major issues that need to be dealt with before the beginning of any data analysis process. The chapter consists of two sections that highlight the need for data pre-processing. To establish the need for data pre-processing and study its effects on the achieved results, three machine learning algorithms named decision tree, Naive Bayes, and artificial neural network were applied to four diverse datasets. The result shows that high accuracy, as well as better data quality, is attained after the application of data pre-processing methods. The solution can be used to solve the problem of data discrepancies, noise, and outliers in different datasets for improved results.

Keywords: Data pre-processing techniques, data analytics, information retrieval, decision tree, neural network, impact of pre-processing

Corresponding author: humanaaz168@gmail.com

Ashok Kumar, Megha Bhushan, José A. Galindo, Lalit Garg and Yu-Chen Hu (eds.) Machine Intelligence, Big Data Analytics, and IoT in Image Processing: Practical Applications, (199–224) © 2023 Scrivener Publishing LLC

9.1   Introduction

In today's world, a huge amount of data is accumulated from multiple sources, which are further used for decision making and information extraction. But, data in the real world is quite inconsistent, noisy, and contains missing values [1]. At times, it also contains some extremely indifferent features known as outliers that need to be removed for achieving appropriate and correct results. As a result, Data pre-processing is the most influential and required step to be performed while accomplishing data analysis tasks. Data pre-processing is the overall process of making data more relevant for data mining. It includes several tasks to transform inconsistent data into quality data [2]. As shown in Figure 9.1, pre-processing of data can be considered as one of the major steps to be performed before data analysis tasks to achieve the expected results as processed or prepared data can only produce better and appropriate results [3]. However, each step has its equivalent importance in the process of Knowledge Discovery in Databases (KDD) but data pre-processing seems to be the most difficult task in the whole mechanism.

The process of converting raw data into an appropriate style for effective analysis is known as data pre-processing, which includes:

- Data collection from diverse sources or repositories
- Outliers and noise removal to manage unambiguity in data
- Assortment of most persuading features for performance upgrading

9.1.1   Tasks Involved in Data Pre-Processing

As evident from the introduction, data pre-processing is required to attain the expected results. Data pre-processing incorporates various tasks that are preferred according to the dataset. This section provides a brief

Figure 9.1 Stages of data pre-processing.

introduction to all the tasks that can be done for data pre-processing. The first step in data pre-processing is data cleaning, which can be done using various techniques available, depending on the type of data. If the targeted dataset contains numerical, nominal, or polynomial values, but not all techniques can process different data types depending on the chosen technique, there is a need to convert the data into a required format accepted by the technique by using data pre-processing techniques. Another important task involved in data pre-processing is data reduction, which requires both domain expertise and technical knowledge. Since the dataset may contain various attributes that may or may not contribute to the expected result [4], at times, there may be redundant values in the dataset, which need to be eliminated for improving the data quality [5]. This can be done using various data reduction techniques, depending on the type of data and algorithm used [6]. Various tasks that are involved in pre-processing of data are shown in Figure 9.2 and further explained below.

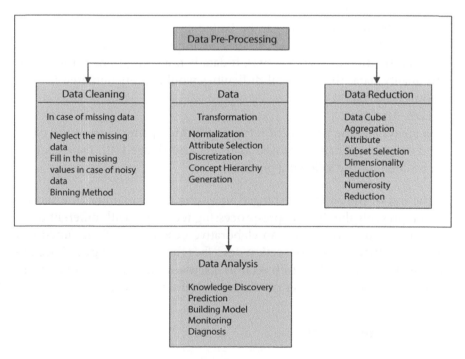

Figure 9.2 Different tasks in data pre-processing [7].

A. Data Cleaning
In missing data case
At times, data collected from different sources can comprise missing information that can result in depreciation of model performance [8], which can be handled by using the subsequent techniques.

    1) Neglect the missing values
    2) Fill in the missing values

B. Data Regression
The data can be smoothened by using linear or multiple regression functions. Regression analysis can be explained as a predictive modeling method that specifies a relationship between dependent (target) and independent variables. The method can be used for diverse applications, such as time series modeling, forecasting, and establishing causal effect relationships.
C. Clustering
Outliers present in the data can be eliminated using the clustering technique because the similar data is clustered in a group and the exceptional data is treated as outliers and removed [6].
D. Data Transformation
The basic limitation of data can be its incompleteness or irrelevancy for which data transformation techniques are required to transform the data into an error-free format. Because, if data is not going to be transformed into quality data, then, several difficulties would be encountered in data analysis tasks [9].
E. Data Analysis
Data analysis is the process of retrieving hidden patterns from the organized data, which can be proved as useful knowledge [10]. Data analysis can be done for the following purposes.

The rest of the paper is arranged in a subsequent manner. The second section presents the distinct pre-processing technique with different datasets. The third section gives an elaborative description of the functional dataset. Further, the fourth section describes the methodology followed to implement the work. The fifth section discusses the Result obtained from the methodology. The paper concludes with the Sixth section.

9.2   Related Work

Numerous studies suggest the use of data pre-processing operations for improving the consistency of data. Also, several studies are available which perform surveys on data pre-processing techniques and their applications. According to Jagdale *et al.* [11], rare event detection can be performed using prediction techniques, and prediction can be done on organized data for which data have to go through various pre-processing stages to remove the inconsistency and vagueness of data. The authors attempted to survey all the technical and practical aspects of the study to provide a complete representation of the importance of data pre-processing and data analysis [11]. A review on image pre-processing has been performed by Sriram *et al.* [12]. A change detection algorithm was being applied to the image dataset, but before the implementation of the algorithm, an image series of pre-processing steps are required. Where image repressing techniques are used for the enhancement of the image quality so that detected objects' efficiency can be improved. All algorithms that induce the change are partitioned into six categories, which combine segmentation, threshold, difference, trajectory classification, statistical boundary, and regression [12]. As the data accumulated is collected from different sources of data, it may contain noisy data, which needs to be eliminated by using data pre-processing operations. Although several tools and techniques are available for data pre-processing, Perez *et al.* [13] have introduced new pre-processing software known as SeqyClean for performing data pre-processing operations. The tool was being validated at both test and production stages by the Institute for Bioinformatics and Evolutionary Studies (IBEST) at the University of Idaho [13].

Tran *et al.* [14] explained the open problems and challenges in the field of data analytics that need to be possessed for improving the problem of class imbalance. Different important areas of research are found in the topic of imbalanced data distribution, which covers the full spectrum of imbalanced data learning, classification, regression, clustering, big data analytics, applications, and data streams. Data pre-processing is considered one of the most vital processes in data analytics or well-known Knowledge Discovery from the Data processes. Nonstructured data that is retrieved from a source will likely have irregularities, noise, and outliers, which means that data cannot be processed directly for any data analytic process. What issues are considered by Mishra *et al.* [15] in their proposed study and the evolving big in recent science, industry, and business applications, are unstructured and need pre-processing for providing good outcomes on any ML algorithm.

The pre-processing algorithm [16] to extract real-time user accessed data and analyze leads to deeper insights into business patterns and trends. Here Banerjee and Pal [17] proposed an efficient pre-processing technique and an innovative hashing technique to identify Distinct Users for Web Usage

Mining. Data preparation tasks are refined and enhanced to the industrial standards by recent authors Imambi and Sudha [18]. A method named "Cross-Industry Standard Process for Data Mining" has been suggested by the author for dealing with noisy and unprocessed data. Data preparation can be of General and Specific types for the epidemiological domain. In the work by Aziz *et al.* [19], noise and outliers are removed by corresponding attraction and density in the data and it is called D-IMPACT data pre-processing which is an iterative technique that yields better results in two-dimensional datasets. Feature selection is performed in various ways either be supervised or unsupervised algorithms often comprising duplicated or associated features. This yields extra cost and removes duplicate values completely which can further make the system vulnerable to district measurement errors discussed by Khoshgoftaar *et al.* [20] and Kang *et al.* [21].

Other researchers, like Mukhopadhyay *et al.* [22] propose a new scheme for feature selection in unsupervised categories labeled as "UFeSCoR" for avoiding the non-significant features and allowing the significant ones. Hence, this approach helps to control the redundancy by comparing around five other feature selection methods. A novel feature weighting scheme "GRW" is provided (Zhang, Y., [23], Nazarzade *et al.* [24]) and proved to support the classification accuracy while testing on a medical data set.

Iterative and multiple partitioning-based filters are tried for data pre-processing in the context of Software Engineering. This has been improving software quality prediction by Nilashi *et al.* [25]. The final results achieved with models built on the datasets based on filtered fitting are more accurate than the models built on the dataset with higher noise levels. Experimental results by Niu *et al.* [26] proposed a novel method that takes less time for training to achieve reasonable success, relative to all the traditional SVM ensembles. The contribution of multi-objective evolutionary algorithms used for data mining is indicated by Palaniappan & Awang [27]. This adds a lighter dimension to feature selection. While the data size becomes voluminous, "crowd sensing" harnesses the power of the crowd by mobilizing a huge group of mobile users attached with various networked devices to gather data with the basic multimodal and large-volume features. Various data mining techniques are applied for modeling by Hung *et al.* [28].

There are diverse methods available that have been applied to different datasets to record the significant improvement in the performance of the model. But the improvement achieved inaccuracy can be enhanced. In this chapter, data pre-processing is done to enhance the performance of the model by applying the equal width binning (data cleaning), Z-score (outlier detection), and SMOTE (class imbalance).

9.3   Experimental Setup and Methodology

The proposed work attempts to show the enhancement in the quality of the dataset using data pre-processing. Four different datasets (PIMA Diabetes Dataset, Cleveland heart disease dataset, Framingham Heart Study, and Diabetic dataset) were analyzed to measure the effects of data pre-processing by comparing the result achieved using three data mining techniques (DT, NB, and ANN). All the datasets were analyzed by applying these three techniques before and after the pre-processing of data. The outcomes on different datasets and their graphical representation have also been shown in this section of the chapter. Also, this section presents the description of the datasets and algorithms. All the experiments were performed using RapidMiner 9.7.0. The RapidMiner tool provides an easy-to-use graphical interface for application data pre-processing techniques and data mining techniques. All the datasets used in this study were taken from the UCI ML repository, which provides open access, free to use datasets for educational and research purposes. All the experiments were performed on Windows 10, i7 processor with 8 GB RAM. The methodology of the proposed work is explained below.

9.3.1   Methodology

1. The initial step is to input the four functional datasets (PIMA Diabetes Dataset, Cleveland heart disease dataset, Framingham Heart Study, and Diabetic dataset).
2. The three data mining techniques were applied on four datasets and results were recorded without applying data pre-processing techniques.
3. Various Data pre-processing techniques as described in section 3.2 were applied on the datasets.
4. The next step is to divide the clean data into training and testing into 70:30 ratios using the holdout method. After that, apply the three data mining techniques (DT, NB, and ANN) as discussed in Section 3.3 on clean data to measure the performance. The results were recorded for further comparison with the results achieved using the same technique without data pre-processing.
5. Validate the model with predicted outcomes using different performance measures such as accuracy, sensitivity, and specificity.

9.3.2   Application of Various Data Pre-Processing Tasks on Datasets

Data Cleaning
As all applied datasets are vast in size, therefore there might be a possibility that discrepancies may exist in such data such as outliers, and missing values and noise. To remove the discrepancies from the applied dataset, the method of equal width binning is implemented to split the attributes with irregularities into equal groups of bins using equation 9.1.

$$G_N = \frac{((A) - min(A))}{N} \qquad (9.1)$$

In the given equation (9.1), G_N indicates the bin sets. Furthermore, A signifies the discrepancy characteristic and N shows the number of equivalent bins sets. After that, the mean binning method has been applied to each set of bins to fill the missing values using equation 9.2.

$$M_{G_N} = \frac{\sum_{i=1}^{n} X_i}{n} \qquad (9.2)$$

where M_{G_N} signifies each bin's mean and n indicates the total number of values in each bin.

Outlier Detection
After filling in the missing values and cleaning the data, it has been examined that the model's performance can be affected by the existence of an outlier in the dataset. Thus, outliers can be detected by applying the outlier Z-score method using equation 9.3. And the standard deviation is calculated using equation (9.4).

$$Z_S = \frac{O - M}{S_D} \qquad (9.3)$$

$$S_D = \sqrt{\frac{\sum (x_i - M)^2}{n}} \qquad (9.4)$$

Handling Class Imbalance Problem

The applied ML algorithms do not perform well with the imbalanced data for training of the model. As per the literature review, SMOTE is one of the most promising techniques to deal with class imbalance problems in datasets in the medical domain. It works by generating new synthetic points for minority classes using equation 9.5.

$$\gamma_{NEW} = \gamma_0 + (\gamma_{0i} - \gamma_0) \times \partial \tag{9.5}$$

Here,

γ_{NEW} shows the new synthetic sample.

γ_0 represents a distinct instance's feature vector of the minority class.

γ_{0i} indicates the ith nearest neighbor of γ_0.

∂ signifies the random number generated amongst 0 and 1.

9.3.3   Applied Techniques

9.3.3.1   Decision Tree

As the name implies, the DT classifier works like a structure of the tree, its internal nodes perform testing on each attribute of the dataset and the final node represents the outcome of the DT. DT verified to be the most basic and efficient classification method for prediction purposes. The attribute with highest weight is considered to be the root of the tree and all the internal nodes are working to apply a decision on the considered root node. DT is a simple and easily interpretable classification problem, although noise and outliers in the training data can cause overfitting in the dataset. When DTs are built using training data, many branches over-expand due to noise, therefore it is essential to offer pre-processed data to DT for better results.

9.3.3.2   Naive Bayes

NB is not a single algorithm but it is a collection of algorithms, which follows the Bayesian classification rule. It represents the association between assumption and actual observations. As shown in equation (9.6).

$$P(X/Y) = \frac{P(Y/X)P(X)}{P(Y)} \tag{9.6}$$

Where P(Y) shows the prior probability of premise
P(X) presents the prior probability of actual value
P(Y/X) present premise probability given X and
P(X/Y) shows the probability of the given value

Typically, the posterior probability of the data object is to be identified. Naive Bayes is the supervised classifier, which means it needs training before performing classification on the data set. On account of this, the classifier needs some training data that contain several instances with the classes in which they are divided. After that, a test set is also required for testing purposes [29]. NB is a classification algorithm that can solve the problem of binary as well as multiclass classification problems. Although, NB performs better with the binary or categorical variable. In this presented work, all functional datasets contain categorical target variables.

9.3.3.3   Artificial Neural Network

Artificial Neural Networks—basically a concept, known from the field of biology where a neural network plays a principal and key role in the human body. A neuron is a unique and special biological cell that is responsible for information processing from one neuron to another with the assistance of chemical change and some electrical change. It is composed and composed of a cell body. And outreaching branches (tree-like) of two types. A nucleus that contains some information regarding hereditary traits and plasma neurons that hold the molecular instruments for supplying the important materials required by the neurons are the constituents of the cell body. An artificial neuron is primarily an engineering approach of a biological neural network or neuron.

9.3.4   Proposed Work

As discussed in the methodology four datasets were selected and the results were recorded by applying three data mining techniques before and after applying data pre-processing techniques [30]. The results were compared to analyze the impact of data pre-processing on various parameters such as accuracy, sensitivity, and specificity as discussed below.

9.3.4.1   PIMA Diabetes Dataset (PID)

The PID dataset was developed at the National Institute of Diabetes, United States, and is freely available on the UCI ML repository for research purposes.

PIMA dataset is related to the PIMA group, which are Native Americans living in Arizona and diagnosed with high sugar levels. The dataset consists of female patients of age group greater than 21 years old of Pima Indian heritage. The dataset samples were collected during the first trimester of pregnancy. This dataset includes 768 instances and 9 attributes. Among the nine attributes, six are continuous, two are discrete and one is a categorical attribute. The attributes of the PID dataset are namely, Number of times pregnant, Plasma glucose concentration 2 hours in an oral glucose tolerance test, Diastolic blood pressure (mm Hg), Triceps skinfold thickness (mm), 2-Hour serum insulin (mu U/ml), Body mass index, Diabetes pedigree function, Age (years) and class. A class label associated with each instance indicates whether the individual is affected with diabetes or not. The target variable identifies whether a person is non-diabetic (represented by 0) or diabetic (represented by 1). The description of attributes in the PID dataset is shown in Table 9.1. Moreover the details related to PIMA dataset including missing values, mean and standard deviation is shown using Table 9.2. Thereafter the prior and post effect of pre-processing on the PIMA dataset is presented using Figure 9.3 and the graphical representation of the same is shown using Figure 9.4.

Table 9.1 Description of PID attributes.

Sr. no.	Predicators	Description of predictors	Unit	Type	Range
1.	**Pregnancy**	Pregnant Number of times pregnant	--	0-17	Discrete
2.	**Glucose**	Plasma glucose concentration 2 hours in an oral glucose tolerance test [32]	Mg/dl	0-199	Continuous
3.	**BP**	Diastolic blood pressure	mmHg	0-122	Continuous
4.	**Skin**	Triceps skinfold thickness	Mm	0-99	Continuous

(Continued)

Table 9.1 Description of PID attributes. (*Continued*)

Sr. no.	Predicators	Description of predictors	Unit	Type	Range
5.	**Insulin**	Applicant's 2-Hour serum insulin	Mm U/ Ml	0-846	Continuous
6.	**BMI**	Body Mass Index/Weight of an applicant w.r.t height in m) ^2	kg/m²	0-67.1	Continuous
7.	**Pedi function**	Diabetes pedigree function that finds attributes used for diabetes diagnosis	--	0-2.42	Continuous
8.	**Age**	Age of participants	--	21-81	Discrete
9.	**Outcome**	Diabetes onset with diabetic and non-diabetic patients	--	0-1	Categorical

Table 9.2 Detailed description of PID attributes.

Sr. no.	Predicators	Missing values	Mean	Std dev
1.	Pregnancy	0	3.845	3.370
2.	Glucose	0	120.89	31.973
3.	Diastolic Blood pressure	0	316.56	1096.927
4.	Skinfold Thickness	0	51.697	88.690
5.	Insulin	0	819.49	3873.732
6.	Body Mass Index	0	60.769	92.015
7.	Diabetes pedigree function	0	0.472	0.472
8.	Age	0	33.241	11.760

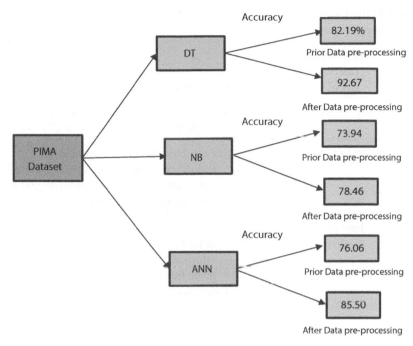

Figure 9.3 Performance of DT, NB and ANN on PIMA dataset before and after pre-processing.

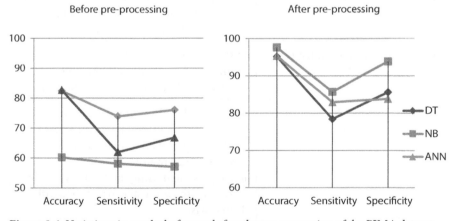

Figure 9.4 Variations in results before and after the pre-processing of the PIMA dataset.

9.3.5   Cleveland Heart Disease Dataset

This dataset consists of 76 attributes and only 14 of them have been used for different studies to predict heart disease as shown in Table 9.3. The after and before effects of pre-processing technique on the Cleveland data have been shown using Figure 9.5, and Figure 9.6 presents the graphical representation of the same.

Table 9.3 The most prominent attributes from Cleveland dataset are extracted using data pre-processing techniques.

S. no.	Attribute name (attribute no from list of 76 attributes)	Description of the attribute
1	Age (attribute no #3)	This attribute presents the age of the participant within a range of 29 to 77
2	Sex (attribute no #4)	Sex attribute displays the gender of individual using binary variable 0 and 1, here 0 signifies female candidate and 1 signifies male candidate
3	Chest pain type (attribute no #9)	Chest pain type is abbreviated as cp in the dataset, it signifies the type of chest pain which an individual is facing using the given format- 1 signifies the typical angina 2 signifies the atypical angina 3 signifies the non-angina pain and 4 signifies asymptotic
4	Trestbp (attribute no #10)	Trestbp is the resting blood pressure of an individual measured in mmHg unit, and it ranges between 94 and 200.
5	cholesterol (attribute no #12)	Display the serum cholesterol of a patient measured in mg/dl unit, the range of cholesterol lies between 126 and 564
6	Fasting Blood sugar (attribute no #16)	Abbreviated as fbs in the dataset. This mentioned field compares the fasting blood sugar of an individual with the value of 120 mg/dl. It is represented using the binary variable - 1. when fbs is greater than 120 mg/dl 0-otherwise

(Continued)

Table 9.3 The most prominent attributes from Cleveland dataset are extracted using data pre-processing techniques. (*Continued*)

S. no.	Attribute name (attribute no from list of 76 attributes)	Description of the attribute
7	Resting ECG (attribute no #19)	This field signifies the resting electrographic of an individual using ternary variable, where-0 represents the normal ECG 1 represents the ST-T wave with abnormality and 2 represents the left ventricular hypertrophy
8	Max heart rate (attribute no #32)	Shows the max heart rate achieved by an individual which ranges between 71 and 202
9	Exercise-induced angina (attribute no #38)	Presented using a binary variable, 0 signifies the yes, and 1 signifies No
10	Oldpeak (attribute no #40)	It shows the ST depression caused due to exercise relative to test and ranges between 0 and 6.2
11	Slope (attribute no #41)	It shows the highest exercise ST-segment using ternary variable where 1 signifies the up sloping 2 show flat and 3 indicate the down sloping
12	No of major vessel (attribute no #44)	It shows the major vessels colored by fluoroscopy, represented using integer and lies between 0 and 3
13	Thal (attribute no #51)	It shows the thalassemia using ternary variable which ranges between three values where, 3 shows normal value 6 indicate fixed defect and, 7 show reversible defect
14	Num-target attribute (attribute no #51)	Present whether the person in dataset is experiencing from heart disease or not: 0 shows the absence of the disease 1, 2, 3, 4- the presence of the disease of an individual

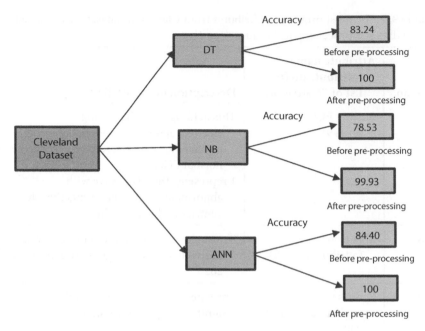

Figure 9.5 Performance of DT, NB and ANN on Cleveland dataset before and after pre-processing.

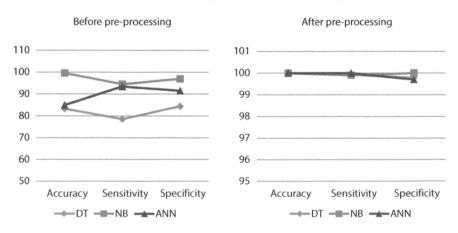

Figure 9.6 Variations in results before and after the pre-processing of the Cleveland dataset.

9.3.6   Framingham Heart Study

FHS dataset was applied for analyzing the Coronary heart study. The reason for choosing the FHS dataset is that the dataset consists of 18 attributes with 4583 instances. There are a number of potential papers available for heart study and published using the FHS dataset. There are 18 attributes in the dataset, significant attributes using pre-processing techniques to predict heart disease shown in Table 9.4. PIMA, Cleveland datasets, and FHS datasets were also tested with before and after effects of pre-processing there shown in Figure 9.7, and Figure 9.8 presents the graphical representation of the same.

Table 9.4 The most prominent attributes from FHS are extracted using data pre-processing techniques.

Name of the attribute	Description of the attribute
Blood Pressure	Blood Pressure is categorized into five groups, i.e., normal, prehypertension, hypertension stage 1, hypertension stage 2 and hypertensive crisis.
Gender	Male and Female
Age	Age of each person diagnosed
Total-Blood Cholesterol	Combination of serum cholesterol exam 1 and exam 2, In that exam 1 cholesterol ranges from 96 – 503 for serum cholesterol (mg/100 ml) and it ranges from 15 to 568 for serum cholesterol (mg/100 ml) in serum cholesterol Exam 2.
Body Mass Index	BMI is body mass index derived from the height and weight of an individual by using the formula: weight(pounds)/ [height(in)] *703
Smoking	"Yes" denotes the cause of death is CA whereas "No" illustrates the unknown cause.
Death from CVD	"Yes" reveals the reason for death due to CVD "No" illustrates the unknown cause.

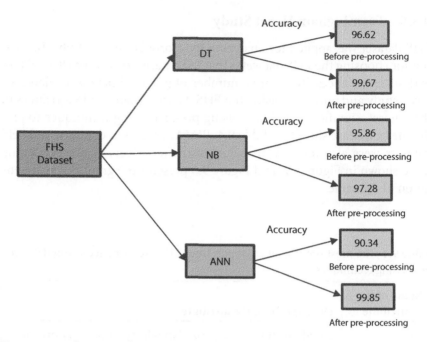

Figure 9.7 Performance of DT, NB and ANN on FHS dataset before and after pre-processing.

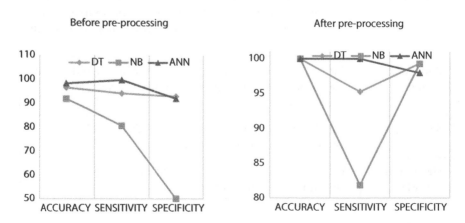

Figure 9.8 Variations in results before and after the pre-processing of the FHS dataset.

9.3.7 Diabetic Dataset

The dataset used for prediction is Diabetic-dataset from Kaggle which has 101766 instances. This particular dataset has numerous attributes which are responsible for the items in an individual. The dataset contains 50 attributes and one class label as shown in Table 9.5. In the diabetic dataset, the impact of the data pre-processing technique is shown using Figure 9.9, and Figure 9.10 shows the graphical representation of the comparison.

Table 9.5 Description of attributes in the database.

Attribute names	Description
Encounter Id	The ID when patient encounters for a test
Patient name	It describes the name of the patient
Race	It describes the ethnicity of the patient
Gender	Male or Female
Age	Age of the patient
Weight	Weight of the patient
Admission_type_id	Type of ID when patient get admitted
Discharge_disposition_id	A person's inherent qualities of mind and character.
Admission_source_id	It depicts the source of the admission
Time_in_hospital	It defines the period of a patient staying in hospital
Payer_code	Electronic code of Patient_ID
Medical_speciality	The specialty of the doctor
Num_lab_procedures	It denotes the number of lab procedures
Num_procedures	It denotes all the procedures
Num_medications	Medications recommended to the patient
Number_outpatient	It is the number of patients who are not admitted to the hospital but came for diagnosis only.

(Continued)

Table 9.5 Description of attributes in the database. (*Continued*)

Attribute names	Description
Number_emergency	Number of patients in emergency cases
Number_inpatient	Number of patients who are admitted to the hospital
Diag_1	Diagnosis number 1
Diag_2	Diagnosis number 2
Diag_3	Diagnosis number 3
Number_diagnoses	Number of total diagnoses conducted
Max_glu_serum	The maximum glucose level in the patient's body
A1Cresult	Assessment of Hemoglobin used to measure the glucose or blood sugar
Metformin	It is the drug used to cure diabetes
Repaglinide	Drug to improve blood sugar control
Nateglinide	Blood glucose lowering drug
Chlorpropamide	Drug for patients with Diabetes type 2
Glimepiride	Drug 1 for patients with Mellitus type 2
Acetohexamide	Drug 2 for patients with Mellitus type 2
Glipizide	Drug to improve control in blood sugar for adults
Glyburide	General Drug to improve control in blood sugar
Tolbutamide	Drug for patients with type 2 diabetes mellitus
Pioglitazone	Drug for Diabetes type 2
Rosiglitazone	An anti-diabetic data
Examide	Drug for Diabetes type 2
Citoglipton	A drug used to lower blood glucose level
Insulin	The hormone used to use sugar from carbohydrates

(*Continued*)

Table 9.5 Description of attributes in the database. (*Continued*)

Attribute names	Description
Glyburide-metformin	The combination used to treat patients with diabetes
Glipzide-metformin	The combination used to treat patients with high diabetes
Glimepiride-pioglitazone	Combination used to control sugar levels in body
Metformin-pioglitazone	Drug for Diabetes type 2
Change	Status change of patient
diabetesMed	Type of medication
Readmitted	Readmission of a patient in the hospital

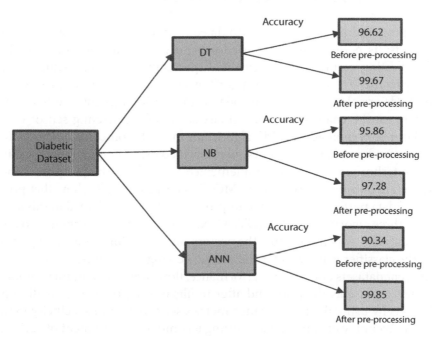

Figure 9.9 Performance of DT, NB and ANN on diabetic dataset before and after pre-processing.

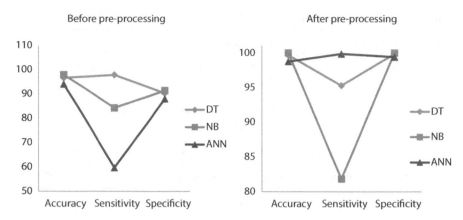

Figure 9.10 Variations in results before and after the pre-processing of the diabetic dataset.

9.4   Experimental Result and Discussion

Our proposed model has been validated on the test data of four distinct datasets; its performance measured using diverse evaluation metrics like classification accuracy, sensitivity, and specificity. The results achieved by applying the DT, NB, and ANN algorithms before and after pre-processing methods are also discussed. In today's era of advancement, the IoT eco-system can be enabled to group smart devices for collecting sensor data and can be combined with ML algorithms for better outcomes [31–33]. In our research work, Experiments are performed using a rapid miner toolkit. Three techniques have been implemented to perform the pre-processing which are binning, z-score, and SMOTE methods. Results show that processed data performs better than unprocessed data as explained in this section. Three algorithms of ML (DT, ANN, and NB) have been applied to the four datasets, which show that the performance of training and testing data on the classifier is better after the pre-processing. After pre-processing on different datasets and earlier performance, flows are shown in this section. Table 9.6 presents the prior and after results of pre-processing on all four datasets by using the performance metrics sensitivity, and specificity [34]. In the Accuracy of applied data mining technique improvement of 12.79% in DT, 4.52% in NB, and 9.54% in ANN has been recorded after using data pre-processing techniques.

Table 9.6 Performance of DT, NB and ANN on four distinct datasets before and after pre-processing.

Applied dataset	Applied classifiers	Before pre-processing			After pre-processing		
		Accuracy	Sensitivity	Specificity	Accuracy	Sensitivity	Specificity
PIMA	DT	82.39	60.21	82.73	97.62	97.62	95.35
	NB	73.94	58.12	62.01	85.71	85.71	82.95
	ANN	76.06	57.07	66.87	93.81	93.81	83.83
CLEVELAND	DT	83.24	99.53	84.99	100	100	100
	NB	78.53	94.53	93.48	99.90	99.90	100
	ANN	84.40	96.97	91.54	100	100	99.71
DIABETIC	DT	96.62	97.86	94.03	99.67	100	98.76
	NB	95.86	81.29	59.70	97.28	84.84	99.88
	ANN	90.34	91.43	88.06	99.85	100	99.42
FHS	DT	96.50	91.90	98.28	99.95	99.95	100
	NB	94.18	80.65	99.72	81.84	81.84	100
	ANN	92.87	50.13	92.26	99.25	99.25	97.99

9.5   Conclusion and Future Work

Data pre-processing is a technique performed on data before data analytics. The main ingredient for this analytics process is the data. The performance or the quality of the results of the analytics or mining process can only be enhanced by the quality of the results in the pre-processing. The underlying data from different sources may have format conflicts, unnecessary attributes and unfilled or blank data. Therefore, data needs to be prepared and preprocessed before applying any kind of data process. In this proposed study, data pre-processing techniques are implemented, which perform data cleaning, outlier detection, and class balancing on four different datasets to improve the performance of the applied machine learning classifiers. Before and after effects of data pre-processing methods are also discussed. The results show that the pre-processing methods show good improvement in the accuracy of applied machine learning algorithms. In the future, we intend to propose a novel pre-processing method by combining the applied pre-processing techniques for better outcomes.

References

1. Duhamel, A., Nuttens, M.C., Devos, P., Picavet, M., Beuscart, R., A pre-processing method for improving data mining techniques. Application to a large medical diabetes database. *Stud. Health Technol. Inform.*, 95, 269–274, 2003.
2. Losarwar, V. and Joshi, D.M., Data pre-processing in web usage mining, in: *International Conference on Artificial Intelligence and Embedded Systems (ICAIES'2012)*, pp. 15–16, 2003.
3. Vokorokos, L., Pekár, A., Ádám, N., Data pre-processing for efficient evaluation of network traffic parameters, in: *2012 IEEE 16th International Conference on Intelligent Engineering Systems (INES)*, June, IEEE, pp. 363–367, 2012.
4. Ma, X. and Jiang, X., Multimedia image quality assessment based on deep feature extraction. *Multimed. Tools Appl.*, 79, 47, 35209–35220, 2020.
5. Kotsiantis, S.B., Kanellopoulos, D., Pintelas, P.E., Data pre-processing for supervised leaning. *Int. J. Comput. Sci.*, 1, 2, 111–117, 2006.
6. Lin, W.C., Tsai, C.F., Hu, Y.H., Jhang, J.S., Clustering-based undersampling in class-imbalanced data. *Inf. Sci.*, 409, 17–26, 2017.
7. Data pre-processing in data mining, 2019, September 9, Retrieved from https://www.geeksforgeeks.org/data-pre-processing-in-data-mining/.
8. Son, N.H., *Data cleaning and data pre-processing*, http://www.mimuw.edu.pl/~son/datamining/DM/4-preprocess.pdf.

9. Bakar, A.A., Othman, Z.A., Shuib, N.L.M., Building a new taxonomy for data discretization techniques, in: *2009 2nd Conference on Data Mining and Optimization*, October, IEEE, pp. 132–140, 2009.

10. Adusumalli, S.K. and Kumari, V.V., An efficient and dynamic concept hierarchy generation for data anonymization, in: *International Conference on Distributed Computing and Internet Technology*, Springer, Berlin, Heidelberg, pp. 488–499, 2013.

11. Jagdale, A.R., Sonawane, K.V., Khan, S.S., Data mining and data pre-processing for big data. *Int. J. Sci. Eng. Res.*, 5, 7, 1156, 2014.

12. Sriram, R. and Malliga, R., Innovative pre-processing technique and efficient unique user identification algorithm for web usage mining. *Int. J. Adv. Res. Comput. Sci. Software Eng.*, 6, 02, 85–91, 2016.

13. Pérez, J., Iturbide, E., Olivares, V., Hidalgo, M., Almanza, N., Martínez, A., A data preparation methodology in data mining applied to mortality population databases, in: *New Contributions in Information Systems and Technologies*, pp. 1173–1182, Springer, Cham, 2015.

14. Tran, V.A., Hirose, O., Saethang, T., Nguyen, L.A.T., Dang, X.T., Le, T.K.T., Ngo, D.L., Sergey, G., Kubo, M., Yamada, Y., Satou, K., D-impact: A data pre-processing algorithm to improve the performance of clustering. *J. Software Eng. Appl.*, 7, 639–654, 2014.

15. Pan, L., Zheng, H., Li, L., A hybrid feature selection approach based on the Bayesian network classifier and rough sets, in: *International Conference on Rough Sets and Knowledge Technology*, Springer, Berlin, Heidelberg, pp. 707–714, 2008.

16. Mishra, D., Rath, A.K., Acharya, M., Jena, T., Rough ACO: A hybridized model for feature selection in gene expression data. *Int. J. Comput. Commun. Technol.*, 1, 1, 85–98, 2009.

17. Banerjee, M. and Pal, N.R., Unsupervised feature selection with controlled redundancy (UFeSCoR). *IEEE Trans. Knowl. Data Eng.*, 27, 12, 3390–3403, 2015.

18. Imambi, S.S. and Sudha, T., pre-processing of medical documents and reducing dimensionality. *Adv. Computing*, 2, 5, 15, 2011.

19. Aziz, A.S.A., Azar, A.T., Salama, M.A., Hassanien, A.E., Hanafy, S.E.O., Genetic algorithm with different feature selection techniques for anomaly detectors generation, in: *2013 Federated Conference on Computer Science and Information Systems*, September, IEEE, pp. 769–774, 2013.

20. Khoshgoftaar, T.M. and Rebours, P., Improving software quality prediction by noise filtering techniques. *J. Comput. Sci. Technol.*, 22, 3, 387–396, 2007.

21. Kang, S., Kang, P., Ko, T., Cho, S., Rhee, S.J., Yu, K.S., An efficient and effective ensemble of support vector machines for anti-diabetic drug failure prediction. *Expert Syst. Appl.*, 42, 9, 4265–4273, 2015.

22. Mukhopadhyay, A., Maulik, U., Bandyopadhyay, S., Coello, C.A.C., A survey of multiobjective evolutionary algorithms for data mining: Part I. *IEEE Trans. Evol. Comput.*, 18, 1, 4–19, 2013.

23. Zhang, Y., Chen, M., Mao, S., Hu, L., Leung, V.C., CAP: Community activity prediction based on big data analysis. *IEEE Netw.*, 28, 4, 52–57, 2014.

24. Nazarzadeh, M., Pinho-Gomes, A.-C., Smith Byrne, K., Canoy, D., Raimondi, F., Ayala Solares, J.R., Rahimi, K., Systolic blood pressure and risk of valvular heart disease. *JAMA Cardiol.*, 4, 8, 788, 2019.

25. Nilashi, M., Ibrahim, O., Ahani, A., Accuracy improvement for predicting parkinson's disease progression. *Sci. Rep.*, 6, 1, 1–18, 2016.

26. Niu, J., An, G., Gu, Z., Li, P., Liu, Q., Bai, R., Du, Q., Analysis of sensitivity and specificity: Precise recognition of neutrophils during regeneration of contused skeletal muscle in rats. *Forensic Sci. Res.*, 7, 2, 1–10, 2020.

27. Palaniappan, S. and Awang, R., Intelligent heart disease prediction system using data mining techniques. *2008 IEEE/ACS International Conference on Computer Systems and Applications*, pp. 108–115, 2008.

28. Hung, A.J., Chen, J., Gill, I.S., Automated performance metrics and machine learning algorithms to measure surgeon performance and anticipate clinical outcomes in robotic surgery. *JAMA Surg.*, 153, 8, 770, 2018.

29. Kumar, A., Sharma, A., Kumar, R., SERVmegh: Framework for green cloud. *Concurr. Comput.: Pract. Exp.*, 29, 4, e3903, 2017.

30. Naz, H. and Ahuja, S., Deep learning approach for diabetes prediction using PIMA Indian dataset. *J. Diabetes Metab. Disord.*, 19, 1, 391–403, 2020.

31. Goel, R., Jain, A., Verma, K., Bhushan, M., Kumar, A., Mushrooming trends and technologies to aid visually impaired people. *International Conference on Emerging Trends in Information Technology and Engineering (ic-ETITE'20)*, February 24-25, 2020, IEEE, Vellore Institute of Technology, Vellore, India, 2020.

32. Mangla, M., Kumar, A., Mehta, V., Bhushan, M., Mohanty, S.N., *Real-life applications of the Internet of Things: Challenges, applications, and advances*, CRC Press, Taylor & Francis Group, 2022.

33. Sharma, S., Nanda, M., Goel, R., Jain, A., Bhushan, M., Kumar, A., Smart Cities using Internet of Things: Recent trends and techniques. *Int. J. Innov. Technol. Exploring Eng. (IJITEE)*, 8, 9S, 24–28, 2019.

34. Verma, K., Bhardwaj, S., Arya, R., Islam, U.L., Bhushan, M., Kumar, A., Samant, P., Latest tools for data mining and machine learning. *Int. J. Innov. Technol. Exploring Eng.*, 8, 9S, 18–23, 2019.

10

Cloud Computing Security, Risk, and Challenges: A Detailed Analysis of Preventive Measures and Applications

Anurag Sinha[1]*, N. K. Singh[2], Ayushman Srivastava[3], Sagorika Sen[1] and Samarth Sinha[4]

[1]Department of Information Technology, Amity University, Jharkhand, India
[2]Department of Computer Science, Birla Institute of Technology, Mesra, Ranchi, Jharkhand, India
[3]Department of Information Technology, Dr. A.P.J. Abdul Kalam Technical University, Delhi, India
[4]Department of Computer Science and Engineering, Central Institute of Technology, Kokrajhar, Delhi, India

Abstract

Cloud computing is an emerging technology for networking and virtual database extension, which is being adopted by several enterprises widely. Nowadays, every single enterprise is mitigating toward virtualization immensely. Several cloud services provide a great range of computer infrastructure tools and technology. Thus, implementing a cloud-oriented environment in industry is very challenging and requires skills to manage its pitfalls. So cloud implementation also brings several challenges, security risks, and threats for data security. Several data security threats in the cloud are very risky and not at all easy to identify, such as Distributed Denial of Service (DDoS) attacks, flooding, data breaching, man-in-the-middle attack, and so on. Therefore, this proposed chapter critically analyzed several risks and issues of data security in a cloud environment, and also analyzed some preventive measures to overcome that as well. This chapter provides a taxonomical solution for data security issues in cloud environments, i.e., steganography can be used for data security in cloud environments. Thus, this chapter provides a complete survey of cloud security risk and its solution for managing the risk effectively.

*Corresponding author: anuragsinha257@gmail.com

Ashok Kumar, Megha Bhushan, José A. Galindo, Lalit Garg and Yu-Chen Hu (eds.) Machine Intelligence, Big Data Analytics, and IoT in Image Processing: Practical Applications, (225–268) © 2023 Scrivener Publishing LLC

Keywords: Cloud computing, security, challenges, analysis, preventive actions, data security, steganography

10.1   Introduction

Cloud computing is moving from business to consumer [1]. According to the recent research 1, a customer is the same every time they examine four other cloud-based applications. Additionally, 41% of commercial products found a huge liability on the mists. The system responsibilities to the cloud; the security of cloud enrolment is under scrutiny [2]. Eighty percent of all computer applications are cloud-based, and 49% of organizations allow cloud-enabled functionality. The novel point of the proposed conveyed figuring is web-based scattered and virtual machine (VM) developments to outfit customers with sensible handling organizations and limit organizations, decreasing the cost of enrolling. Various customers' benefit remarkably from the New York Times had rented Amazon's cloud preparing organizations to more than 11 million stories in a solitary day to convert electronic files for customers [3], the cost of the traditional responses for one of modest bunches of course even hundreds development, the as of late settled association under tremendous measure of money frank hypothesis venture assets through cloud figuring, for instance, the getting cost of the laborers, programming, figuring limit augmentation is, moreover extremely invaluable, speedy. Each cloud expert association needs to take care that the client has data to the cloud, and there should be some confirmation concerning the permission to that data that it will be basically limited to the endorsed admittance [4]. The client's data security and real practices and assurance approaches are to be supported to ensure the cloud customers of the data security and insurance. Security and assurance is an indispensable point of view also similarly as with the creating development moreover ends up bearing cost for that. As all the cloud affiliations need a web relationship as the web transforms into the method of data transmission its unapproved nature can cause a commotion to the reliability of the data. Today, there are various intruders and developers that endeavour to interfere with the different firewalls and take or fetter customer's data. Disseminated figuring in the current circumstance is a making and rapidly creating advancement that is all around by and large used all through the planet. It utilizes the power of Internet-based enrolling and here the data, information and various resources are given to the customer through computer or device on demand. It is another start that uses virtual resources for sharing data. Hooray or Gmail are some fitting occasions of disseminated registering. Various activities that fuse clinical benefits, banking, and preparing are gliding toward this advancement, all because of the capability given by the system which is fuelled by using models, and subsequently, it manages the bandwidth, data improvement, trades, and limit data. In continuous years, conveyed processing has been an emerging enrolling model in the information technology (IT) business. Various enormous associations are throwing resources into it. It gives useful preparation by bringing together storing, memory taking care of and moving speed. Accepting appropriate registration can achieve both positive and unfriendly outcomes on data security. This chapter presents find out with regards to the challenges of disseminated figuring. It includes the novel sorts of risks and what their existence can mean for the cloud customers. Different challenges are being raised for embrace of the cloud, including data security, openness, and protection [5].

It shows how important it is to use the flow of inscriptions as it is today, without proximity or archival correlation. The importance of enlisting in security is an immediate consequence of the different organizations offered by the cloud. One of these associations restricts customers to their information as a Cloud Information Tracker licence community. The downside to this assistance is that it poses a security concern due to the availability of sensitive information. Customers want the world to be a great place to store their information in the cloud. Therefore, it needs models that will innovate in information security. Image steganography prevents access to information without authorization. Image steganography covers client information on the cover. In this article, System examines a fragment of a proposed new practice for obtaining information about clouds using image steganography. The evaluation they use; the most basic team of benefit and barrier assessment forms. Second, model relation linked to the characteristics of steganography [5, 6].

In this chapter, an overview of preventive measures for cloud computing security risk is discussed as past research reads have generally described cloud security challenges across the accompanying arrangements: data access, network issues and virtualization characterizations. Regardless, past research has not zeroed in on human mix-ups, i.e., cloud security challenges that happen in view of the cloud modellers' oversight when organizing, arranging and passing on cloud organizations or mix-ups as a result of customer's inconsiderateness. Considering that, it added another characterization called human error to plan. This proposed system by finding compositions that talked about security challenges that were possibly a direct result of human lack of regard [7]. The second significant focal point of this chapter is that it gives an itemized depiction of the suggested protective measures for every one of the difficulties that have been distinguished and portrayed. Besides, it likewise portrayed a planning arrangement of the cloud security difficulties to their safeguarding efforts. In the later part, it introduces several techniques of cloud data security.

This chapter highlights the focal point of security and loopholes in data transmission and data security aspects in cloud computing with identifying major challenges in cloud data security and thereby discussing the taxonomical projection of solution, preventive measures to be adopted for cloud data security.

10.2   Background

10.2.1   History of Cloud Computing

To get what distributed computing is and is not, it is vital to see how this model of processing has advanced. As Alvin Toffler notes in his popular book, development has advanced in waves (three of them to date: the main wave was farming social orders, the second was the modern age, and the third is the data age). Inside each wave, there have been a few significant subwaves. In this postmodern data age, it presently works toward the start of what many individuals feel will be a period of distributed computing [22]. Rapid change from legacy to cloud system is described below:

- Before the advent of cloud computing, client/server computing was mainly about product applications, information and controls are stored on the server side.
- If a single client needs to search out to clear progression or run a program, the person should initially associate with the server and get approved admittance prior to continuing [8].
- Then after, disseminated processing came into picture, where every one of the computers are organized together and offer their assets when required.
- Cloud figuring ideas emerged because of the previously mentioned registering and were subsequently executed.

Cloud computing implies that as opposed to having the entire product and equipment on a computer or on an employer's network, it might get everything from one more organization as an assistance and access it through the Internet, normally in a totally consistent way. To the client, it does not make any difference where the equipment and programming are arranged or how they work—simply it is some place up in the undefined "cloud" that the Internet addresses [11].

- Salesforce.com started circulating applications to clients through a basic site in 1999. The applications were circulated to organizations through the Internet, bringing the fantasy about computing as a utility nearer to the real world [21].
- Amazon started Amazon Web Services in 2002, offering administrations like storage, compute, and even artificial intelligence [10].
- Google Apps started offering cloud computing undertaking applications in 2009 [15].
- The entirety of the central parts is engaged with the cloud computing development, some sooner than others [9].

From Figure 10.1, in spite of the fact that the cloud security service model of distributed computing that is alluded to as Software-as-a-Service (SaaS), there is a significant contrast in how these administrations are

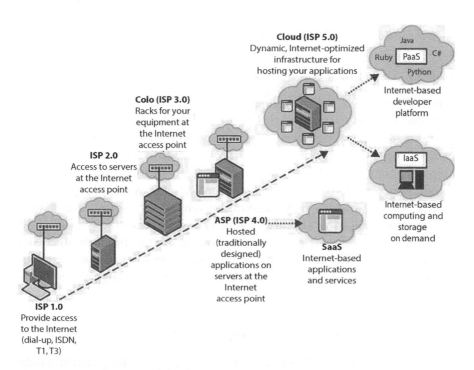

Figure 10.1 Beginning era of cloud.

given, and in the plan of action. Despite the fact that ASPs as a rule offered types of assistance to various clients (similarly as today), they did as such through committed frameworks. That is, every client had its own committed case of an application, and that case as a rule ran on a devoted host or server. The significant distinction between SaaS suppliers and ASPs is that SaaS suppliers offer admittance to applications on a common, not devoted, framework. In reality, IT organizations large and small follow the standard approach to IT development. Cloud computing data primitives are taken care of by a particular set of data security specialists, which can be understood by following example i.e. in this staff room there are data scientists; Mail employee sorting firewall; switches; modem switches. This high speed system needs to spend a lot of money to build such a system establishment [33]. To conquer this mound of issues and to reduce the IT system cost, Cloud Computing appears. A registering model in which outside purchasers are given colossally versatile (and flexible) IT-related abilities over the Internet. Cloud suggests an association or the web. Cloud computing permits users to get to IT assets like hardware and software on request over the web. It offers a minimal expense IT foundation arrangement. Since the product should not be introduced locally on the computer devices, cloud computing gives stage freedom. Subsequently, cloud computing permits us to make business applications more portable and community [9].

10.2.1.1   Software-as-a-Service Model

From Figure 10.2, conventional strategies for buying programming included the client stacking the product onto his own equipment as a trade-off for a permit charge (a capital cost, known as CapEx). The client could likewise buy upkeep consent to get patches to the product or other help administrations. The client was worried about the similarity of functional frameworks, fix establishments, and consistency with permit agreements. SaaS empowers programming merchants to control and restrict use, forbids replicating and dispersion, and works with the control of all subsidiary forms of their product. SaaS unified control frequently permits the seller or provider to lay out a continuous income stream with various organizations and clients without preloading programming in every gadget in an association.

10.2.1.2   Infrastructure-as-a-Service Model

In the conventional facilitated application model, the seller gives the whole foundation to a client to run his applications. Regularly, this involves

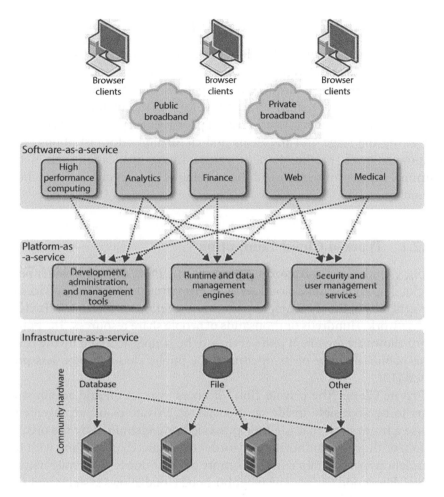

Figure 10.2 Types of cloud services and architecture.

lodging devoted equipment that is bought or rented for that particular application. IaaS model likewise gives the framework to run the applications, yet the distributed computing approach makes it conceivable to offer a compensation for every utilization model and to scale the help contingent upon request. From the IaaS supplier's point of view, it can assemble a foundation that handles the pinnacles and box of its clients' requests and add new limits as the general interest increases. Essentially, in a facilitated application model, the IaaS merchant can cover application facilitating just, or can reach out to different administrations (for example, application support, application advancement, and upgrades) and can uphold the more exhaustive re-appropriating of IT.

10.2.1.3   Platform-as-a-Service Model

In the Platform as a Service (PaaS) model, the seller offers an advancement climate to application engineers, who foster applications and proposition those administrations through the supplier's foundation. The supplier regularly creates tool stash and norms for advancement, and channels for dispersion and instalment. The supplier regularly gets an instalment for giving the stage and the deals and circulation administrations. This empowers quick proliferation of programming applications, given the minimal expense of section and the utilizing of laid out channels for client obtaining.

10.2.2   Types of Cloud Computing

Public Cloud: Google, Amazon, Microsoft, and more companies offer public cloud options. Public cloud services give infrastructure and services to the general public; the company secures a portion of that infrastructure and network. Hundreds or thousands of people share resources. The public cloud allows arrangement and services to be simply accessed by the complete public. Because of its openness, the public cloud may be less protected [10].

Private Cloud: The private cloud licences association and administrations to be accessible inside an association. Private cloud arrangements utilise a framework that is either possessed or constrained by the organization, or they may authoritatively order that a merchant dealing with the foundation meet certain rules. It is more popular due to its private nature.

Neighborhood Cloud: The neighborhood cloud allows systems and services to be available by a group of organizations.

Hybrid Cloud: Hybrid cloud systems combine public and private cloud resources. Because the organization must manage several platforms and select where data is housed, this is a more complicated cloud solution. An enterprise that wishes to keep secret information protected on their private cloud but create more generic, customer-facing material on a public cloud is an example of a hybrid cloud solution [11].

10.2.3   Cloud Service Model

Cloud computing is based on service models. These are categorized into three basic service models, which are shown in Figure 10.3.

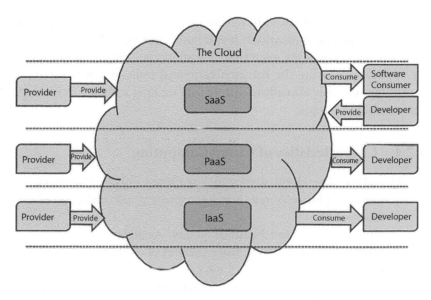

Figure 10.3 Cloud computing services.

Software-as-a-Service (SaaS): It is the least complex sort of distributed computing. Albeit the client does not approach an outsider turn of events or assets, SaaS applications can give incredible components directly to the internet browser. The SaaS worldview empowers end-clients to burn-through programming applications as a help.

Infrastructure-as-a-Service (IaaS): IaaS, or "Foundation as a Service," is the second kind of cloud administration. This cloud platform is gen-erally thorough and for the most part utilized by full-time engineers and enormous scope venture customers. While SaaS and PaaS permit to utilise cloud applications and create applications, IaaS gives framework to cre-ating, executing, and putting away programming in cloud settings. The benefit of IaaS is that designers have basically limitless capacity and com-putational limit without keeping any actual equipment on-site. IaaS gives admittance to fundamental assets including workers, organizations, virtual systems, and virtual stockpiling [12].

Platform-as-a-Service (PaaS): It makes available a request runtime atmo-sphere, as well as development and deployment tools. Platform as a Service, the third type of cloud service, gives developer's access to proprietary APIs that allow them to create applications that run in a specific environment.

While a developer is allowed to design whatever app they want, the app is bound to the platform on which it was built. This way of developing apps is low-cost (in some cases, even free) and allows building or migrating current applications using the infrastructure and tools of a well-established cloud company. This also allows immediate making software available to a large number of people.

10.2.4   Characteristics of Cloud Computing

1) Agility: The cloud works in a scattered enlisting environment. It splits resources among customers and works incredibly speedy.
2) High openness and steadfast quality: Availability of workers is high and more strong since chances of establishment frustration are immaterial.
3) High Scalability: Means "on-demand" provisioning of resources for an immense extension, without having engineers for top weights.
4) Multipopulation: With the help of distributed computing, various consumers and requests can effort even more profitably with fee lessen by sharing emblematic structure.
5) Appliance and position autonomy: Cloud figuring enables the clients to obtain structures using a web program paying little brain to their space without a doubt contraption they use for instance computer, cell thus forth [13].
6) Low Cost: By using distributed computing, the cost will be reduced; the IT association needs not to set its own system and pay-as indicated by usage of resources.

10.2.5   Advantages of Cloud Computing

There are a few benefits of distributed computing innovation [14].

1) Lower cost computer for clients: In the cloud, it need not bother with an incredible (and in like manner costly) COMPUTER to run distributed computing on the web applications since applications run on cloud not on workspace.
2) Lower IT establishment cost: By using distributed computing, it does not require placing assets into greater amounts of the relative multitude of more noteworthy workers.

Furthermore, I need not to require the IT staff for managing such stunning specialists.

3) Fewer support costs: The upkeep cost in distributed computing fundamentally diminishes both gear and programming upkeep for affiliations, taking everything into account.

4) Low rate of software; this reduces the cost of this item because it does not have to buy separate programming groups for each system.

5) Instant program update related benefit in distribution calculations is that customers do not have to choose between old plans and repair costs. If the app is an electronic device, the prelaunch is normal and the customer can get it again when they enter the cloud.

6) Increase the preparation power; the distance for workers from the cloud is significantly higher. It is estimated that the application is very fast.

7) Unlimited storage space; The Cloud gives a bigger limit, like 2000 GB anyway.

10.2.6   Challenges in Cloud Computing

Theft of Data: With the passage of time, internet penetration has increased. However, as the number of people using the internet grows, so does the risk of data theft. There is a large population of attackers and hackers among internet users [15].

Customers should be aware of the provider's encryption and backup procedures. Many cloud email providers, for example, encrypt emails so that their workers cannot read them. Customers should review these precautions on a regular basis to determine whether they are adequate for their needs.

Load on the Network: Cloud network load can also have an impact on the cloud computing system's performance. If the network load exceeds a certain threshold, the computers become unresponsive. This is a threshold limit. Due to high volume data movement between the disc and the computer memory, computers and servers break. When the threshold is exceeded by more than 80%, the suppliers defend their services and pass the loss on to their customers. One of the many reasons for brief outages is this [16].

Sanitization of data: Data sanitization is the technique of masking sensitive information in test and development databases by overwriting it with data that appears to be real. This method is used to verify the data protection. When a storage device is taken out of operation or transported to be stored, the data is lost. For security reasons, a sanitization process is used for that data. This method is also used for the backup copies that are used in the recovery process One of the sanitization strategies employed is masking [17].

Loss of Authority: When utilizing cloud foundations, the customer should surrender power over an assortment of safety related issues to the cloud Provider. Simultaneously, the system may not give a responsibility with respect to the cloud supplier to supply such administrations, leaving a security hole. The client surrenders information administration and possession. In the event that information possession rights are not tended to in agreements, cloud specialist organizations might utilise information for auxiliary reasons.

Immaturity of standards: In the cloud, standards are immature, and things change at a breakneck pace. Different technologies and standards are used by different IaaS and SaaS providers. Amazon's storage system is distinct from that of a normal data centre. The Azure storage engine does not use a traditional relational database, and neither does Google's App Engine. As a result, it cannot merely move apps to the cloud and expect them to work. Migrating an application to the cloud requires at least as much effort as moving it from an existing server to a new one. There is also the matter of personnel capabilities; employees may require retraining, and they may hate and dread a move to the cloud [18].

10.2.7   Cloud Security

Cloud security, generally named cloud computing security, is a bunch of safety efforts pointed toward shielding cloud-based foundation, applications, and information. These shields guarantee client and gadget validation, information and asset access to the board, and information security assurance. They likewise assist with information security and administrative consistency [16].

10.2.7.1   Foundation Security

While surveying host security and evaluating chances, it ought to consider the setting of cloud administrations conveyance models (SaaS, PaaS, and IaaS) and arrangement models (public, private, and mixture). In spite of the fact that there are no known new dangers that are explicit to distributed computing, some virtualization security dangers, for example, VMs escape, framework arrangement float, and insider dangers via frail access control to the hypervisor-convey into the public distributed computing climate. The unique nature (versatility) of distributed computing can bring new functional difficulties from a security the board viewpoint. The functional model spurs quick provisioning and passing cases of VMs. Overseeing weaknesses and patches is along these lines a lot harder than simply running an output, as the pace of progress is a lot higher than in a customary server farm [22].

10.2.7.2   SaaS and PaaS Host Security

As a general rule, cloud service providers (CSP) do not freely share data connected with their host stages, have working frameworks, and the cycles that are set up to get the hosts, since programmers can take advantage of that data when they are attempting to interrupt into the cloud administration. Consequently, with regards to SaaS (e.g., Salesforce.com, Workday.com) or PaaS (e.g., Google App Engine, Salesforce.com's Force.com) cloud administrations, have security is murky to clients and the obligation of getting the hosts is consigned to the CSP. To get affirmation from the CSP on the security cleanliness of its hosts, it requests that the seller share data under an agreement or just interest that the CSP share the data through a controlled evaluation system, for example, SysTrust or ISO 27002. From a controls affirmation viewpoint, the CSP needs to guarantee that suitable preventive and criminal investigator controls are set up and should guarantee the equivalent by means of an outsider evaluation or ISO 27002 sort appraisal system [33]. Both the PaaS and SaaS stages are dynamic and conceal the host working framework from end clients with a host deliberation layer. One vital contrast among PaaS and SaaS is the openness of the deliberation layer that conceals the working framework benefits the applications consume. On account of SaaS, the reflection layer is not noticeable to clients and is accessible just to the designers and the CSP's activities staff, where PaaS clients are given backhanded admittance to the host deliberation layer as a PaaS application programming connection point that thus communicates with the host reflection layer. So, a SaaS or a PaaS client, depending on the CSP to give a solid host stage on which the SaaS or PaaS application is created and conveyed by the CSP separately.

10.2.7.3   Virtual Server Security

Clients of IaaS have full admittance to the virtualized visitor that are facilitated and detached from one another by hypervisor innovation. Consequently clients are liable for getting continuous security on the board of the visitor system. This framework when arranged properly, can give flexibility to assets to develop or recoil in accordance with responsibility interest. The powerful life pattern of virtual servers can bring about intricacy if the interaction to deal with the virtual servers is not computerized with appropriate systems. From an assault surface point of view, the virtual server (Windows, Solaris, or Linux) might be open to anybody on the Internet, so adequate organization access moderation steps ought to be taken to confine admittance to virtual occurrences. Regularly, the CSP impedes all port admittance to virtual servers and suggests that clients utilise port 22 (Secure Shell) to direct virtual server occasions. The cloud executives id add one more layer of assault surface and should be remembered for the extent of getting virtual servers in the public cloud. A portion of the new host security dangers in the public IaaS include:

- Taking keys used to get to and oversee has (e.g. private keys)
- Assaulting unpatched, weak administrations tuning in on standard ports (e.g., file transfer protocol, etc)
- Seizing accounts that are not as expected got (i.e., powerless or no passwords for standard records)
- Assaulting frameworks that are not as expected got by have firewalls
- Sending Trojans inserted in the product part in the machine or inside the configuration itself.

10.2.7.4   Foundation Security: The Application Level

Application or programming security should be a basic component of a security program. Most ventures with data security programs still cannot seem to establish an application security program to address this domain. Planning and executing applications focused on arrangement on a cloud stage will expect that current application security programs rethink current practices and guidelines. The application security range goes from independent single-client applications to complex multiuser internet business applications utilized by a large number of clients.

10.2.7.5   Supplier Data and Its Security

Notwithstanding the security of own client information, clients ought to likewise be concerned regarding what information the supplier gathers and how the CSP safeguards that information. Explicitly with respect to client information, how metadata treats the supplier about information, how it is obtained, and what access the client has. As volume of information with specific supplier increments, so does the worth of that metadata. Moreover suppliers gather and should safeguard an enormous measure of safety related information. For instance, at the organization level, suppliers ought to gather, observe, and safeguard firewall, collision detection framework, and security aspects and switch stream information. At the host level, suppliers ought to gather framework log files, and at the application level, SaaS suppliers ought to gather application log information, including confirmation and approval data.

10.2.7.6   Need of Security in Cloud

Maintaining a solid cloud security act assists partnerships with receiving the rewards of cloud computing, which incorporate lower forthright expenses, lower progressing authoritative and the board costs, simplicity of scaling, expanded dependability and accessibility, and a brand novel method of working.

Cloud figuring security is progressing in lockstep with dangers, which are regularly observed to be past the point where it is possible to stay away from mishaps. Cloud computing represents a novel and genuine risk to all members because of its troublesome nature, convoluted engineering, and utilized assets. Understanding the danger and sufficiently relieving it is urgent for all partners and entertainers. As far as a possible danger, security should be inserted into each level of a cloud computing stage by accepting prescribed procedures and creating innovation. To adequately ensure the cloud computing stage, shoppers, suppliers, specialists, transporters, examiners, and every other person should make the necessary strides against risks.

10.2.8   Cloud Computing Applications

Distributed computing applications are mainly associated with convey ability of the customer, this may incorporate worldwide travel. The customer needs a sea wreck joining of organizations using various devices consistently from various regions. The customer needs solace and to pass on the base proportion of gear. Loss of gear accumulating contraptions may in like manner be an issue. In this manner, applications that cook for these needs have a phenomenal charm. Security applications, cataclysm mitigation, swarm figuring including long reach casual correspondence is prime competitor for gathering of flexible disseminated registering applications. Both Apple and Google give cloud applications, for instance,

Apple's iCloud and Google's compact email organization. Major foreign enterprises—delivery and plan stage (App Engine) and usefulness instruments (Google Apps). Minimal expense equipment scalable software system, imaginative applications, simple to develop, keeps up with and scale. The more cloud center customers there are, the more prosperity will be. Online businesses are business-based SaaS and PaaS models. Fuse complete web have system CloudEx Computing Service, Cloud security organizations, data support for individuals and dares to the Internet. A few more use of cloud are displayed in Figure 10.4.

Business applications rely on specialized teams in the cloud. Today, every affiliate needs a cloud business application to grow their business. It provides similar commercial services to customers 24*7 [19].

Data storage and application distribution (data, reports, images, sounds and accounts) are stored in the cloud and this information is obtained using a web affiliation. The cloud provider is responsible for security; hence, they provide a one-stop media recovery application to recover lost data.

The cloud enrollment in the school app guide area is amazing. It offers informative offers for different levels of online distance education and students who want to learn. The potential for using the cloud in the tutoring

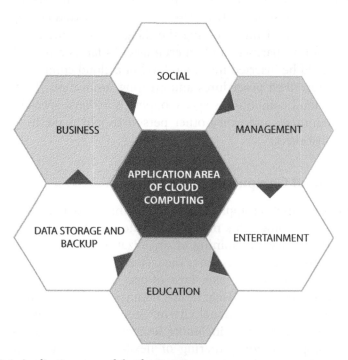

Figure 10.4 Application areas of cloud computing.

arena is that it can be used in solid virtual homeroom scenarios; Ease of use Secure data collection; Flexibility; these are essential hardware requirements for students to be more accessible and usable. Diversion applications media outlets use a multicloud method to interface with the ideal vested party. Circulated processing offers distinctive redirection applications for instance, web games and video conferencing. The executive applications cloud processing offers distinctive cloud the board gadgets which assist chairmen to manage a wide scope of cloud works out, similar to resource game plan, data consolidation, and disaster recovery. These organization gadgets similarly give administrative control over the stages, applications, and establishment.

10.3   Literature Review

This article assesses appropriate handling of information security issues, counting tile security of information transmission, putting away security and the heads of prosperity. Zero in on expansive information the board swayed cloud security evaluation, and called attention to that a leap forward in the progress of this spread figuring, attempting to list the differentiating systems and significant length movement bearing. Last part is an outline and standpoint about future movement of passed on enlisting security issues [20]. Through a creating study, its purpose is to safeguard philosophies that are utilized by specialists to battle these security difficulties and it gave an organizing of the difficulties to their protecting undertakings. Their disclosures reveal that there is, no ifs, ands or buts, a case for human blunder to be joined as a class in the solicitation for the security challenges experienced in cloud selecting, and if cloud oriented networks and their clients are completely instructed on the security challenges experienced in the cloud, the two players can thoroughly advantage from the benefits this model of enrolling offers [21]. This region will at first spotlight on the fundamental musings and separate the basics of information security issues relating to disseminated processing [22]. Then, by then it fosters each issue by talking about its tendency and existing plans if available. In this article, the system performed an assessment of the security assessments offered for the cloud selection steps. The document discusses the expected risks, it presents the security issues at the stages of the cloud project and a comprehensive assessment process carried out in these areas. Basic cryptography security calculations; for example Double Encryption Standard (DES), Advanced Encryption Standard (AES) were examined and the identities of these evaluation indicators were considered and their results introduced. Some of the frameworks used to implement cloud learning security have completely disappeared compared to others; they are analyzed in depth and provide an overview of the safety aspects by examining the constraints of the current detailed plans. Finally, it provides us with different conformations and provides a cloud security architecture that sees the levels of dependency between them. The system encounters cloud security opportunities negotiated with five orders. It reports nine different general attacks in the neighborhood and assesses rationality. This article or section needs sources or references that appear in credible third-party publication, describes the appropriate registration requests including qualifications and issues [10–15].

10.4   Cloud Computing Challenges and Its Solution

There are different security issues and troubles in cloud computing since it incorporates various advancements like associations, databases, working structure, virtualization, resource arranging, trade the board, synchronous control and memory the chiefs [8]. This is fundamental because the cloud expert community should ensure that the customers are not managing any significant issue like data disaster and data theft which may cause a phenomenal mishap depending upon the affectability of the data set aside in the cloud. A harmful customer may profess to be the genuine customer and sullying the cloud [20].

There is a great deal of safety issues to be examined:

Security issues: Information exceptionally still is the huge issue in appropriated figuring since customers may store all their ordinary, private, or even fragile data in the cloud which can be gotten to by anyone wherever. Data forgery is a very typical issue that is being looked at by the cloud expert associations nowadays. In addition, some cloud expert centres even do not give their own laborer because of the cost reasonability and flexibility. There are moreover events like data disasters which might be also a significant issue for the customers. For example, the specialist is unexpectedly shut down and causes data loss of the customers. Also, disastrous occasions may moreover cause data to be hurt or spoiled. Thus, real data region can be seen as one of the security issues in conveyed registering [19].

Information issues: The conveyed registering expert community ought to approve their own game plans to ensure the security of the data customers set aside in their cloud model. They ought to guarantee that they comprehend who is truly moving to the data set aside in the cloud and simply the supported individual can stay aware of the cloud organization model [8]. The security of appropriated registering should be done on the provider side and besides the customer side. Cloud expert community should give a good layer of security protection for the customers while the customers should not modify the other customer's data. The appropriated processing is a respectable strategy to reduce the expense and give greater limit if and just if the security is done by both provider and customer [9]. Guaranteed that regulatory change is crucial to guarantee tricky data in the cloud since one of the most troublesome points of view in appropriated processing is to ensure that the client has trust in insurance and security of their data. Application issues: Checking and upkeep should be done by the cloud expert association a large part of an opportunity to ensure that the cloud is secure and not tainted by the malicious code that have been moved to the cloud by the developers not set in stone to take sensitive information or regardless, hurting the information of explicit customers.

In Figure 10.5, various threats in cloud environment area as follows:

A. Network Load: Cloud network burden can likewise influence the presentation of the distributed computing framework. The system becomes slow off chance that the organization load is in excess of a limit. This limit

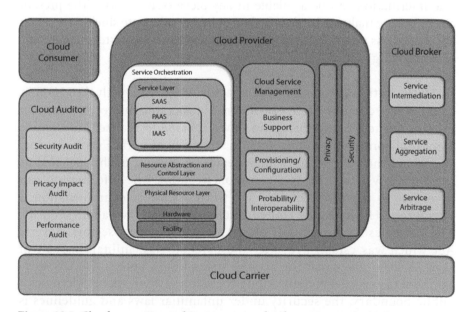

Figure 10.5 Cloud computing architecture as per cloud-server communication.

is the threshold. The systems and the workers collide due to high volume movement of information between the plate and the computer memory. When the limit surpasses 80%, the sellers secure their administrations and give the corruption to clients. This is one of the many reasons that give transitory blackouts [21].

B. Data decontamination: Information filtering is the most common way of masking delicate data in test and advancement data sets by overwriting it with information which resembles reasonable yet bogus information of a comparable type. This method is utilized to guarantee the security of information. At the point when a capacity gadget is eliminated from administration or moved somewhere else to be put away, the information disinfection procedure is applied to that information for security purposes. This method is likewise applied on the reinforcement duplicates that are utilized for recuperation. Covering is one of the procedures utilized for data filtration.

C. Data Location One of the most widely recognized consistency issues confronting an association is the information area. In distributed computing the information is dissipated in the workers present in various pieces of the world. The client does not know about the area of his/her information. The information can be available in any piece of the world. The present circumstance makes it hard to find out whether adequate shields are set up and regardless of whether lawful and administrative consistency prerequisites are being met.

D. Data location constraint: The enactment changes with the country. Every single country has its own constitution and laws of privacy. On looking at the US and European Laws, it came to realize that there is a colossal contrast in the protection strategies of both. Europe is worried about the protection of their residents in light of the fact that as per their laws, security is a common freedom. The information can be moved to the third country just when assurance of information is guaranteed. However, the United States has various laws about security. The strategies are profoundly affected by the business sectors. Several Act which was established after the psychological militant assault of Sept 11, 2001 gives that the arrangements would permit the U.S. government to get private data. Along these lines, when the data crosses a public boundary, the security under unfamiliar laws and guidelines is not ensured [22].

E. Lack of Cyber Crime Laws: There is absence of digital wrongdoing laws in numerous nations. Around 52 nations have reacted to the digital wrongdoings by authorizing the laws. Out of them, just ten nations including Australia, Japan, and the United States have completely or significantly refreshed laws. Nine nations including Brazil, Chile, and Poland soon have to some extent update laws. In any case, 33 nations have no refreshed laws. Thirteen nations have shown some advancement. Such nations do not make severe moves against digital violations. In such circumstances if the information is moved to such a nation, where the completely refreshed digital wrongdoing laws are not there, it can be hazardous for the security of information.

F. Availability of Service: However there are structures intended for high help dependability and accessibility yet the clients experience blackouts and lull in exhibitions. There is no 100% unwavering quality of administration. The help can be hindered because of specific reasons like more traffic, gear disappointment; cataclysmic events and so forth. There are a few instances of such outages. Amazon S3 blackout for An AppEngine incomplete blackout for 5 hours [30] on 17 June, 2008 and Gmail site was inaccessible for 1.5 hours on 11 August; 2008 [42]. This makes the client as a result of inaccessibility of administration [42].

G. Permanent Loss of Data: The bother is not restricted to the brief blackouts, there can be extremely durable loss of information because of specific reasons like liquidation or office misfortune. It is feasible for a specialist co-op to encounter significant issues, similar to insolvency or office misfortune which can cause total shutdown. There is an illustration of Texas where it struck assistance places and held onto many workers if there should arise an occurrence of extortion charges in April 2009. Due to this, the help was upset and numerous misfortune their information who were utilizing that processing habitats.

H. Data Segregation: In the event that information of the client is available in the common mode, the client should ensure that information is secure and cannot be gotten to by different clients. Since information might consist of individual data, for example, financial balance numbers, secret phrases and so on, the cloud supplier should deal with the security of the information which can be given by the information segregation. Encryption of the information is a protected method to give information isolation. The client should ask the specialist organization whether the execution of encryption was tried by the specialists. Since the mishaps happen because of the disappointment of the execution of encryption techniques. Prior to putting away all the data of a client, encryption should be done on the information of the client and afterward it is put away. Prior to recovering the information by the client, decoding is performed on the information of that specific client and afterward the data is given to the client. It is likewise to discover who approaches the unscrambling keys [23].

I. Insider Access: Information put away on a cloud supplier's worker might conceivably be gotten to by a representative of that organization, and the standard staff powers over those individuals. This is likewise a negative mark of cloud computing. The specialist co-ops may not uncover the entrance of workers to the information of the clients of the cloud. This degree of access chooses the security of information. The representative might approach the private data of the clients and workers can abuse such data. This will not just influence the client yet additionally the brand picture, monetary usefulness of the specialist co-op. Along these lines, the clients of the assistance should think prior to putting away the secret data into the cloud and ought to get some information about the workers who will approach information and oversee it.

J. Long-Term Viability: Specialist organizations should guarantee the information wellbeing in changing business circumstances like consolidations and acquisitions. Clients should guarantee accessibility of the information in these circumstances. The supplier's terms of administration may not uncover the genuine proprietor. The purchaser could be an administration organization, an unfamiliar office, or an Internet news administration. On the off chance that an administration office possesses the supplier, terms of administration that permit imparting to members could bring about the entirety of the client's data being gotten by investigators or knowledge offices minus any additional notification or interaction. Some chapter 11 issues are additionally there. Insolvency is a danger to the information of the client. The specialist co-op should cautiously structure the agreements with the financial backers and loan bosses. Another arrangement can be the protection contracts which can give the assets to run the foundation for a period after bankruptcy [24].

10.4.1 Solution and Practices for Cloud Challenges

The cloud environment has become better known in light of the fact that various customers start to comprehend its benefits. It allows the customer to helpfully withdraw the movement and moreover help to save cost. In any case, with the extended gathering speed of the cloud organization, the security issues and risk have been extended moreover. To make circulated processing a better option than to construct the customer accumulating breaking point and save their ordered information securely, there are relatively few game plans and practice that has an effect [25].

Shortcoming protecting: The cloud expert community should additionally foster the fix of the board. They should check the shortcomings of their cloud organization, generally and reliably revive and stay aware of the cloud to confine the possible path and diminishing the risk of attack of the cloud by the developers. The cloud expert centre may in like manner use the Intrusion Detection System to guarantee the cloud organization is secure and safe.

Use cloud organization commendably: The data set aside in the cloud should be private and shockingly the cloud expert association should not move toward that information [23]. The data set aside in the cloud should be mixed throughout to ensure the security of the customers' information. Any person who needs permission to the data in the cloud should demand the assent of the customers preceding doing accordingly.

Security really takes a gander at events: The customers should have clear concurrence with the cloud expert community so the customers can ensure any incidents or breaks of the delicate data set aside in the cloud. The customers ought to have clear simultaneousness with the cloud expert association before using the cloud organizations given by that particular cloud expert association. The customers should ensure that the cloud expert community gives adequate bits of knowledge concerning fulfilments to ensure break remediation and specifying probability [26].

Data storing rules: The plan of the cloud environment is a critical point to ensure the security of the data set aside in the cloud. The customers ought to appreciate the possibility of the data accumulating rules which the cloud expert association follows. Cloud expert associations that outfit security plans reliable with rules, for instance, data protection laws are presumably the best choice.

Workplaces for recovery: Cloud expert community ought to expect the risk to recover the data of the customers in the event that there is any data mishap in light of explicit issues. Cloud expert community ought to guarantee that they have proper support and can recuperate and recover the private data of the customers that might be extreme. Plus, the cloud expert communities can similarly do the going with deals with any consequences regarding ensure data recovery.

   i. Utilizing fastest plate development in event of catastrophe for replication of data at genuine danger.

   ii. Changing the dingy page limit.

   iii. Forecast and replacement of hazardous contraptions.

Escapade system: The customer must get the data that they need to keep in the cloud structure. The cloud expert centre should give a structure that gives work for the customers to present and plan hardware parts like firewalls, switches, laborer and mediator specialist.

Access control: The cloud expert association should set up the data access control with rights and the customers who access the data should be affirmed by the cloud expert association come what may. The cloud expert association should ensure that fundamentally the supported customers may move toward the data set aside in the cloud. The method can help with diminishing the risk of the data access by the unapproved customers and as needs be given a much secure environment to store fragile data. In addition, outcast assessing can similarly be one of the decisions to ensure data genuineness of the limit in the cloud. Regardless, the looking into technique should have the going with properties:

- Confidentiality: Auditing shows should keep customer's data grouped against the inspector.
- Dynamic inspecting: Auditing convention should uphold updates of information in the cloud.
- Batch reviewing: Auditing convention should support bunch examining for various clients and mists.

10.5   Cloud Computing Security Issues and Its Preventive Measures

Before moving on to security issues, understand the definition and order of clouds. Scattered calculations are much stronger that can be updated on demand. It can be done in any way that really matters. With the community of masters, it can use the Internet in a personal way without any income required. The cloud is about innovating and interacting with things, Provides better proposal ideas for reflection and evaluation. As noted in the cloud computing Reference Architecture. There are five great performers nearby who are entertained by his thoughts of safety. This report focuses on the risk and information fragments of customers and cloud providers as shown in Figure 10.5 [27].

Cloud consumer: a cloud provider who has a relationship with an individual or an organization: Cloud provider: individual; Answers to help an alliance or a participant; Cloud Auditor - Part of cloud associations; Responsibilities relating to the data framework; Free evaluation of cloud management performance and security; Cloud Broker - One Entity It fosters execution and development, and fosters integration among cloud users. Carrier-A is a hub that integrates individuals from cloud providers to cloud teams.

10.5.1   General Security Threats in Cloud

As the challenges of cloud security guide social interactions, it increases the demand on offer. The goal is to understand the audit and integrate the past cloud security challenges of the study, and to make the cloud intelligence challenges more comprehensive through the challenges faced in the New Year. During our evaluation, Related reasons; Forms of communication; Network issues; Information possibilities; Cloudless dangers; virtualization issues; the affairs of the board of directors. It found a collection of articles that included content transfers.

Types of Cloud Security Issues and dependencies are part of this job. It has conducted the best analysis of current cloud security issues and class security plans. The 28 security issues are listed in five categories (Table 10.1). A comprehensive overview of current security plans and high-level counterattacks are provided [29]. The proposed work projected that the safety concerns described in Groups 1 be summarized in accordance with the Five Commandments. Cloud security concerns cover a few game plans and four illustrations.

(C1) The security standard governs managing authorities and regulators who demonstrate cloud security philosophies to ensure a dust-free workplace. This involves team-level approaches; Different strategies between evaluation and clients; Local know-how and various accessories.

(C2) Network class refers to an intermediate group of cloud frameworks and clients for optimal assessment. It selectively selects programs, Provides network connectivity and data exchange.

(C3) Access Control is a client management class that provides access to documents, involved in recommendations and support issues.

Table 10.1 Taxonomy of challenges in cloud security.

Trend	Explanation
Encryption [25]	a. In traditional IT, encryption has turned into a mainstay of information security, and it will keep on being so in the distributed computing worldview. Information security and information trustworthiness are two key assurance builds that encryption handles. Encryption and access control have long become significant apparatuses of cryptography for getting information on the two sides of the edge. b. However cryptography can assume a huge part in numerous information security issues, cryptographic strategies are intrinsically tedious. Accordingly, any symmetric encryption plan or technique created to address the issues should represent actuated dormancy, which will affect the accessibility of information and administrations facilitated in the cloud.
Data Lineage [11]	Logging or following programming is utilized in a standard IT climate to catch data about a party's activities, for example, gadget use, data set tasks, program execution follows, network bundle follows, and working framework occasions. Bookkeeping and bookkeeping, gadget checking, computer observation, IT review, criminology, and peculiarity identification are largely programs that might profit from such logs.
Data At Rest [26]	Data very still alludes to information that is put away on an optional capacity unit or in a cloud. Nonstop stockpiling includes cloud-based progressed storerooms also with respect to occasion stockpiling for each occurrence. While some stockpiling suppliers give profoundly adaptable, flexible, and secure capacity, occurrence stockpiling can be wasteful and is deleted until the occasion is ended. In any case, in such circumstances, information is put away in the cloud, far away from the client's system.

(Continued)

Table 10.1 Taxonomy of challenges in cloud security. (*Continued*)

Trend	Explanation
Data in Progress [40]	The dynamic condition of information, otherwise called information in progress or information in handling, is the point at which it is being utilized. In this condition, information can be put away predominantly on a gadget like RAM. What is more reserve recollections, and may likewise be put away in CPU logs. To process encoded information through the CPU, a large part of the security in the information very still should be encoded and subsequently convert cipher text to plaintext design. Information are along these lines generally delicate in their lifetime while in this state.
Data in Transit [33]	The state of data in transit, also known as file Transfers, is when it is travelling or is moved from one place to another through a network. Data may be transferred from a cloud client's local storage to a CSP as cloud storage, or it may be moved from cloud storage to local storage in the cloud computing environment. Data can pass between different clouds as well as within a single cloud.

(C4) The integration of cloud infrastructure integrates security issues within SaaS, PaaS and IaaS, particularly in relation to the virtualization climate.

(C5) The data class describes information security and complaints issues [30].

   a. shared Technology inadequacies—extended impact of resources gives the aggressors a specific nature of attack, which can cause hurt disproportional to its importance. A framework of course of action progress is a hypervisor or cloud alliance [31].
   b. Information Breach—with data protection moving from cloud client to cloud ace connection, the risk of unconstrained, poisonous, and deliberate data break is high.

c. Record of Service traffic getting—maybe the best advantage of cloud is access through Internet, yet for all intents and purposes indistinguishable is a risk of record compromise. Losing approval to maintain a record might mean loss of affiliation.

d. Denial Of Service (DoS)—any denial of affiliation attack on the cloud provider can influence all essentials

e. A Vindictive Insider not actually settled insider can find more ways to deal with oversee attack and cover the track in a cloud circumstance.

f. Web protocol—various insufficiencies of brand name, for instance, caricaturizing, Address resolution protocol flooding, Cache Poisoning are real risks.

g. Imbuement vulnerabilities—deficiencies, for instance, database injection, and implantation at the affiliation layer can cause monstrous issues across different cloud buyers.

h. Programming interface and browser susceptibility—Any inadequacy in cloud supplier's system or Interface tends to a tremendous danger, when gotten together with friendly arranging or program based assaults; the wickedness can be fundamental.

i. Changes to business model—dispersed processing can be a colossal change to a cloud customer's game plan. IT division and business needs to change or go up against receptiveness to peril.

j. Extra use—certain components of circulated figuring can be used for hateful harass purposes, for instance, the use of trail season of usage to dispatch zombie or DDoS attacks.

k. Malevolent insider—a noxious insider is reliably a huge risk, regardless, a dangerous insider at the cloud provider can make basic mischief for various purchasers.

l. Availability—The probability that a structure will work as per need.

In Table 10.1, taxonomy of cloud threat is discussed.

According to this stage, it perceives evolution with seven classes of cloud security issues: meeting/hacking cookies - sessions and benefits are typically used to guarantee the credentials of clients in the program. If any of them are threatened, they can be used to complete a critical attack, or the aggressor can be used to give misleading testimony and turn the client into other help in which the accident falls. To carry out this attack, the aggressor uses a flaw in the structure of the cloud or malware in the client program; some major security threats are shown in Figure 10.6 [32].

XML signature part binding - An XML certificate is a means of delivering messages from the various sources of information used in the cloud [16] in real time, A different, reliable and authenticated encryption system. An attacker could hijack Simple Object Access Protocol messages; it enhances the messages sent by the acknowledged sender by incorporating the transmissions and information used to move information to the cloud and the host computer. This will usually trick the customer into thinking they do not understand the confusing information.

Entering database—Here the attacker includes the revenge code in the SQL code model, and the attacker receives the returned information using some unprecedented characters in the SQL configuration. They allow an intruder to obtain unconfirmed consent to favorite data and retrieve data from an illuminator assortment, as the embedded information will be misinterpreted by the expert as client information as intended, allowing the programmer to reach the SQL employee [5]. The database and SQL Wizard are the core bits of the cloud framework, and must ensure that their core is protected against this attack.

Attacking a man in a mid-attack a man in the middle of an attack by an attacker receives correspondence between the sender and the finder of the item. An attacker takes private data that is transferred between two players, declaring it to be one of the social events [15]. This attack uses advanced information generation, using tools such as Packet Creator, Ettercap, Dsniff, to possibly modify lists [33].

Denial of service—Denial of Service (DOS) attacks Attempts to associate with a service through memory or through data transmission. DOS

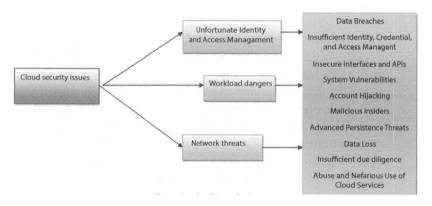

Figure 10.6 Cloud security issues.

attacks should be designed in the branch offices of each division. At the interaction level, the intruder can fill the expert network interface card with unnecessary packages, which encourages the unit to provide real-time access to the client. DOS can be used in the vehicle layer using the system Flood framework. In this situation, the attacker was unable to complete the manual operation of the request wizard using the operation method [34].

Application vulnerability—A cloud association client is using a web application to access the cloud framework. To support customers and protect information from dangerous or unauthorized customers, web applications typically cannot perform XML signing or XML encryption to ensure the security of web associations. Therefore, this vulnerability can cause the aggressor to use customer statements destructively to connect to the cloud worker.

Scripting between sites - this is an attack in which an attacker injects toxic substances into the information entry points of the site. Even if this is done, it may require the expert to execute scripts to be executed by external measures, which may later lead to the disclosure of key information. This attack is carried out in any legal proceedings; continuously capturing retaliatory codes in material used by a web application or displaying malicious code to a customer.

10.5.2　Preventive Measures

Weaknesses and dangers of clouds are documented everywhere. Depending on their assessment, every cloud specialist and cloud customer needs to take action and respond to the risks to mitigate the risks. In either case, the delay is likely to be acceptable procedures in response to considerations and controls [34–38]:

a. End-to-end encryption—the information in a cloud conveyance model may navigate through numerous geological areas; scramble the information start to finish.

b. Scanning for malignant exercises—start to finish encryption while energetically suggested, prompts new dangers, as scrambled information cannot be perused by the firewall. Accordingly, have suitable controls and countermeasures to relieve chances from vindictive programming going through encryption.c. corroboration of cloud buyer—the cloud

Table 10.2 Explanation of major security issues.

Challenges	Types of security threat	Major security issues
Dynamic environment: The flexible idea of conditions on the cloud, makes ideal continuous perceivability of virtual occurrences troublesome. Safeguarding of such conditions require a ceaseless disclosure, security appraisal and proactively make moves to safeguard them.	Unfortunate Identity and Access Management	a. Account Hijacking: If Cloud merchant console or Programming interface qualifications are lost, a malevolent entertainer outside of an association can assume responsibility for the cloud climate. b. Data Breaches: Poor access the executives of item stockpiling pails and information stores make touchy data be unveiled, which has been one of the significant reasons for information breaks on the cloud. c. Malicious insiders: Malicious insiders attempting to take administrator/root honors, can bring about loss of delicate information and frameworks. d. Abuse and Nefarious utilization of cloud assets: Account seizing of a cloud record can bring about the malevolent client to utilize the compromised assets to send off DDOS, spam and phishing efforts leaving the association inclined to lawful risk. e. Insufficient Due-industriousness: Organizations which handle information and fall under administrative consistence regulations, need to have an unmistakable arrangement to move to the cloud, else this represents a security danger and lawful risk.

(Continued)

Table 10.2 Explanation of major security issues. (*Continued*)

Challenges	Types of security threat	Major security issues
Edge definitions: Cloud responsibilities are frequently divided across a few different geo-areas and conditions, making it hard to midway oversee resources	Workload dangers	a. Advanced industrious threat : Malware and Advanced constant dangers once enter a climate, adjust to the safety efforts and over the long run gain a traction in the climate and engender itself along the side and when it arrives at the expected objective, it will exfiltrate delicate information. These dangers are hard to distinguish and remediate. b. Vulnerabilities: With Cloud administrations being multi-inhabitant, weaknesses that incorporate honor acceleration and limit bouncing can cause information breaks and leave the applications and responsibilities powerless. c. Insecure application administrations: Insecure API's serving different assistance of an application can leave the application helpless against known assaults and result in application personal time or information breaks.

(*Continued*)

Table 10.2 Explanation of major security issues. (*Continued*)

Challenges	Types of security threat	Major security issues
Loss of control on actual security: As associations let go completely over actual security, the obligation of safeguarding information and responsibilities at travel and rest falls into the lap of the client. Virtualized and multi-occupant nature of public cloud make it important that an association is consistently up to speed with the most recent weaknesses and make remediation moves when fundamental	Network dangers	a. DOS and DDOS assaults: Poor organization division and firewall the board lead to cloud assets being focused on by DOS and DDOS assaults which bring about unfortunate application execution and even application vacation. b. Data infiltration: Poor outbound firewall controls prompts information infiltration endeavours by compromised jobs.

supplier needs to play it safe to monitor the cloud procurer to forestall significant components of cloud being utilized for malignant assault purposes.

c. protected Interface and APIs—the interfaces and APIs are critical to carry out robotization, organization, and the board. The cloud supplier needs to guarantee that any weakness is alleviated.

d. Insider assaults—cloud suppliers should play it safe to screen workers and workers for hire, alongside fortifying interior security frameworks to forestall any insider assaults.

e. Secure utilized assets—in a common/multioccupancy model, the cloud supplier has secure shared assets, for example, hypervisor, organization, and checking instruments [39, 40].

f. Business Continuity plans—business cogruence plan is a course of archiving the reaction of the association to any occurrences that cause inaccessibility of the entire or part of a business-basic cycle.

In Table 10.2, the prominent security issue of cloud threat and challenges is enumerated for each type of threat category.

10.6   Cloud Data Protection and Security Using Steganography

Steganography is one way to solve this problem. These methods will increase the security of sensitive data as it allows hiding other things, such as video. This system is talking about image steganography to deliver cloud data [41]. Figure 10.7 discusses the process of image steganography in a cloud environment. In this article, current cloud data security methods have been identified for image steganography. This article is consistent as follows: Region II provides a reasonable calculation plan. Provide a steganography program in Zone [28].

Overview of this segment offers a steganography sketch, its aprons, and targets. Steganography is the study of hiding inside information in an interactive media document. Steganography as a word is a combination of two Greek words "Sregano" and "Grafie" and means "expression of coverage".

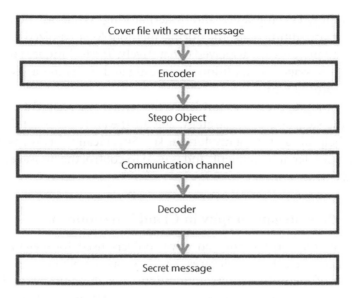

Figure 10.7 Steganography process.

10.6.1   Types of Steganography

- Text steganography: Use text documents to conceal privileged information.
- Picture steganography: Hide privileged information in a cover picture.
- Sound steganography: Use a sound document to cover privileged information.
- Video steganography: Hide privileged information in a video document.
- Image-based steganography: Employ haphazardness of image to implant privileged information.

Steganography is the method involved with implanting data in a transporter to secure the mystery in the terms of text, music, video and pictures. Steganography is the specialty of concealing data. Typically data is implanted in pictures; it stays imperceptible in most business picture data sets, like Getty (gettyimages.ie) or iStockPhoto (istockphoto.com). In this way the upside of utilizing steganographic strategies for data security is that the presence is safe in terms of precise algorithmic function. In this chapter, it presents the vital ideas and the portrayal of the steganography region is graphically and numerically shown. Differentiations between steganography, cryptography and watermarking as far as method and expectation are summed up. The normal methodologies utilized for installing data into pictures are displayed exhaustively. The techniques applied for concealing messages are additionally investigated. Steganography apparatuses are featured [25].

In a steganographic correspondence, the sender and recipient concur on utilizing the steganographic strategy, and for the anticipation of data location, it is helpful to utilise a mysterious key. Specifically, this key is divided among the sender and recipient, and is used to control the message stowing away and extraction.

10.6.2   Data Steganography in Cloud Environment

Pictures are the most renowned cover objects used for steganography. In the space of electronic pictures a wide scope of picture report plans exist, most of them for express applications. For these unmistakable picture report plans, various steganographic computations exist as shown in Figure 10.8 [32].

1. Least Significant Bit Algorithm

Stage 1. Select a cover image of size M*N as a data.
Stage 2. The message to be stowed away is embedded in the part of an image.
Stage 3. Use a pixel assurance channel to get the best districts to disguise information in the cover picture to get a predominant rate. The channel is applied to Least Significant Bit (LSB) of every pixel to cover information, leaving cipher text which is hard to scramble [41, 43].

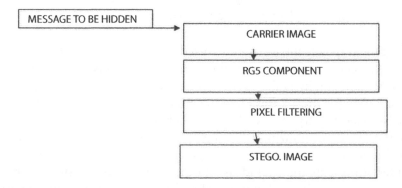

Figure 10.8 Steps of steganography data encryption.

10.6.3 Pixel Value Differencing Method

The Pixel Value Differencing strategy (PVD) can successfully provide an elevated place in the boundary and unexpected impalpability for the stego-images. The PVD technique partitions the cover image into non covering blocks containing two interfacing pixels and alters the pixel distinction in each square (pair) for information installation. To appraise the number of mystery pieces inserted into pixels, the biggest distinction esteem between the other three as well as four pixels near the objective pixel is determined as displayed in figure. PVD is planned so that the pixel change does not disregard dark scale range span. The determination of the reach stretches depends on the qualities of human vision: affectability to dark worth (0-255) shifts from perfection to differentiate [43, 44]. It depicts the pixel value distribution for equalizing the hash value in four different segments. Likewise, in image processing, the image compression the features and attributes of an image is subdivided into different pixels and using cross validation of these features the labelling is done using a hashmap of ciphered image. The PVD method comprises the four different matrices, such as G(x-1, y-1) To left pixel, G(x-1,y) Top pixel, G(x, y-1) Left pixel, G(x, y) Target pixel.

It gives a simple method to deliver a more intangible outcome than basic LSB substitution techniques. The installed secret message can be separated from the subsequent stego-images without referring to the first cover picture. Additionally, to accomplish mystery insurance of stowed away information a pseudo-arbitrary component might be utilized. In the event that privileged information is put away haphazardly it is hard to comprehend by the interloper. PVD installation is utilized for edged regions to build picture quality. It is likewise used to conceal messages into dark scales just as in shading pictures [45].

2. Least Significant Bit Technique The most every now and again utilized steganography strategy is the procedure of LSB replacement. In a dim level picture, each pixel comprises 8 pieces. One pixel can subsequently show 28 = 256 varieties. The essential idea of LSB replacement is to insert the private information at the furthest right pieces (bits with the littlest weighting) so the implantation technique does not influence the first pixel esteem extraordinarily. The numerical portrayal for LSB strategy is [46]:

$$XY + (A-Y) = AY \qquad (10.1)$$

in Equation (10.1), x'i addresses the Ith pixel worth of the stego-picture, xi addresses the first cover-picture, and mi addresses the decimal worth of the Ith block in classified information. The quantity of LSBs to be subbed is signified as k. The extraction cycle is to duplicate the k-furthest right pieces straightforwardly. Numerically the removed message is addressed as:

$$XY + AB = X\text{-}1 \tag{10.2}$$

Moreover, the essential difference reduced gives the principal private data. While using a 24-bit picture, a bit of all of the red, green and blue concealing parts can be used, since they are each addressed by a byte and each byte can address 256 particular tones. With everything taken into account, one can store 3 pieces in each pixel. A 800 × 600 pixel picture, would in this way have the option to store a total amount of 1,440,000 pieces or 180,000 bytes of embedded data. Cover the twofold worth 1011011 into 24-bit picture 24-bit picture using LSB as shown in Figure 10.9 [46].

This part introduced a foundation of Steganography and an exhaustive investigation of some applications in distributed computing. Steganography as a data security framework can have some helpful applications, as other apparently related frameworks (cryptography). Steganography targets making undercover channels for mystery or private correspondences. Every one of the strategies under the picture space techniques is centre which generally works around the most un-huge pieces of the pixel esteems. The device for estimating the nature of a picture subsequent to implanting is the hidden value. The determination of the cover pictures impacts the security

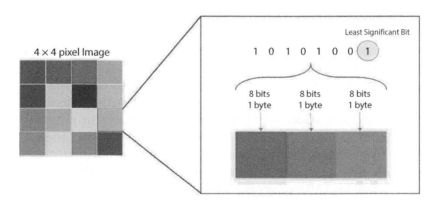

Figure 10.9 Color bit representation.

Table 10.3 Survey discussed in existing literature.

Reference	Survey
[21]	Cloud computing security issues
[32]	Classification of the security challenges
[11]	Security threats nature
[22]	Preventive measure
[8]	Analysis of the proposed countermeasures
[29]	Data related security challenges

of steganography frameworks. Benefits and disservices of steganography technique are shown the accomplishment of this review is to recognize the dependable steganography approaches in computerized pictures [42].

10.7   Related Study

In Table 10.3, the significant literature and their main focus is highlighted which have been used for analyzing the taxonomy.

10.8   Conclusion

In this chapter, a critical security concern of cloud computing is discussed including data privacy, localization, server, communication, traceability in various models of cloud services, like IaaS, PaaS, and so on. This chapter gives an introductory overview of various kinds of data security challenges faced by this service model, like Ddos, identity theft, breaching, man-in-the-middle attack, insider threat, etc. The taxonomical discussion of this chapter highlights the major solutions to these issues like data filtration, authentication, security protocols, like secure socket layer and transport layer security services. Various kinds of security mechanisms were introduced, like encryption standard, cryptographic algorithms, to enhance the data security with data security with the help of steganography. In the future, this chapter gives a pathway for research and development of various algorithms for cloud data security.

References

1. Bal, P.K., Mohapatra, S.K., Das, T.K., Srinivasan, K., Hu, Y.-C., A joint resource allocation, security with efficient task scheduling in cloud computing using hybrid machine learning techniques. *Sensors*, **22**, 3, 1242, 2022.
2. He, Q. and He, H., A novel method to enhance sustainable systems security in cloud computing based on the combination of encryption and data mining. *Sustainability*, **13**, 1, 101, 2020.
3. El Sibai, R., Gemayel, N., Bou Abdo, J., Demerjian, J., A survey on access control mechanisms for cloud computing. *Trans. Emerg. Telecommun. Technol.*, **31**, 2, 16–35, 2020.
4. Kunal, S., Saha, A., Amin, R., An overview of cloud-fog computing: Architectures, applications with security challenges. *Secur. Privacy*, **2**, 4, 6–14, 2019.
5. Samarati, P., De Capitani di Vimercati, S., Murugesan, S., Bojanova, I., Cloud security: Issues and concerns, in: *Encyclopedia of Cloud Computing*, pp. 205–219, John Wiley & Sons, Ltd., Chichester, UK, 2016.
6. Kamara, S. and Lauter, K., Cryptographic cloud storage, in: *Financial Cryptography and Data Security*, vol. 6054, pp. 136–149, Springer, Berlin Heidelberg, 2010.
7. Ahmad, W., Rasool, A., Javed, A.R., Baker, T., Jalil, Z., Cyber security in IoT-based cloud computing: A comprehensive survey. *Electronics*, **11**, 1, 16, 2021.
8. Al-Ruithe, M., Benkhelifa, E., Hameed, K., Data governance taxonomy: Cloud versus non-cloud. *Sustainability*, **10**, 1, 95, 2018.
9. Yang, P., Xiong, N., Ren, J., Data security and privacy protection for cloud storage: A survey. *IEEE Access*, **8**, 131723–131740, 2020.
10. Kaufman, L.M., Data security in the world of cloud computing. *IEEE Secur. Priv. Mag.*, **7**, 4, 61–64, 2009.
11. Jagtap, S., Bader, F., Garcia-Garcia, G., Trollman, H., Fadiji, T., Salonitis, K., Food logistics 4.0: Opportunities and challenges. *Logistics*, **5**, 1, 2, 2020.
12. Wang, Q., Su, M., Zhang, M., Li, R., Integrating digital technologies and public health to fight COVID-19 pandemic: Key technologies, applications, challenges and outlook of digital healthcare. *IJERPH*, **18**, 11, 6053, 2021.
13. Angel, N.A., Ravindran, D., Vincent, P.M.D.R., Srinivasan, K., Hu, Y.-C., Recent advances in evolving computing paradigms: Cloud, edge, and fog technologies. *Sensors*, **22**, 1, 196, 2021.
14. Abdulsalam, Y.S. and Hedabou, M., Security and privacy in cloud computing: Technical review. *Future Internet*, **14**, 1, 11, 2021.
15. Fernandes, D.A.B., Soares, L.F.B., Gomes, J.V., Freire, M.M., Inácio, P.R.M., Security issues in cloud environments: A survey. *Int. J. Inf. Secur.*, **13**, 2, 113–170, 2014.
16. Angelopoulos, A., Michailidis, E.T., Nomikos, N., Trakadas, P., Hatziefremidis, A., Voliotis, S., Zahariadis, T., Tackling faults in the industry 4.0 era—A

survey of machine-learning solutions and key aspects. *Sensors*, **20**, 1, 109, 2019.

17. Bhatia, T., Verma, A.K., Sharma, G., Towards a secure incremental proxy re-encryption for e-healthcare data sharing in mobile cloud computing. *Concurr. Comput. Pract. Exp.*, **32**, 5, 2020.

18. Habib, S.M., Ries, S., Mühlhäuser, M., Varikkattu, P., Towards a trust management system for cloud computing marketplaces: Using CAIQ as a trust information source: Towards a trust management system for cloud computing. *Secur. Commun. Netw.*, **7**, 11, 2185–2200, 2014.

19. Chang, V. and Ramachandran, M., Towards achieving data security with the cloud computing adoption framework. *IEEE Trans. Serv. Comput.*, **9**, 1, 138–151, 2016.

20. Liu, J.K., Liang, K., Susilo, W., Liu, J., Xiang, Y., Two-factor data security protection mechanism for cloud storage system. *IEEE Transactions on Computers*, **65**, 6, 1992–2004, 2016.

21. Visconti, P., Capoccia, S., Venere, E., Velázquez, R., de Fazio, R., 10 Clock-periods pipelined implementation of AES-128 encryption-decryption algorithm up to 28 Gbit/s real throughput by Xilinx Zynq UltraScale+ MPSoC ZCU102 platform. *Electronics*, **9**, 10, 1665, 2020.

22. Hasib, A.A. and Haque, A.A.Md.M., A comparative study of the performance and security issues of AES and RSA cryptography. *2008 Third International Conference on Convergence and Hybrid Information Technology*, pp. 505–510, 2008.

23. Patil, P., Narayankar, P., Narayan, D.G., Meena, S.M., A comprehensive evaluation of cryptographic algorithms: DES, 3DES, AES, RSA and blowfish. *Proc. Comput. Sci.*, **78**, 617–624, 2016.

24. Patil, P., Narayankar, P., Narayan, D.G., Meena, S.M., A comprehensive evaluation of cryptographic algorithms: DES, 3DES, AES, RSA and blowfish. *Proc. Comput. Sci.*, **78**, 617–624, 2016.

25. Oswald, E., Mangard, S., Pramstaller, N., Rijmen, V., A side-channel analysis resistant description of the AES S-box, in: *Fast Software Encryption*, vol. 3557, pp. 413–423, Springer, Berlin, Heidelberg, 2005.

26. Zheng, Q., Liu, N., Wang, F., An adaptive embedding strength watermarking algorithm based on shearlets' capture directional features. *Mathematics*, **8**, 8, 1377, 2020.

27. Akkar, M.-L. and Giraud, C., An implementation of DES and AES, secure against some attacks, in: *Cryptographic Hardware and Embedded Systems — CHES 2001*, vol. 2162, pp. 309–318, Springer, Berlin Heidelberg, 2001.

28. Chowdhary, C.L., Patel, P.V., Kathrotia, K.J., Attique, M., Perumal, K., Ijaz, M.F., Analytical study of hybrid techniques for image encryption and decryption. *Sensors*, **20**, 18, 5162, 2020.

29. Joshi, N., Wu, K., Karri, R., Concurrent error detection schemes for involution ciphers, in: *Cryptographic Hardware and Embedded Systems - CHES 2004*, vol. 3156, pp. 400–412, Springer, Berlin Heidelberg, 2004.

30. Kumar, P. and Rana, S.B., Development of modified AES algorithm for data security. *Optik*, **127**, 4, 2341–2345, 2016.

31. Sai Srinivas, N. S. and Md Akramuddin. FPGA based hardware implementation of AES Rijndael algorithm for Encryption and Decryption. *2016 International Conference on Electrical, Electronics, and Optimization Techniques (ICEEOT)*, 1769–1776, 2016.

32. El Batouty, A.S., Farag, H.H., Mokhtar, A.A., El-Badawy, E.-S.A., Aly, M.H., Improvement of radio frequency identification security using new hybrid advanced encryption standard substitution box by chaotic maps. *Electronics*, **9**, 7, 1168, 2020.

33. Abikoye, O.C., Haruna, A.D., Abubakar, A., Akande, N.O., Asani, E.O., Modified advanced encryption standard algorithm for information security. *Symmetry*, **11**, 12, 1484, 2019.

34. Kumar, A., Kumar, R., Sharma, A., A swarm intelligence based quality of service aware resource allocation for clouds. *Int. J. Ad Hoc Ubiquitous Comput.*, **34**, 3, 129, 2020.

35. Kumar, A., Kumar, R., Sharma, A., Energy aware resource allocation for clouds using two level ant colony optimization. *Comput. Inf.*, 37, 76, 2018.

36. Kumar, A., Kumar, R., Sharma, A., EQUAL: Energy and QoS aware resource allocation approach for clouds. *Comput. Inf.*, 37, 4, 781, 2016.

37. Kumar, A., Sharma, A., Kumar, R., SERVmegh: Framework for green cloud. *Concurr. Comput.: Pract. Exp.*, 29, e3903, 2017.

38. Ambrose, J.A., Parameswaran, S., Ignjatovic, A., MUTE-AES: A multiprocessor architecture to prevent power analysis based side channel attack of the AES algorithm. *2008 IEEE/ACM International Conference on Computer-Aided Design*, pp. 678–684, 2008.

39. Zhang, X. and Parhi, K.K., On the optimum constructions of composite field for the AES algorithm. *IEEE Trans. Circuits Syst. II*, **53**, 10, 1153–1157, 2006.

40. Elminaam, D.S.A., Kader, H.M.A., Hadhoud, M.M., Performance evaluation of symmetric encryption algorithms. *IJCSNS International Journal of Computer Science and Network Security*, 8, 561–581, 2008.

41. Razaque, A., Frej, M.B.H., Alotaibi, B., Alotaibi, M., Privacy preservation models for third-party auditor over cloud computing: A survey. *Electronics*, **10**, 21, 2721, 2021.

42. Reddy, M.I.S. and Kumar, A.P.S., Secured data transmission using wavelet based steganography and cryptography by using AES algorithm. *Proc. Comput. Sci.*, **85**, 62–69, 2016.

43. Messerges, T.S., Securing the AES finalists against power analysis attacks, in: *Fast Software Encryption*, vol. 1978, pp. 150–164, Springer, Berlin Heidelberg, 2001.

44. Feldhofer, M., Dominikus, S., Wolkerstorfer, J., Strong authentication for RFID systems using the AES algorithm, in: *Cryptographic Hardware and Embedded Systems - CHES 2004*, vol. 3156, pp. 357–370, Springer, Berlin Heidelberg, 2004.

45. Parrilla, L., Castillo, E., López-Ramos, J., Álvarez-Bermejo, J., García, A., Morales, D., Unified compact ECC-AES co-processor with group-key support for IoT devices in wireless sensor networks. *Sensors*, **18**, 1, 251, 2018.
46. Rewagad, P. and Pawar, Y., Use of digital signature with Diffie Hellman key exchange and AES encryption algorithm to enhance data security in cloud computing. *2013 International Conference on Communication Systems and Network Technologies*, pp. 437–439, 2013.

11

Internet of Drone Things:
A New Age Invention

Prachi Dahiya

Computer Science Department Delhi Technological University (DTU),
New Delhi, India

Abstract

The Internet of Drones (IoD) has attracted the major commercial sector by its technological advancements and utility functions and capabilities. It is a layered network that is designed mainly for unnamed aerial vehicles, which are used in a controlled airspace. It provides various kinds of navigation services and several drone applications. There are a lot of applications related to drones technology, like food delivery and packaging systems, search and rescue operations, traffic surveillance systems, home and office security systems, etc. The IoD architecture concepts and features can be implemented in the practical world and can be beneficial in a lot of ways. With a lot of applications come the security issues in Internet of Drones Things (IoDT) systems also. The following chapter introduces the IoD technology at an advanced level along with its architecture, features, application areas, and the security issues regarding the system. The chapter will help in getting detailed information regarding the IoD systems.

Keywords: Internet of Drones, aerial vehicles, surveillance systems, machine learning, artificial intelligence

11.1 Introduction

Drones are one of the most emerging sectors in the field of Internet of Things (IoT) smart devices and appliances. They fly high in the sky with all the network connecting capabilities as they are backed by IoT [1], and they

Email: prachidahiya_2k20phdco11@dtu.ac.in

Ashok Kumar, Megha Bhushan, José A. Galindo, Lalit Garg and Yu-Chen Hu (eds.) Machine Intelligence, Big Data Analytics, and IoT in Image Processing: Practical Applications, (269–304) © 2023 Scrivener Publishing LLC

use all types of latest technologies, like cloud computing, wireless communication, wireless sensor networks, machine learning, smart computer vision, high-end security methods, etc. Smart drones are coming-of-age devices with sensors and actuators as embedded mechanisms and have high cognitive capabilities. They recognize the objects far away and track them to where they are going as when the drones are flying above the concerned object, it becomes easy for the drones to track and recognize the designated objects [2]. Drones also solve the problem of the users who try to track and observe these devices from a remote position as their work is now done by the drones. A power-constrained environment is needed for the drones to function properly and to provide the best analysis and results for its users.

In today's world, drones are used everywhere, and their usage is increasing in diverse fields as they are easy to use and provide several functionalities to the users [3]. There are several application areas where drones are being used, such as in agriculture, where the drones can help the farmers in knowing when a wild animal comes to their farm and they can help in spreading manure and chemicals over large areas of fields, which is very difficult to do manually. Drones are used as Internet of Medical Things (IoMT) [4] in identifying where the patient is and it can carry and serve the medical care products to the patients in the time of need when the doctors cannot reach the patient. Drones are used as industrial equipment as they help in performing day-to-day activities and carry out several tasks in the industries, such as inspection, surveillance, security, etc. They are used to perform several government activities and private organizations also. From smart cities to the monitoring of the rural areas, drones are being used everywhere. All the applications mentioned above use drones in one way or another and hence with the implementation of drones all of these existing smart environments become more intelligent and can perform the activities of security, surveillance, monitoring the smart devices more easily. IoDT is just a mixture of smart sensors and actuators, big data, machine learning (ML) and artificial intelligence (AI), and several other technologies and they just make the functioning of already existing smart environments more easy and simple [5].

There are a large number of advantages of working with Internet of Drone Things (IoDT), but there are some security challenges also. The drones are present in large numbers so they work in a coordinated manner and collect the data in an efficient manner. The main issue is the handling of the unstructured data that comes with the drones. This big data needs to be processed in real time and for that a large number of mining techniques and machine learning methods are required to solve this problem. Hence, several servers are also required to coordinate these drones and the technologies are also required. Due to all these things, the implementation and maintenance of drones become an expensive thing.

11.2   Unmanned Aerial Vehicles

Unmanned aerial vehicle is the other name for drones, which is like an aircraft but on autopilot and without any human pilot or any passenger on board. The unmanned aircraft system (UAS) contains several components out of which one is unmanned aerial vehicles (UAVs). UAS contains ground-based controllers, which mainly maintain the flight and speed controls of the drones. It also contains the system of communication with the drones when in the air. These flights of drones or UAVs are basically controlled through a remote that is further controlled by a human operator and then it works as a Remotely-Piloted Aircraft (RPA). RPA contains several levels of autonomies where there is autopilot assistance provided to the drones or UAVs and then there is a level of full autonomous aircraft where there is no provision for the human intervention.

UAVs [6] were developed in the twentieth century mainly for military purposes. They were mainly used for the spy operations that were too dangerous for the humans to perform and by the twenty first century, they became the most essential assets for carrying out the military operations. With the advancements in the technology of drones, the application areas of the drones also increased toward the non-military platforms. Some of the uses of drones include aerial photography, in agriculture, the product deliveries, monitoring and surveillance systems, infrastructure inspections with aerial views, drone racing competitions, etc.

UAVs and UASs have become very popular in a very short time as they were developed in the twentieth century due to their advancements and implementations in the diverse fields along with their flexible capabilities in the real time operational activities and very low acquisition costs. Hence, drones when combined with different technologies easily become IoDT and can be accessed by anybody as this technology has made it easy for the people to understand it. Now, the work is done toward optimizing these UAV systems [7] with working here and there to reduce the production costs, weight reduction techniques are also used for drones, battery life improvisation, smart sensor and actuator techniques, increasing the camera resolution attached to the drone and its integration with the ground controller, etc. The main advancement is the fusion of this technology with latest developments [8] in the field of computer science like ML [9], Deep

Learning (DL), soft computing, AI, and these all technical fusions try to bring a list of operations that cannot be performed by drones on their own.

The term UAS was adopted from the Federal Aviation Administration (FAA), which has various terms like "Remote Plane," "Flying Robot," "Pilotless Aircraft," and then the main word "Drone" was coined by the military. It is defined as an aircraft, which is operated without any human intervention as it is operated autonomously. A UAV is defined as the smart aircraft system [9], which is remotely controlled by a human or an on board computer system, which is mainly used for monitoring and surveillance purposes and for capturing photos and videos or for other real time operations that can be performed by the drones at the time of flight. There are various components of UAS which are controller, sensors, and actuators, communication system, camera, and several mechanical components.

The end users of drones are increasing across the globe mainly in the fields of entertainment and media, security systems, agriculture, oil and gas, E-commerce industries, in hospitals and at several other places. According to a research, the UAV market is increasing at a fast pace as in 2016, the market had reached around $8 billion and in 2021, the sales of the drones have crossed the $12 billion mark and theses sales are up by 7.6% from $8 billion in 2016 [10] to $12 billion in 2021. Hence, one can say that the market growth of drones and UAVs is increasing at a fast pace and is expected to increase much more in the market in the coming years. The main categories in which the production of drones is increasing are consumer drones, which have crossed the 30 million mark in 2021, then there are the enterprise drones, which have crossed the 8 million mark in 2021, and then in the last are government drones, which are increasing at around 50% rate with every coming year.

With the advancements in different technologies and different fields like wireless sensor networks (WSNs), ubiquitous computing, machine-to-machine communication (M2M), etc., IoT provides the ability to sense the data and understand the things [11] and then action is taken by the actuators according to the environmental conditions and necessary requirements. IoT helps the drones to act according to the situations and conditions, also, they help in deploying the necessary smart sensor systems and provide better communication capabilities. IoT consists of "things" that communicate with each other with one method or another. The main objective of IoT is to identify and address [12] the devices such that they can exchange the necessary and the confidential information with each other with the least possible human intervention. Same approach applies on the drones where they make communication with the on ground systems by detecting a secure pathway for communication.

The drones are controlled remotely by the ground systems with the help of the microcontrollers that are embedded inside the drones, and their flights are software controlled through these microcontrollers. This whole working happens in a coordinated manner with the help of sensors, actuators, Global Positioning System (GPS) technology, advanced communication systems, etc [13]. Originally, they were used via hot air balloons in World War 1 and 2. And today, they are remarkably improved along with the advanced technologies and hence, they are used in varied forms, combinations, configurations along with complex implementations [3] according to the requirement of the conditions and the type of work assigned to the drones. Today, these drones are given different names, such as Nano Air Vehicles (NAV), Micro or Miniature Air Vehicles (MAV), Low Altitude (LASE), Vertical Take-Off and Landing (VTOL), Medium Altitude, Long Endurance (MALE), Low Altitude, Long Endurance (LALE), etc. So, one can see that according to the functions performed by the drones, they are given different names.

The usage of these UAVs or drones is skyrocketing in the previous few years as the researchers are finding this new technology very interesting and hence, they work in improving and optimizing this technology as much as possible. According to a study, the sales of drones in the year 2017 was around 3 million and it will cross the 11 million mark in 2020 and it is increasing day by day even in 2021 [12]. Many multienterprise companies are investing in this product with varied functions provided to the customers. The next-generation future, which is termed as "IoDT" is being shaped with the fusion of different technologies [14] with the drones like fog computing, cloud computing, etc. Smart or intelligent sensors that are being used by drones with the help of IoT are also giving very good results. The term "IoDT" will change the future of drones in the coming years as this existing technology of drones will now have broadened missions, with a wide scope toward the application areas, improved collaborations with latest problem solving methods, operational research methods, faster connectivity, advanced data analysis techniques, etc.

IoDT is the facelift of the drone technology, which will help drones to reach greater heights and will help the drones to work in the challenging conditions like some natural calamity situation saving lives, providing with food and water supplies, rural area monitoring and operations related to the farming and agriculture, underwater monitoring—sea and ocean cleanliness initiatives and protecting the underwater wildlife, monitoring forest fires with the help of WSNs, monitoring the underground oil and coal mines, etc. [15]. Hence, the abovementioned places are hard to be monitored manually and the monitoring is almost impossible. IoT is not confined to smart cities, smart homes connectivity, it has broadened its application areas with the help of drone technology and now it can easily transform and enhance the drone technology not only in the air but also in underground environments like underwater monitoring, etc.

11.2.1   UAV Features and Working

There are a lot of features present in the proper functioning of the drones. First, the voice commands are sent over a bluetooth device or through WiFi for the purpose of voice recognition. Then there are some applications present on the device that control some components of the drone which can control it. These applications also contain the neuro speech services, which can detect any kind of language or speech. Hence, between the speech recognition services and the applications of the drone, the speech transaction and recognition is a constant process. The flight of the drone is controlled through a flight controller. In Figure 11.1, the data is sent and then received by the remote controller to the flight control system of the drone [9]. The flight system controls all the flight operations of the drone in a secure way. So, in this way, the features of the drone are used and then one can easily perform all the actions using a drone. All the applications related to drones find it convenient to work with the drones and fulfill their goals.

For a drone to be remote controlled and operational, a certain number of things need to be taken care of. The user must see to it that the drone is powered on and the remote controller is present with the user, the

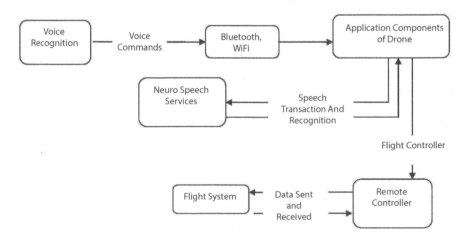

Figure 11.1 Working of UAV.

application device is connected with the remote controller and the drone, there can be a Bluetooth headset that is also connected with everything. The user can easily talk over the bluetooth headset and the application [11] present in the device is in a constant listening mode for whatever the user talks over the headset. Now, this voice is sent over the speech recognition mechanisms, the speech is recorded in the natural language of the user's choice. Once the speech is recognized irrelevant of the background voices and noise present there as they don't contaminate the original voice of the user. The speech that is taken as an input is matched in the speech recognition systems with the already existing speech present in these systems. Then the input speech is matched with the already existing speeches and then the output is shown on the screen. The speech is converted into a series of regular expressions [16] and these expressions are designed in such a way that the user does not need to memorize the instructions present for the pattern matching. The regular expression helps to model the language by using and creating a platform independent mechanism and then the implementation is done in a particular programming language.

11.2.2   IoDT Architecture

This section of the chapter will discuss in detail about the architecture of the IoDT system. The purpose of the architecture is to provide a diverse range of generic services to a large number of applications, such as surveillance systems, navigation systems, location detection, and communication services in a coordinated and an efficient manner. A navigation control is required for the proper functioning and operation of a drone. Several applications are enabled in this architecture, which make the mobile drones to perform the local tasks for the customers [17]. If the talk is about the pool drones, they are much more efficient than the individual drones that work alone. Hence, the pool drones are mainly responsible for performing [6] the local tasks, and the local drones are notified about the completion of the tasks. Providing the location of the drones for communication purposes is an important aspect of the applications of the architecture.

The IoDT architecture provides an abstract design along with the feature requirements of the customers of what they want for the function and features that need to be implemented on the IoDT system [8]. The IoDT system provides the concrete system protocols [18], which include different kinds of algorithms and interfaces that implement the features for the IoDT architecture. There is a possibility that there can be several IoDT systems implemented in different applications that are based on the same core IoDT architecture with several advantages and disadvantages. Not all architectures are viable in the engineering designs and applications, and hence, slight changes are needed to be done in the original IoDT architecture [9]. To prove that the IoDT architecture is viable, then the system needs to implement this architecture for its correct functioning.

Describing the architecture will require a certain set of concepts and definitions, which have special meanings in the architecture. Airspace is the term used by the drones to mention the limited area that is used by the drones for their flight purposes of surveillance and other investigation methods. The airspace is used as a roadway network [19] for the drones in the cities. The drones are allowed in certain paths, such as airways, which are similar to roads and intersections that are formed by the two airways and the last nodes, which are pinpoints of interest that are reachable with alternate sequences of both the airways and the intersections. These airspace structures have static features, which help in avoiding collisions over the airways. The movement of the drones and the direction in which they are going over an airway or intersection is regulated and when they are inside the node then they are in the free flight mode.

The airspace is divided into several zones and hence, each zone contains its own set of airways, nodes, and intersections. The adjacent zones can easily be accessed by the inbound and the outbound gates, which act as the intersections at the border of the zones and hence, they belong to both the zones. Hence, drones can take these intersection roads and then can go from one zone to another. Airways are not allowed the kind of access that is given to the intersections [20]. For the airways to access both the zones, they need to be segmented at the border into two parts, one of which will be in one zone and the second one will be in another zone. The two airways get separated at the gate of the two zones.

One can also form a graph by using these nodes and intersections as the vertices, and it also includes gates and the airways which act as the directed edges, and this kind of graph is called a zone graph. Pathway is a path that is present in the zone graph. Element word refers to the intersections, airways and nodes present in the zone graph. Each and every element has a global address in order to be reachable. This is similar to the global addresses of the host present online on the internet. When the gates are taken as the vertices and are connected over the cozones, then they are known as the transits, and then, the resulting graph is basically known as the interzone graph. Transit cost is the cost of traveling of the drones in between any pair of two gates. This transit is calculated in the terms of distance, time, speed, etc. The path of this zone graph is known as a route.

The portions of the zone graph are divided into the public and the private section. The elements that are present in the public and the private section of the graph are in turn called as the public and the private members of the graph, respectively. The private elements have a certain number of access rules for the drones which are specified in the meta data, such as whether the drones are allowed to access the metadata or not. The last level of abstraction contains the points or the nodes that are present in the airspace. A coordinate system, which includes latitudes, longitudes, altitudes help in identifying the points present in the airspace. The geometry of the airspace is learned through these points. Trajectory is formed through the path followed by these points.

The architecture of the IoDT in Figure 11.2, the system contains the UAV application layer [5] which contains a lot of interfaces that are used by the application layer in order to communicate with other layers in the architecture. It mainly contains the UAV interface, the cloud interface and the web services. The UAV interface is used to connect the UAV device with the ground base station control. A communication channel is provided by the UAV interface such that the messages can be passed in a secured way by the drone to the ground systems. The cloud interface connects the IoDT system with the cloud services in order to store the confidential data that is connected by the drones. The cloud is a virtual platform and provides a

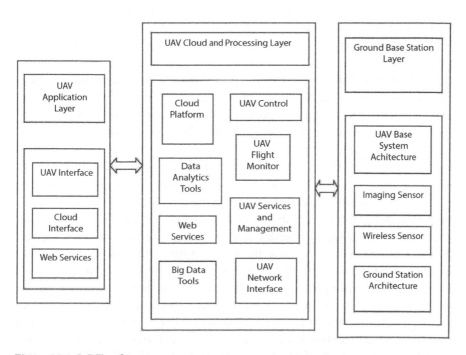

Figure 11.2 IoDT architecture.

set of services for storing the data in an efficient manner such that no data gets lost. Web services are provided in such a way that the proper functioning of the drones takes place over the web. Whenever any device transfers information over the web then the web services are used.

The next layer is the cloud and processing layer, which contains a lot of analytics tools, methods and mechanisms which help the drone system. Other than this, the controls and the management systems related to the UAVs are also present in this layer. The cloud platform provides a safe space for the drones to store their important information. Several data analytics tools are present in the system which help in managing the data. The data is collected, segregated, the duplicate data is eliminated, refinement of data is done and then the data is stored and processed further. Data analytics tools also analyze the data and derive some useful insights from the collected data. Similarly, big data tools are also used over here, which are used to handle the large amounts of data and helps with the problems of information overload and the overflow of the information. Hence, the big data tools help with the management of the big data tools.

The UAV cloud and processing layer also contains several aspects of the UAV system as it includes UAV control mechanisms, UAV flight monitoring systems, UAV services and management methods and the UAV network interface. UAV control mechanism contains a set of methods which look into it so that the system is able to work in a proper format. In the UAV flight monitoring system, the flights that the drone takes in its lifespan. The monitoring system monitors the proper functioning of the drone while it takes a flight for its own purposes. UAV services and management methods see to it if the drones want any kind of help while performing the tasks to them and hence, the proper management becomes necessary for that. The UAV network interface contains different network channels for the communication purposes of the drones. It provides a secure gateway for the passing of the confidential information and to prevent his information from the hackers by using a certain number of security protocols.

The information navigation in the drone architecture is done with the help of zonal graph format in any two elements that are designated inside the zone. The license to transfer information in a specified zone is given by the authorities who manage the whole system. The governing laws are needed to be followed by the drones regarding the intersections, airways and the nodes either public or private. The UAV system softwares is implemented easily with the help of zonal service providers for the proper functioning of the drones. There are several service provider companies which provide services to the zones as Zone Service Providers (ZSPs) similar to the IoDSP Internet of Drone Service Providers (IoDSP) [3]. The gates coordinate with each other with a handoff, whenever a drone crosses the border into another zone and then the responsibility gets transferred to the new ZSP. The ZSP is able to command the drone into performing various functions, such as hovering, landing position, takeoff, making some patterns, these actions are called the grounding and the holding positions, respectively.

Layers of the IoDT [9] architecture provide several benefits to the drones, such as maintainability, scalability of the devices, flexibility of the device layers with very minimal changes that are needed by the other layers. The fundamental goal of the IoDT architecture is to enable the drones to perform with several applications with minimal advancements and changes with the provision of some common and generic goals and services for the applications. Basically there are two goals that the maker of the architecture keeps in mind. The first is to provide a way or guidance to the drones to travel from a source node to the target node. This is done when all the drones are coordinated in the airspace. The second task of the architecture is to provide an extensible platform for the future services that are easily needed by the applications for the purpose of the delivery of messages that are to be sent to the pool of the worker drones for the purpose of the specification of the applications.

The navigation of the drones is a very important aspect. The drone traverses a path in the inter zone graph from the source of the zone to the destination present in the zone. The drone also has to traverse each and every zone and also, the intersections and the airways present in the zone graph. The trajectory of the drones is already decided such that the drone travels inside the boundaries of the airspace, airways, the intersections and the nodes of the zone also. Separate layers deal with separate tasks and difficulties faced by the drones. Dividing the work among different layers is an efficient way of tackling the navigation of the drones with a single giant system which has a map as well as the map of airspace where the drones travel and this all becomes very complex as well as unsustainable at times. Now, these bugger problems can be converted into smaller problems and then all these problems become tractable and solvable. Hence, the optimal solution is the best solution which is tractable.

There are some features [7] that are help and control the drones while they are in airspace:

(a) Broadcast and Track: While in airspace, the drones broadcast periodically their three dimensional coordinates as well as their trajectories and where they will go during the flight. Hence, path planning for the future flights of the drones becomes easy and calculating the progress of the drones becomes easy.

(b) Planning Trajectory: The ZSPs need to plan the trajectory that is to be followed by the drone. There is a need to see to it that the drones remain inside the boundaries of the intersections, airways in a planned pathway.

(c) Precise Control: The airspace should have a precise control feature and the ZSP needs to be notified about the specific maneuvers of the drones, such as hovering, takeoff, landing, etc. Hence, this is a reasonable feature for the architecture.

(d) Collision Avoidance: There can be collisions when the drones are flying with various dynamic objects, such as birds, other drones, which may be obstructing the airways or intersections. Drones must try to avoid the collisions by seeing the trajectory points beforehand and taking rapid decisions during the flight of the drones. Standard protocols need to be used for the coordinated maneuvers for the collision avoidance.

(e) Weather Conditions: The weather conditions like rainfall, snowfall, fast winds should be known earlier and before the flight of the drones. This needs to be done such that there is a safe flight for the drones during the trajectory.

11.3   Application Areas

Rapid implementations and innovations are helping IoDT in diversified fields and industries and hence, they are gaining strong attention from innovators and researchers. There are endless possibilities that have emerged in the past few years for IoDT and this means a large number of real time application areas in Figure 11.3 [7]. Smart wireless connections are coming up which are enhanced by the sensors and actuators embedded inside the drones. Through smart integration, several data analytics tools are used in IoDT along with fog computing and cloud computing, big data analytics, etc. These aerial systems are finding their way with the help of strong collaboration from the IoT, and this also helps in the advanced technologies and opens doors for a bulk of real time opportunities. With the help of IoT, these drones can be easily deployed at high altitude regions as well as underwater seas and oceans which means now drones can be deployed at the extreme locations also.

Figure 11.3 IoDT applications.

A. Aerial Photography
Drones have made aerial photography easier and efficient as they can easily capture the footage, which is rather very expensive and difficult for helicopters and cranes [6]. Various action film sequences are shot with the help of drones and hence, it makes the cinematography skills easier. Many autonomous flying devices are used in real estate as well as for sports photography. The journalists also use these drones for collecting important footage as well as for the live broadcasts, these are also used for the military purposes to carry out the string operations, etc.

B. Shipping Delivery
Many major e-commerce companies like Amazon.com, UPS, etc. favor drone delivery to their customers [7]. This saves a lot of manpower and also avoids major road traffic and this unnecessary traffic is shifted to the air. Several small delivery packages over a small distance can be transferred via drones, these packages include food packages, letters, beverages, medicines, etc. Hence, small luxury delivery packages are transferred easily. Amazon.com has already deployed around 300 IoDT drones for package deliveries in the USA.

C. Geographic Mapping
Several unattainable geographic locations can easily be captured by drones which are very difficult for people to reach like mountainous regions, islands, coastlines, etc. [8]. 3D maps are images formed by the footage

captured by the drones and they actively contribute to crowd sourced mapping applications.

D. Disaster Management

Whenever a man-made or a natural disaster occurs, then drones provide quick relief measures to the affected sites and to the injured people. The drones gather the information of the affected sites and then navigate the debris and the rubble in order to look for the injured people [9]. Very high resolution cameras, sensors and actuators, radars provide the rescue teams with all the important information like providing access to a higher field view in turn saving the time and money of the manned helicopters and other vehicles. In such cases, the small size drones are much more efficient than the larger aerial vehicles as the smaller drones provide a close up view for the affected sites.

E. Precision Agriculture

Agriculture is the largest field which provides a bunch of application areas for IoDT. The farmers always try to find some cheap solutions and effective methods for monitoring their crops and farms [10]. The crop health can be monitored by the infrared sensors present in the drones which enable the farmers to improve their crop health by taking the necessary measures like using fertilizers, insecticides, etc. Threat identification of crops from wildlife, fires, etc. is done by IoDT as the farmers get the early alerts about all kinds of threats. With the help of IoDT, farmers can get precise information regarding the soil quality, soil moisture levels, crop maturity, etc. which may help farmers in making better water and fertilizer adjustments and hence, making the crop yield better.

F. Search and Rescue

Thermal sensors that are present in the drones help in giving night vision and provide a powerful tool for surveillance. It becomes easy for the drones in discovering the location of the victims and the lost persons in the harsh conditions and in the challenging terrains [11]. Apart from these, drones can help in dropping necessary food supplies to the disaster hit areas and to the unreachable sites. So, one can say that the drones perform multiple tasks as a GPS locator, medicine and food supplier, providing water and clothes supplies. Drone based cameras, retrieval of information and location of drowning survivors, searching for victims in case of water emergency situations. The delivery of healthcare products with the help of IoDT is easy and feasible. Hence, the healthcare industry can get some huge advantages from drones.

G. Weather Forecast

The drones are designed to reach the most difficult and dangerous places where humans are not able to. Some drones are developed to monitor very dangerous and unpredictable weather conditions. The cost of developing the drones is very low, hence, one can send them in very harsh conditions like hurricanes, tornados, volcanic eruptions, etc. [12]. New insights are gained by the scientists and weather forecast departments and then they predict the behavioral developments and trajectories in the weather at particular conditions. When drones are equipped with smart sensors, they are able to collect important data, weather parameters, detailed data points, etc., which may help in determining the early mishaps.

H. Wildlife Forecasting

The drones have helped the wildlife protectors by keeping the poachers away. Several drones systems are installed for protecting the animals like rhinos, big cats, elephants which are mainly the favorite targets of the poachers [13]. With advanced cameras and smart sensors, the drones are capable of taking flights in the night and the night vision cameras help in taking good pictures of the incidents that occur in the forests. Monitoring and researching purposes on the wildlife without causing any kind of disturbances and they also provide regular insights, patterns, animal behavior statistics, etc. to the researchers.

I. Law Enforcement

Law enforcement activities are also carried out by drones. Large crowded areas are guarded by drones in order to ensure public safety. These drones also assist in monitoring illegal and criminal activities [14]. Several other illegal activities like smuggling of migrants, drugs, fire investigations, illegal weapons transportations via sea borders, which are monitored by the border patrol police with the help of drones.

J. Entertainment

Drone fighting is performed to entertain the players. The cage matches are played where the two contenders and their drones put up a fight against each other. As the technology gets more advanced, the drones become more robust and can fight for multiple hours along with heavier loads [15]. Artificial drone intelligence can be used to capture the videos and the photographs of the fight scenes as well as the selfies. Hence, the drone entertainment industry comes with a lot of immense opportunities and several businesses. So, there is a need to discover more opportunities for drone technology and the businesses dealing with drones need to build the required infrastructure and test the drones and their services over there.

11.3.1   Other Application Areas

There are some other application areas which need to be focused related to drone technology.

A. Aquatic Life Monitoring

The drones work well with the aerial viewpoint but they work the same way when it comes to the undersea aquatic life monitoring systems. IoDT based drones are capable of clicking high resolution images underwater [16]. With the pilot feature provided by the IoDT technology, the drones help in monitoring aquatic life with ease and high quality and these drones can go down very deep in the sea. The researchers are able to explore the unexplored areas underwater for marine research. Several other underwater observatory tasks like mooring the lines inspection, predator net damaged submerged infrastructure inspection, etc.

B. GIS Mapping

The traditional drone systems or UAVs took time to operate, set up and the pilot feature helps in autonomous flying, collision detection and the Geographical Information System (GIS) mapping [17]. GIS mapping is a conceptualized framework which is backed up by drones which provide the ability to capture and then analyze the geographical locations. Other than this, IoDT network boosts crop productivity and boosts the yield and enhances the profit. The robotic drones are used to enhance the revolution in the shipping industry, undersea researchers and fisheries.

C. Undersea Infrastructure Maintenance

Several government corporations, private companies and several service providers get the help from the drones that use enhanced high resolution cameras to get the inspection checks regarding the underwater drainage pipe system or the sewerage pipes, etc. [18]. The drones keep the pipe in check, if any of the pipes is severed or damaged and then the proper functioning of the is resorted.

D. Border Drone Surveillance

Cross border threats, intrusions, illegal trespassing have increased in the past few years and the countries are trying to combat this situation at the border lines for a long time with the help of several smart drones [18]. With the help of IoDT, the military forces can easily fly drones over the border for a certain period of time and the forces can pinpoint the location where the trespassing has occurred. Hence, with the help of IoDT, the military can monitor the intrusion area and can send a task force to prevent any harm done by the intruders. Hence, drones can help the border security forces in a lot of ways.

E. Early Threat Detection
With the help of IoDT, now drones can easily monitor the sensitive areas and can give an early warning to the military forces about an attack regarding the enemy vehicles like fighter jets, trucks and arsenal like missiles, machine guns, etc. [19].

F. Safety of Military Soldiers
The drones can help in getting the clear images of injured army soldiers, war personnels, and can pinpoint their location such that the soldiers are ensured to be safe during the complex missions and operation [19]. Monitoring the sensitive areas is done by the drones and the military gets the intelligence data such that they can perform the strategic missions and operations.

11.4   IoDT Attacks

The components that comprise the security protocols of the IoDT system are mainly IoDT devices, sensors and actuators, IoDT network and the communication links, etc. The security breach in an IoDT system can lead

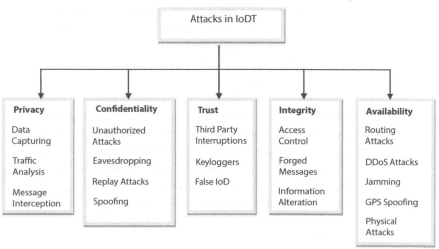

Figure 11.4 IoDT attacks.

to loss of confidential information, their resources, availability of devices, the most important thing- trust in the IoDT system by the customer. According to the threats and the vulnerabilities, the IoDT attacks [8] are divided into around 5 categories as shown in Figure 11.4.

A. Privacy

Privacy [2] is the main concern for the IoD systems which are basically data oriented. When the data is processed in the IoD system, the threats and the vulnerabilities are on the rise and the attackers try to target this confidential data in order to gain the sensitive information by following several approaches. The following are the attacks that play a vital role in breaching the confidentiality of the information carried by the IoDT system.

> (a) Data Computing: Through traffic analysis, a huge amount of data is collected from the IoD. Data forensics attack the data by extracting the confidential information from the collected data. One has to formulate some solutions for the counter attacks in order to prevent the information breach in the case of the mechanisms based in forensics research that are applied for attacking the IoD.
>
> (b) Traffic Analysis: The useful information about the traffic is obtained by the IoD systems which analyze this information present in the network and during the communication process. The information is shared in the form of packets in between the drones and the ground control systems. These packets contain very confidential and sensitive information. This information is basically device locations that are connected with IoD, sensor information and the captured data from the smart sensors.
>
> (c) Message Interception: Message interception is done when the attacker takes control over someone who regularly checks on the functioning of the network communication. Capturing the intruder in such a situation becomes very difficult sometimes as in sensitive missions, the drones contain very important and sensitive information. The tracking and monitoring of the intruders in the IoD system can be dangerous at times for the companies which specially carry out such missions.

B. Confidentiality

The confidentiality [4] keeps the private information in check that it is not leaked to any of the illegitimate users or to the unauthorized devices. The deficiency in the security leads to several attacks against the IoD system and the ground control systems. When the illegitimate or the unauthorized users get access to the IoD system, the confidentiality agreement is breached and there can be severe consequences to this. From that point, the retrieval of the information becomes problematic. Now, there are a lot of ways in which the confidentiality gets breached:

(a) Unauthorized Access: Here, the attacker gets the access to the IoD devices and gets an access into the IoD network such that it starts sending or spamming the authorized users with unnecessary emails. The services of the IoD network can be gained by hacking into someone else's account by duplicating the IDs or by using the public and private keys. These kinds of attacks lead to the unauthorized disclosure of the confidential information of the IoD services.

(b) Eavesdropping: Eavesdropping is the unauthorized real time interception of the messages that are exchanged among the IoD devices over the network, hence eavesdropping over the communication system. Eavesdropping is a very dangerous attack regarding the important information as it allows the attacker to take over the system very easily once he is inside the communication network and retrieve the important data. The availability of the unencrypted data and lack of authentication protocols in every connected IoD device can lead to such attacks.

(c) Replay Attacks: Here, a malicious user sniffs on the IoD network and then by overcoming the security mechanisms by continuously replaying the requests over the IoD network. These kinds of replay attacks can be carried out in a lot of different ways as the attackers have found a lot of ways to infiltrate the IoD network and communicate among the IoD devices. These attacks can be prevented with the authentication mechanisms, they should use the concept of fresh message requests in a very secure manner.

(d) Spoofing: Here, the attackers masquerade as a legitimate user in order to get the private information from the IoD network with the help of a spoofing ID of a particular legitimate user and hence, the attacker can gain access into the IoD network and communications links. The only solution for these kinds of attacks is to make the encrypted IDs and the one time usable IDs which can be the efficient solution against these attacks.

C. Trust

Trust [12] plays an important role with the risks associated with the latest technologies and it depends and influenced by the measurable as well as the non-measurable properties of the technology like the strength, goodness, availability, reliability and the ability. Privacy and security [11] are the two pillars of the trust. Several organizations and companies are deployed which build a trust relationship with the IoD system which is very important. Limited security mechanisms present at the corporations lead to the various threats to the trust between the two parties. Breach of trust also happens due to the wrong configuration of the messages also.

(a) Third-Party Interruptions: There are some applications which involve the trust of the third parties, and this is done through IoD communication formats like public and private keys, digital certificates, the management authorities, etc. Whenever the trusted parties release or leak the confidential information of the deployed devices of IoD and hence, this information leak leads to the loss of finances and the intellectual properties thus, leading to the breach of the trust between the parties.

(b) Keyloggers: Keyloggers come under the category of the internal threats. Keyloggers work in the software of the system and are embedded during the deployment and development of the software system of IoD. The confidential and the sensitive information is forwarded by the keyloggers to the attacker.

(c) Falsified IoD: Sometimes, the legitimate IoD devices are replaced with the fabricated ones and hence, they start sending the important information to the attackers and eavesdroppers which then get easy access to the IOD network and communications. Hence, in this way, the IoD devices can be falsified which may create trouble for the whole IoD network and can leak sensitive information.

D. Integrity

Integrity [14] contains three basic rules for the confidentiality of data which are consistent, accurate and trusted. The transmission of the information should not be altered in any possible way by any kind of attacker or intruder. Protection mechanisms used over here for the data integrity part are the hash functions, checksums, etc. There are several attacks that affect the integrity of the data.

(a) Access Control: There are some rules and policies that keep the devices in check that how the data is to be communicated with other devices in the IoD system and how the users can access this data in the system. Now, this access control of the confidential information acts as the mind for the IoD system. When the attacker gains the access controls of the information then these attackers can change the permissions, authorizations, privileges for the sharing of information.

(b) Forged Messages: Sometimes, in order to get access to confidential information, the messages are forged by the attackers. They forge the method during the authentication protocols and then change the authentication protocols according to their needs. After this, the attackers can easily modify and then retransmit the message to the users by forging them.

(c) Information Alteration: Information alteration means that the information transferred to the users in the systems gets fabricated and changes and the original meaning of the information gets changed. Modifications, alterations, fabrications, substitutions, data injections are some terms and methods that change or modify the information that is used in the communication in an IoD system. These changes just misguide the users with the modified information.

E. Availability

Availability [15] of the services that get initiated immediately whenever there is a need to maintain the proper functioning of the IoD system is required. The availability of the information is required which ensures that only the legitimate users can get access to the IoD network and be a part of the communication network of the IoD system. When the IoD network is operated in the mission oriented applications then the availability of major IoD resources, devices and information plays a major role in ensuring the security of the IoD system. The availability of the IoD network can be affected in the following ways:

(a) Routing Attacks: Several kinds of network routing attacks are there in the IoD network, such as node isolation, node capture, flooding, phishing, location disclosure attacks, etc. Node capture is where a particular node is overpowered by an attacker and the attacker takes the node as hostage and whatever information is passed through the node, it is infiltrated by the attacker. Flooding is an attack where the inbox of the host is flooded by the attacker with unnecessary emails.

(b) DDoS Attacks: Denying the functions performed in a network, accessibility to the resources, prevention of the unauthorized users accessing the important information, etc. is known as the Denial of Service attack. To prevent such an attack, IoD needs a proper communication system to send and receive the message signals. Sometimes, the attack is in the form of flooding which may lead to the resource unavailability as the network is infiltrated and then interrupted.

(c) Jamming: The main objective of jamming is the disruption of the communication network of the IoD system intentionally. These vulnerabilities can affect the IoD network in several ways. Jamming can affect the IoD network with the unavailability of the resources and cause the collision of the IoD network.

(d) GPS Spoofing: GPS is basically used to determine the position of devices or vehicles in the IoD network or the drones that fly in the sky toward the designated target. The attackers can easily modify the information of the GPS signals or even they can generate the spoofing signals through the GPS signals of generators. Due to the interruption through the spoofing signals, the GS signals get delayed and this can cause big loss to the IoD network and this can break the coordination between the devices and the IoD network and hence, in the end, there are collisions. There can be a list of countermeasures for the GPS signals, such as there should be a proper use of authenticated, as well as the encrypted signals, routine checkup of the strength of the GPS signals with some absolute values and the relative values. There can also be checking of time intervals while receiving the GPS signals.

(e) Physical Attacks: Physical attacks are basically on the hardware devices of the IoD network. The hardware components are attacked the most with the physical attacks. Hardware components are very expensive, difficult to manage and hence, the protection of the hardware components of the IoD system against all the physical attacks can be a serious issue.

11.4.1 Counter Measures

Drones would be the face of advanced security systems in the coming years. The military equipment will also be powered by the IoDT technology. Hence, almost in all the fields whether the commercial sector or the business sector or commercial sector. The drone industry has been giving positive and promising results in the technology based industries which will enhance the security and the efficiency of several businesses [5]. FAA have established the requirements for the remote identification of various drones which have now installed the safety standards where the drones are used in the line of sight and case to case basis. Widespread home delivery packages are sent via drones.

There is a profitable opportunity for the drone industry in the coming years and the FAA estimates that there will be around 1,25 million personal drones by 2023 [7] in the United States and there will be around 4 lakh commercial drones in the country. As the commercial businesses are adopting the UAV system for their deliveries, inspections and several other uses, these commercial drones will be around 8 lakh in total by 2023 which will be double than the current amount of commercial drones. The drone industry will be creating thousands of jobs in the coming years and hence, the cyber security systems will be using drones for the security purposes. The major applications of drones in the coming years will be the construction businesses, utility industries, oil and gas industries and the biggest of them all, the security industry will be benefiting the most from the drone industry for surveillance purposes and several other things. Monitoring the parameters of the buildings, construction sites will soon be done by drones.

With the immense number of uses and applications of drones, there are still some vulnerabilities of drones. Sometimes, with no timely notice to people, the drones can endanger the general public as well as drones can crash into each other and that will be a huge problem. Also the drones can easily be compromised by the attackers just like any other IoT device from a mile away. Security is a very big issue for drones that needs to be worked upon.

Whenever the attacker can locate the drones in the air then it can be hacked like any other normal device, the attacker can get access to the information that the drone must be sharing or sending to the base control station. The drones are connected to the base station with some connection networks like WiFi and if a drone is connected over a network then the drone is in a vulnerable situation and needs to be well secured. Single factor authentication is carried out by the IoDT networks with the use of weak passwords that can easily be guessed or cracked with the absence of the encrypted devices and resources. Hence, the attackers can intercept a particular café or a building.

Putting patches at the drone security systems whenever their security is breached can be a good way to for the drone security. Hence, these patches should always be installed in order to counter the emerging security threats. Virtual Private Network (VPN) [8] also uses the patch system when the hackers try to infiltrate the communication systems over the Internet. In the coming years, the data of the drones will be stored on the cloud and hence, the cloud should also be encrypted and that is the only way to protect the data from the attackers.

There are a lot of counter measures that can protect the security of the drones in the IoDT systems against the drone hacking. Some of the methods are:

(a) Regular firmware updates: The firmware of the drone needs to be updated regularly and the manufacturers issue the security patches whenever some new security threats emerge. That's why there is a need to regularly update the different firmware and software of the drones such that they are a step ahead from the hackers. The security patches are issued when there is some unwanted activity in the system like unauthorized access of information by the hackers, access to the videos and photographs, map views of certain locations in real time, etc. Even after these kinds of security breaches, some clients refuse to install the new patches and hence, giving the hackers a potential access to all the confidential data.

(b) Using strong password: One must use very strong passwords because with weak passwords, it becomes very easy for the hackers to guess the password. Proper use of special characters, symbols, combinations of numbers and digits is advised which may deter the attackers from guessing the passwords. Hence, strong passwords will avoid the hacking of the drone signal.

(c) Keep the connected devices secure: All the devices connected in the IoDT network can be hijacked by the hackers. There was a case of malware that attacked the US military drones as one of the operators used the connected drone's computer to download the video game. There must be an antivirus installed in the drones and the users should not download any dodgy programs or visit ant dodgy sites.

(d) Return to Home mode: Every IoD device must have the return to home mode such that the drone can return to home in case the drones loses the signals or if the signals are jammed or when the battery is down and needs recharging. The return to home feature will help the drones to get prevented from a hostage or hijacking situation. There must be a proper functioning of GPS in order to

(e) Use of VPN: There is a need for the drone network to subscribe to a VPN which will stop the hackers from accessing the private information and getting access to the network communications over the internet. A VPN makes a secured path or a gateway to the internet and then it encrypts the connection so that the hacker cannot get any idea what the information is as it becomes difficult to decrypt the information.

(f) Limit the connected devices: There should be a limit to how many devices are allowed to get connected with the IoD network. This will prevent the hijacking of the devices as the signals can easily control a limited number of devices.

So, in the previous section, there were some countermeasures for the proper functioning of the drones and to ensure the drone safety. The data of the drones has become more mobile due to the WiFi and the loud services, which makes it difficult to protect the data. IoDT has made it possible along with Radio Frequency Identification (RFID) to enable the access and flow of the data in the entire IoDT network and in smaller devices, such as goods tags, security cameras. pallet tables, etc. Physical access restrictions present in Bluetooth, RFID, WiFi, and several other technologies have made it possible to prevent hacking but due to the mobility of data, hackers easily get access to the data. There are some malicious drones that are programmed by the attackers and hackers in order to get access into an IoD network. Such drones are installed in buildings and offices which exploit the vulnerabilities of technologies like WiFi, Bluetooth, RFID, etc. These drones can easily steal the data from the IoD devices or they can even hijack the system peripherals like mouse and keyboard once they enter the IoD network. Keyloggers are used to steal the information in such a format.

The drones come under the FFA as the UAVs or as Unarmed Aircrafts (UAs) and hence, they cannot be shot down by anyone. They cannot be harmed in any physical manner and no one can interrupt the functioning of these drones. The signals cannot be intercepted whenever there is communication between the drones and the ground control system. So, the defense mechanisms must focus on the protection regarding the data and the space. There are some counter measures that can help in monitoring the malicious drones and to protect data from them.

Geofencing

Geofencing [16] is a way of dealing with the malicious drones menace. Whenever the IoD network is using the GPS or RFID tags then the geofencing creates a virtual border such that no unwanted devices can get access to the information being passed in the drone system. It generates a response whenever the unauthorized drone enters the protected zone and then the geofencing system prevents these commercial drones from entering or flying or taking off into the geofenced areas. Big drone making companies have installed geofencing at their vulnerable sites, such as the power stations, airport buildings, prisons, etc. Some attackers have found their way around the geofencing software. However, the hacking of the geofencing software is very easy and one can find a method for attacking this over the internet. One way to block the geofencing is by attaching a tinfoil around the drone and folding it such that the GPS of the drone is blocked.

Geofencing is a costly software and hence, not available for most of the consumers. If the unauthorized drones cannot be blocked then one must detect these malicious drones but no method is 100% effective. Hence, there is no reliability on these methods but one must try to find a way. Radar is one of the methods that are used for drone detection but again, they are not 100% reliable as sometimes, they can detect a bird as a drone. There are acoustic sensors that must be used to detect the unwanted drones as they are mainly programmed to recognize the minute sounds, vibrations, signatures of the drones and can detect whether they are unwanted or not.

Radio Frequency (RF) scanners are also designed to detect the drones which mainly detect the electromagnetic waves over the spectrum and they recognize the transmissions of the drones over the electromagnetic field. With the advanced drones which use the GPS signals do not use the electromagnetic waves to transfer the information or to navigate. Hence, drones using GPS cannot be detected by RF sensors. Another way is through thermal imaging that mainly detects the heat that is emitted by the objects or in this case drones. Now, the drones can be detected through their thermal footprint. This technique has a very high rate of the false positives for detecting drones. From all the methods discussed above, it is confirmed that detecting the malicious drones is very difficult and hence, most of the consumers are using their home security methods, cyber security and the wireless network security methods.

There can be airspace attacks on the drones present in the IoD network. So, there are a lot of ways by which this data present in the drones can be protected. There are various ways to do this:

(a) Using a VPN whenever working over a WiFi network can be an optima option and this ensures that the communications over the network cannot be hacked. There must be ways which can protect the home or office whenever one uses the public WiFi hotspots and hence, the VPN system is a good option to save the information of the drones.

(b) The security of the IoT devices which are present at homes or office and then their confinement should be up to the guest network. This will prevent the hacker from entering the secured network through any device or drone.

(c) Regular changing of the username and the passwords must be a priority. Sometimes users don't change the default usernames and passwords and it becomes very easy for the attackers to guess the passwords for the network. Regular changing of the passwords leads to the protection of the router of the network. The attacker will not be able to guess what type of router one is using or what kind of connection is used by the customer.

(d) Never use the identical passwords for different devices or even for different networks as it makes the task of the hackers easy as they can easily access the information of the different devices. These hackers can get access to the whole digital life and can get access to a camera also.

11.5   Fusion of IoDT With Other Technologies

The integration of the drone technology and other advanced technologies have created a lot of potential for the other application softwares and organizations from the security to the logistics and from agriculture uses to the industries based on drones and hence, IoT offers the interactivity [10] and the connectivity. To shape up the technologies and get the solutions in the form of IoT drones. There are a lot of technologies that are the key players in shaping the world.

Cloud computing
Cloud computing [15] is a key concept of the IoDT as cloud computing heavily relies on the sharing of the sources and is the primary requirement of the IoT platform. The primary characteristic of the IoT system is to be location independent and the end users can get access to the cloud services from any remote location of the devices and then they are connected to the Internet via the IoT platform. Rapid elasticity and the dynamic stability and scalability of the resources and hence, the fusion of the cloud and IoT has a wide scope of the diverse applications that are based on the real life time applications of the real world. The two major approaches of cloud computing and IoT include the cloud centric IoDT and the IoT based cloud. All the communications that take place are through the IoT architecture and hence, Quality of Service (QoS) becomes a major concern. Hence, to facilitate the IoDT operations via the cloud computing platform, the service providers give the various cloud computing services and hence, they provide the best QoS to gather all the information from clients on a daily basis. Hence, it becomes easy to share the data with the help of IoT centric cloud platform, latencies can also be eliminated along with the high traffic road and the least hop count with a very less overhead.

Fog computing
Fog computing [17] is another technology that when fused with the IoDT technology can provide wonders to the IoT industry in terms of processing the information in an efficient manner. Now, with the help of fog computing, a lot of issues in the IoDT system are resolved, such as Internet connectivity, real time processing, QoS requirements, and real-time processing in the remote areas where the internet connectivity is poor. Drone applications perform the computations with very approximate solutions for the problems which rely on the ground base stations for the computational intensive tasks of the systems and thus, improvise the scalability, stability, accuracy and the reliability of the system. IoDT has benefitted

with the help of fog computing like low latency rate, services and computing operations in the real time, performing the tasks, without any kind
of loss in the areas with poor connectivity, and in this way large numbers
of drones can be deployed in an integrated network. Heterogeneity is an
important aspect in this where adaptive streaming sites are used, such as
parallel successive refinement methods, based on streaming, etc.

Smart agriculture
Agriculture [15] has used the IoDT technology in various formats and still
the agriculture sector deals with the large monitoring areas where there
is limited connectivity. Hence, a precision solution is required for monitoring the fields and the crops for the farmers, the moisture levels of the
soil, their fertility, pH levels, pesticides quality, detection of animals in the
farms, monitoring the cattle and monitoring the agricultural equipment.
Wireless Sensor Networks (WSN) is a good solution for the monitoring
of the farms and the sensors are deployed in limited areas and sometimes
in real time they are prone to some frequent failures due to bad weather
conditions and other factors. Agricultural farm monitoring is used by the
IoDT system but still it has precision issues. Cattle monitoring systems and
sensors, data acquisition bots, weather monitoring bots come to the rescue
of the farmers. With the implementation of IoDT and the deployment of
the drones, certain problems are removed, such as the animal intrusion
cases, border of the farms can be kept in check easily, and this can be highly
expensive with the cost and maintenance of the drone deployment and the
sensor deployment or the camera installation. The cost for maintaining
everything can be a big task and cost a big time.

   With the help of IoDT, the cattle monitoring is done at a precision level.
Sensors are deployed at every animal present in the farm so that monitoring
can be done. With the surveillance of the fields with the help of drones, timely
actions are taken before any threat. Any kind of unusual activity that takes
place related to animals like animal health, cattle going outside the boundaries of the farms which can be prevented with timely triggered actions. Live
data can be acquired by the IoDT systems from the sensors at regular intervals. Agricultural bots are used for cultivating farms which keep the water
levels in check and also monitor the weather. SO, the complex problems of
agriculture can be solved through IoDT in order to yield better crops.

Smart cities
The monitoring of the smart cities [1] has also become a very necessary
aspect of the advanced technologies and that's why IoDT plays an important role in monitoring of the smart cities. The deployment of the IoD

system in the urban areas is easy where the internet connectivity is good but the deployment in the rural areas is very difficult because the connectivity is a big issue in the rural areas. So, the internet connectivity in rural areas is a big drawback. The government organizations also take less interest in deploying their drones in such areas as the expenditures are also very high in order to deploy the latest technologies. The IoDT system provides new opportunities and new business ventures can be undertaken by the organizations which may help in crime detection, implementation of smart cities, smart monitoring, search and rescue operations, object and people tracking, etc. Hence, the existing smart systems are able to be transformed into smart cities at the next level.

The main issue with the smart cities is to monitor the coverage of the entire city ends that do not facilitate the smart internet connectivity and it is not able to update the data at regular intervals or even synchronize the data and this is the reason the efficient monitoring Is not facilitated. Also the emergency systems are not highly reliable in the existing network infrastructure. Cameras and sensors are deployed at every nook and corner of the system, this process is very expensive as there is a need to cover the entire city. Hence, there is a very high cost of infrastructure that includes the smart sensors and the HD cameras. This problem exists with the mobile sensors also while deploying the GSM network like 4G, 5G, etc. that also tries to cover the entire city for better communication purposes. Also, the response teams and also it takes a lot of time to take decisions and actions. Hence, the deployment of sensors for the IoDT system can help the existing problems in the smart cities.

With the help of IoDT, the monitoring of the entire smart city becomes easy and the regular monitoring of each and every nook and corner of the city can be done. Also, the data provided by the sensors present in the drones is sent to the base ground station and is updated at regular intervals. This makes the monitoring process even more easy as live traffic feed is recorded, weather of the city is monitored and predicted for the next few days, etc. Electric power companies can perform their task by reading the electricity bills through automation with the help of Smart Grids. Manual reading is a cumbersome and a time consuming task and smart grid makes it possible to read the bills through automation. RFID tags are used by smart homes for the purpose of smart reading, etc. The drones are deployed by the companies for smart reading as well as to back up the data and for the electricity billing purposes. Natural disasters, checking of the pollution levels, monitoring the environment through keeping a check on the gases, etc. can be done through the IoDT system.

11.6   Recent Advancements in IoDT

Everything can be transformed in the near future with the help of drones and this is due to the reason the IoDT is a fusion of a lot of advanced technologies like smart sensors, cloud computing, big data applications, etc., and these technologies help transform the infrastructures in major sectors. IoDT has advanced monitoring capabilities in performing the complex missions and their response time is also very low for undertaking the necessary actions [15]. The drones are designed in such a way that the users get a safe, reliable and an optimized set of secure results. Several terminologies are discussed here related to the IoDT system which deal with the real time applications, technologies empowered by IoDT, dealing with serious security threats to the confidential data.

Infrared thermography
Infrared thermography [2] is a thermal imager and it detects radiation that is coming from an object which converts this heat into temperature that is shown in the form of image. This image shows temperature distributions and that's why they are called thermograms and these objects can be detected then because they may be invisible to the naked eye. Thermal imagers get the infrared wavelengths regardless of the light they are producing. There are various examples for this like the night goggles which easily detect the objects in the dark. There are various applications also like building moisture diagnostics, machine condition monitoring, chemical and earth science imaging, plant maintenance, locating gas and liquids, monitoring electrical and mechanical conditions of motor, medical imaging for diseases, metabolism, etc. The main aim of infrared thermography is to see that the machinery is working properly, abnormal heat patterns in objects and defects.

Spot infrared thermometers are used to detect and then measure the temperature at a specific spot on a particular surface. They are ideal for measuring the thermal radiation on the assets that are present in the extreme conditions. They have applications like checking water leaks, fluid handling systems, monitoring the electrical rooms, etc. Infrared scanner systems are able to scan very large areas and are used in manufacturing plants. Conveyer scanning belts are the largest applications of the infrared scanner systems. Infrared thermal imaging cameras are an advanced type of radiation thermometer that can measure the temperature at multiple points and create two dimensional thermographic images. They can be hooked up to a specialized softwares for a better evaluation with accuracy and efficiency. Other features of the infrared cameras include color alarm, pictures blending, color blending, etc. From all these features an important information related to the object under scanner is obtained.

Hyperspectral imaging

Hyperspectral Imaging (HSI) [1] is a new advanced technology coming up in the field of IoDT. Here, optical spectroscopy is used as an analytical tool which when combined with a visualization of two-dimensional objects that is obtained from the optical imaging can benefit the drones a lot. Chemical materials exposure, facial recognition, food processing, astronomy, pesticide applications, mineralogy, etc. are the major applications of hyperspectral imaging. These applications can benefit from the hyperspectral imaging and hence, they will get much more imaging views for the objects. HSI is an analytical technique that is majorly based on spectroscopy. Hundreds of different imaging wavelengths are collected by the HSI technology of the same spatial area of the object. An average human eye has the three color receptors namely green, red and blue, the hyperspectral imaging basically measures the continuous spectrum of this light and for each of the pixel of the object or scene must have a fine wavelength resolution which is not only visible but also with the help of near infrared scanners. The collected data of the object is called the hyperspectral cube in which the spatial extent of the objects is described from the two dimensions seen and the third one is the spectral content.

The multispectral mapping is used to provide the continuous and the discrete portions present in the spectral range of the objects. The hyperspectral imaging is used for the creation of the hypercube which has a comparatively larger number of the contiguous spectral points and bands. The values of the complete spectrum of each pixel observed and calculated, then recorded and sent to the imaging centers. Each and every material possesses some of the specific spectral values and spectral signatures that are used as the unique identification in the form of fingerprints. The hyperspectral imaging has a multiple number of applications in the field of remote sensing and monitoring, hence the nondestructive and the label-free capabilities are required in recognizing the components of the matter. The hyperspectral imaging is deployed in different fields namely agricultural, biomedical imaging, astronomy, molecular biology, food processing, cultural heritage, environment, surveillance capabilities and monitoring systems.

Hydrogen fuel cells

A hydrogen fuel cell [2] basically uses the chemical energy of the hydrogen or some other fuel which provides the electricity efficiently. If only the hydrogen is used as a fuel then the products obtained will be heat, water and electricity. Fuel cells have several potential applications which provide power for the large systems in the utility power stations. Fuel cells have applications in multiple sectors, such as transportation, commercial sector, industrial, residential buildings, large energy storage, as well as some grid reversal systems. The fuel cells have several benefits over the combustion based technologies that are still used by vehicles and power plants. Fuel cells convert the chemical energy directly into the chemical energy and can be upgraded to 60%. These fuel cells only emit water that addresses the climate challenges and see to it there are no carbon dioxide emissions. The fuel cells are like batteries that don't get over or need any kind of charging. They can produce the heat and electricity as long as the fuel is supplied to the system. Now, the fuel cell consists of two things an anode and a cathode which are sandwiched together as an electrolyte. Now the fuel which is hydrogen is fed to the anode and into the air is fed to the positive cathode. Now, the catalyst present at the anode breaks down the hydrogen fuel into protons and the electrons which in turn take different paths to the cathode line.

There are several kinds of problems that the scientists deal with in the hydrogen fuel cells production, such as cost, performance, and durability of these systems. The research and the development part of the industry is trying to reduce the overall cost of the hydrogen fueled cells with advanced technologies and the development of the low cost fuel cells. The main emphasis is on increasing the activity as well as the utilization of more reduced content of the catalysts along with the long term applications. To improve the performance of the hydrogen fueled cells, the R&D works toward these innovative ideas, reduced cost, efficiency, durability, innovative materials, integration strategies, membrane electrolytes, etc. Adequate performance is required for several fuel cells applications that are maintained over a long period of time. The robustness and the system reliability is required for carrying out several operations like for dynamic operations and harsh conditions.

Optoelectronics

Optoelectronics [12] is basically the study of the light emitting or the light detecting devices. This technology is widely considered as the application or a branch of photonics. Photonics is the study of the physical science of light which is quickly becoming an emerging technology in the field of drones. It consists of the electronic devices that source their detection toward the control of light. These devices all have a lot of applications like the automatic control systems, military services, medical equipment, telecommunications and much more. The devices that are considered to be a part of optoelectronics are very high in amount which include the LEDs, pick up devices and elements, optical storages, information displays, remote sensing systems and medical equipment and several communication systems. The optoelectronics are used in devices like the telecommunication laser, blue laser, solar cells, optical fiber, photodiodes, etc.

Lidar

Lidar stands for Light Detection and Ranging [15] which is a remote sensing method that is used to examine the different surfaces of the earth. Lidar generated programs are used to examine both natural and manmade environments. The lidar data that is collected supports the activities, such as storm surge modeling, inundation, emergency response systems, hydrodynamic modeling, hydrographical surveying, and various analysis systems. This method uses the light in the form of the pulse laser that measures the variable distances or range to the earth. It generates precise, three dimensional information about the size and shape of the earth and several other surface characteristics. A lidar equipment includes a scanner, specialized GPS receiver, a laser, etc. It is most common in airplanes and helicopters which need to cover more surface area platforms.

There are basically two kinds of lidars, namely topographic lidar and the bathymetric lidar and they are used in various applications. A topographic lidar basically uses a near infrared laser that maps the land while the bathymetric lidar uses the green light which is water penetrating and also measures the seafloor and the riverbed elevations. The lidar systems help the scientists and the professionals to examine the natural and the manmade environments with precision, accuracy as well as the flexibility. The lidar driven apparatus produces better elevation models, accurate shoreline maps for the geographical information systems, then assisting the emergency response teams and various other applications.

11.7 Conclusion

IoDT is an emerging technology and many people can benefit from this advancement and hence, the continuous evolving of the data from the heterogeneous sources toward the new era applications that are connected to the internet. The attacks are also a serious problem in IoDT in hacking the data which lead to problems in data confidentiality, integrity, authenticity, availability, etc., also physical attacks on UAVs are there. The security as well as the vulnerable threats are present here which can be dealt with many cryptographic solutions and mechanisms as well as control signal mechanisms. Hence, in the coming future there is a need to provide the threat prevention mechanisms and several vulnerability assessment methods for the IoDT protection.

References

1. Sudhriti, D.K., Smart cities using Internet of Things: Recent trends and techniques. *Int. J. Innov. Technol. Exploring Eng.*, 8, 24–28, 2278–3075, 2019.
2. Mangla, M., Kumar, A., Mehta, V., Bhushan, M., Mohanty, S.N., *Real-life applications of the Internet of Things: Challenges, applications, and advances*, Apple Academic Press, New York, 1st Edition, pages 536, 2022.
3. Goel, R., Jain, A., Verma, K., Bhushan, M., Kumar, A., Mushrooming trends and technologies to aid visually impaired people, in: *International Conference on Emerging Trends in Information Technology and Engineering (ic-ETITE)*, IEEE, pp. 1–5, 2020.
4. Wazid, M., Das, A.K., Rodrigues, J.J.P.C., Sachin, S., Youngho, P., IoMT malware detection approaches: Analysis and research challenges. *IEEE Access*, 2, 1–18, 2019.
5. Verma, K., Bhardwaj, S., Arya, R., Islam, M.S.U., Bhushan, M., Kumar, A., Samant, P., Latest tools for data mining and machine learning. *Int. J. Innov. Technol. Exploring Eng.*, 8, 9S, 18–23, July 2019, Available: https://doi.org/10.35940/ijitee.I1003.0789S19.
6. Gharibi, M., Boutaba, R., Waslander, S.L., Internet of Drones. *IEEE Access*, 4, 1–15, 2016.
7. Ngo, Q.-D., Huy, N., Le, V.H., Nguyen, D.-H., A survey of IoT malware and detection methods based on static features. *Science Direct, ICT Express*, 6, 1–8, 2020.
8. Chao L., D.H., Neeraj K., K.R.C., Alexey V., X.H., Security and privacy for the Internet of Drones: Challenges and solutions. *IEEE Commun. Mag.*, 2018.
9. Shubhani, M.S., Neeraj, K., Mauro, C., A new secure data dissemination model in Internet of Drones. *International Conference on Communications*, IEEE, 2019.
10. Hall, R.J., An Internet of Drones. *IEEE Internet Comput.*, 2016.
11. Gaurav, V.S., Takshi, J.K., Ilsun, Internet of Drones (IoD): Threats, vulnerability, and security. *The 3rd Int. Jour. on Mob. Internet Sec. (MobiSec'18)*, 37, 1–13, 2018.
12. Tiwary, A., Mahato, M., Chidar, A., Chandrol, M.K., Shrivastava, M., Tripathi, M., Internet of Things (IoT): Research, architectures and applications. *Int. J. Future Revolution Comput. Sci. Commun. Eng.*, 4, 23–27, 2018.
13. Tewari, A. and B.B.G., Privacy and trust of different layers in Internet-of-Things (IoTs) framework. *Future Gener. Comput. Syst.*, 108, 909–920, April 2018.

14. Zhi-Kai Z., M.C.Y., Chia-Wei W., C.H., Chong-Kuan C., S.S., IoT security: Ongoing challenges and research opportunities. *IEEE Int. Conference on Service-Orntd. Comp. & App.*, 2014.

15. Teng, X., James, B., Miodrag, P., Security of IoT systems: Design challenges and opportunities, IEEE, 3, 1–7, 2017.

16. Umesh, B. and Sudeep, T., Secure data dissemination techniques for IoT applications: Research challenges and opportunities, *Journal of Software Practice and Experience*, 51, 2469–2491, January 2020.

17. Sharma, S., Nanda, M., Goel, R., Jain, A., Bhushan, M., Kumar, A., Smart cities using Internet of Things: Recent trends and techniques. *Int. J. Innov. Technol. Exploring Eng.*, 8, 9S, 24–28, July 2019, Available: https://doi.org/10.35940/ijitee.I1004.0789S19.

18. Suri, R.S., Dubey, V., Kapoor, N.R., Kumar, A., Bhushan, M., Optimizing the compressive strength of concrete with altered compositions using hybrid PSO-ANN. *4th Int. Conference on Info. Systems & Mgmt. Sci. (ISMS 2021)*, Springer, 2021.

19. Kholiya, P.S., Kapoor, A., Rana, M., Bhushan, M., Intelligent process automation: The future of digital transformation. *10th Int. Conference on Sys. Mod. & Advan. in Rsrh. Trends (SMART)*, pp. 185–190, 2021.

20. Samant, P., Bhushan, M., Kumar, A., Arya, R., Tiwari, S., Bansal, S., Condition monitoring of machinery: A case study, in: *6th Int. Conference on Sgnl. Process., Computing and Control (ISPCC)*, IEEE, pp. 501–505, 2021.

Computer Vision-Oriented Gesture Recognition System for Real-Time ISL Prediction

Mukul Joshi, Gayatri Valluri, Jyoti Rawat* and Kriti

School of Computing, DIT University, Dehradun, Uttarakhand, India

Abstract

Sign language is a nonverbal type of correspondence used to transfer information unlike speech, i.e., the verbal type of correspondence that utilizes oral articulations through gestures. This chapter aims to bridge the communication barrier between individuals with vocal and hearing disabilities and non-sign language speakers by creating a recognition model. A system is created using real-time motion analysis where the pre-processing of the dataset is performed and converted to a gray scale. Subsequently, HSV conversion is applied along with skin masking and Canny-edge detection applied for hand tracking and detection. Sign language prediction is performed with the help of a 2-dimensional Convolutional Neural Network (CNN) used mainly in image processing applications with the result being translated and stored into a database such that the signed gesture is recognized and translated to text. The overall accuracy achieved via classification through the 2-D CNN, where 33 labels are detected successfully out of 36 labels is 91%.

Keywords: Visual recognition system, sign language, convolutional neural network, Indian sign language

12.1 Introduction

Sign language is a language incorporating non-verbal correspondence that takes its origin in visual cues and hand gestures. Within the language, there

---

*Corresponding author: drjyotirawat19@gmail.com

Ashok Kumar, Megha Bhushan, José A. Galindo, Lalit Garg and Yu-Chen Hu (eds.) Machine Intelligence, Big Data Analytics, and IoT in Image Processing: Practical Applications, (305–322) © 2023 Scrivener Publishing LLC

are fully established hand movements with their grammatical significance and inbuilt formation systems. Sign language is composed of various gestures formed by the shapes of the hand, its movements, and orientations. Presently, sign language is the fundamental language [1] of people who have a hearing disability and is adopted by people who can hear but are unable to speak. Sign is a complex language that comprises and is not limited to hand gestures, facial expressions, and body movements. Sign language, although not practiced universally but is specific to countries that have their native sign language having their own set of rules, grammar, and word orders. The issue derives when the differently able correspond with signing with the people who are unfamiliar with this language. Thus, there is a necessity to develop an automatic and interactive interpreter interface to bridge the communication gap. Often, people with hearing and speech impediments seek the help of an intermediary sign language interpreter, a person proficient in both sign and regular speech, to translate their thoughts to common people and vice versa. However, this way turns out to be very costly and does not work out all the time. Thus, the need to introduce a system that can automatically recognize sign language gestures becomes a priority. Introducing such a system would significantly bridge the gap between differently-abled and people who use speech as a form of communication in society. Often, the sign language in use is different depending upon the culture and language [2].

Particularly, Indian Sign Language (ISL) is used by people who have difficulty in speaking and hearing in India. It is an accepted and advanced form of communication for the differently-abled people in India and people communicating in the English language. Various symbolic representations are involved for different characters present in the ISL.

An effective solution to this problem is one involving a computer vision-oriented gesture recognition system that uses image processing techniques. Sign languages are fully functional yet, are not universally used nor understood, and the easiest way to perceive them is by creating a computer recognition model that through video sequence captures spatial movements and records them, predicting the translation system for sign through Convolutional Neural Networks (CNN).

Further, the chapter is divided into different segments: Literature Review, System Architecture, Implementation and Results and finally the Conclusion. The proposed model is a CNN network, i.e., model which is trained on pre-processed sign image data set without any complex pre-processing wherein a live video stream can directly be input with a camera by a signer and fed to the network.

12.2   Literature Review

Multiple techniques for Sign Language Recognition (SLR) model classification are implemented by various authors, where each method has its convenience over other methods including self-proposed methods for the development of their SLR models. The objective of this segment is to review the approaches to various SLR systems and find an efficient method in comparison that has been adopted by researchers in terms of affordability and accessibility. Hidden Markov Model (HMM) is a widely used method for SLR processing. The HMM algorithm and another algorithm are used in combination for SLR. According to the Wang *et al.* [3] have used the light HMM model for Chinese SLR, in which the keyframes can be selected with the help of low-level approximation and consecutively distinguish the states hidden in HMM such that the keyframes are a scaled-down and precise estimation of the hidden state can be made. In a similar work by Zhang *et al.* [4], another approach called the Tied Density-HMM uses the technique similar to continuous HMM, where the computations are reduced by tying identical Gaussian compound components.

Another research method by Ekbote *et al.* [5] used Artificial Neural Network (ANN) for recognition using shape descriptors, combined with Histogram of Oriented Gradients (HOG) techniques for feature extraction and ANN with the help of Support Vector Machine (SVM) classifiers for classifying the gestures. Table 12.1 specifies the comparative study of accuracies obtained per method considering the hardware implications and number of images for each method. Rekha *et al.* [6], Fatmi *et al.* [15] also did the same task in their work. Another, unsupervised neural network method used in SLR, proposed by Tewari *et al.* [7] is Self-Organizing Map (SOM) using 2D Discrete Cosine Transform (2D-DCT) to create compressed images and SOM or Kohonen Self Organizing feature map (SOFM) simulated in MATLAB [7].

The proposed processing method implemented is a CNN based on the research presented by Yang *et al.* [8], a Chinese Sign Language (CSL) recognition model based on CNN and Long Short-Term Memory (LSTM) network.

This model is created in which CNN takes images as input as it is which prevents the complex hand gesture segmentation and feature extraction, exploiting as much information as possible. Since there are various dynamic signs in CSL, it is more practical to have a continuous recognition model. Similarly in another CNN-LSTM approach [9], a CNN model known as Inception-V3 for feature extraction that is trained on approximately a

Table 12.1 The comparative analysis of different studies in literature for SLR.

Article	Method	Image count	Hardware	Accuracy (%)
[3]	Light Hidden Markov Model	7000 video frames.	Microsoft Kinect sensor, Camera and Color gloves	83.60
[4]	Tied Density Hidden Markov Model	1756 training and 439 test Samples.	Microsoft Kinect sensor, Camera and Color gloves	91.30
[11]	HMM	320 samples sign images and 10 images for each sign.	Microsoft Kinect sensor, Camera and Color gloves	96.87%
[5]	ANN and SVM	Self-created database of 1000 images.	Multiple cameras	(SIFT + SVM=96.2), (HOG + SVM=93), (Combined 99.00)
[7]	Kohonen Self Organizing feature map	A dataset of 35 images.	High specification camera	80.00
[8]	CNN and LSTM	320 corresponding gesture videos, 256 for training and 64 for testing.	Video camera, Web camera	98.43 Input vector of size 1024.
[9]	CNN and LSTM	1100 Videos.	Video camera, Web camera	81.00

million images from the Image Net database where a Recurrent Neural Network (RNN) architecture classified the gestures and translated them to text. In the processing method by Subha *et al.* [11], the images are subjected to the method of Grid-based feature extraction that depicts the hand's gesture as a feature vector. The hand gestures are further classified via the k-nearest neighbors algorithm. For the proposed model, a 2D-CNN directed approach is used which is identical to ANN [12, 13]. It consists of nodes or neurons attached through weighted links further producing an output to the given input.

The 2-D CNN implementation is performed by extracting a layer to feature map via the kernel window also known as the local receptive field, minimizing free variables and increasing the generalization capacity of the network. The test accuracy using HMM model by [3] is obtained as 83%, for [4] the test accuracy is 91.30% for 439 test samples using Tied-Density HMM. In the case of ANN-SVM, [5] the combined accuracy obtained is 93%. For the Kohonen SOM approach by [7], the test accuracy obtained for a dataset of 35 images is 80%. Lastly for the CNN-based approach using LSTM by [8] using 64 test samples the accuracy obtained is 98.43%. Similarly, for [9] the accuracy obtained upon testing the dataset formed using 1100 videos is 81%.

12.3   System Architecture

The proposed framework comprises model building phase and the development environment phase for the input of live feed.

12.3.1   Model Development Phase

1) Data Acquisition: The model building phase is initialized with the acquisition of training data from multiple resources. For the proposed model the Indian sign language dataset is acquired from Kaggle [10] with approximately 40,000 images. The length of the dataset obtained is 34153, where each input image is passed through a series of convolutional layers.

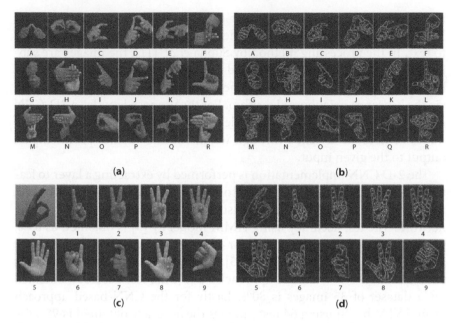

Figure 12.1 (a) Characters Dataset [A-R] before pre-processing, (b) Characters Dataset [A-R] after pre-processing, (c) Numeric Dataset [0-9] before pre-processing, and (d) Numeric Dataset [0-9] after pre-processing.

2) Image Processing: The segment is applicable for various image pre-processing techniques that are important for feature extraction when the model is trained over the data set and for increasing the quality of the image. The sample dataset before and after pre-processing is shown in Figure 12.1.

The pictorial representation of the architecture for the proposed model for ISL prediction is given in Figure 12.2.

3) Data Resizing: All the images that are used in training the model should be uniform for the application of the image resizing technique.

4) Data Augmentation: It is an approach that allows practitioners to automatically increment the data available for training models, without any collection of the new data [13] by using various methods as rotation, scaling, normalization, and horizontal flipping.

5) Feature Extraction: A CNN model is considered as a combination of two components consisting of feature extraction and

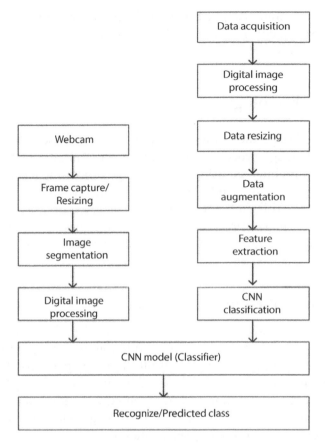

Figure 12.2 System architecture for the proposed ISL Model.

the classification component. The initial layers of CNN perform feature extraction (convolution layer + pooling layer).

6) CNN Classification: The last few layers (fully connected layer + output layer) of the CNN perform the classification part for the proposed model [14].

12.3.2   Development Environment Phase

Live feed is taken as an image input through a web camera or a mobile phone and the frames from the live feed are extracted and resized. The extracted images are passed through the image segmentation process, which furthers to more important features in an image. The same image processing technique is applied again in the model building phase for prediction improvement. The processed output is then classified with the help of the developed CNN model.

12.4   Methodology

The proposed ISL prediction system executes in two phases. The first phase is the digital image processing phase, where the important features are extracted and highlighted using several image processing techniques, and the second phase, or the model building phase involves the classification of the hand gesture using a trained classifier.

After performing several pre-processing techniques, the images used for sample training are further fed to the classifier model during the training process.

12.4.1   Image Pre-Processing Phase

Image frames are extracted from the video such that if the frame per second rate is beyond 70 then the frame extraction process is commenced if not then upcoming process is halted till the frame per second rate limit is exceeded. Frame extraction is followed by image processing techniques [16–18] for feature extraction and frame enhancement via performing the grayscale conversion, HSV image conversion, edge detection, noise reduction and region of interest (RoI) segmentation.

1) RoI segmentation: A specific region in the frame is initialized by providing dimensions using the rectangle method, highlighting the RoI [14], of the frame where only the object within the RoI is processed.
2) Grayscale conversion: The conversion of input color image into a grayscale image is viable due to the lack of useful information available for the identification of a gesture in case of a colored image, thus working on a single-color channel decreases the complexity of the system at an initial level, subsequently maintaining a faster processing speed.
3) HSV image conversion: The input color image is converted to HSV color space to build a skin-mask of intensity ranging from the human skin color so as to highlight specific portions of the image i.e., the hand in this case.

4) Noise reduction: It's a process of reducing minimal to moderate levels of noise while preserving important features. Median blur [19] is performed on the proposed model for noise reduction by processing the edges while removing the noise, maintaining the kernel size of 5 for the process.

5) Edge detection: The processing technique accounts for boundary detection within images and works by detecting discontinuities in brightness where in the proposed model a canny edge detector with threshold values of 60-60 is applied.

12.4.2   Model Building Phase

A deep learning approach is applied to the proposed model based on a 2-D CNN [20–24] for building a classifier. The CNN model used for the ISL prediction is shown in Figure 12.3.

In the initial, convolutional layer C1 takes input images and outputs a stack of 16 feature maps. All the convolutional layers earlier defined. Defining two or more each of which doubles the depth of the output until you get a layer with a depth of 64. It will begin with an image depth of 3 then moving to 16 then 32 and finally 64. Each of these layers uses a convolutional kernel of 3×3 and has a padding of one. There is a max pooling layer, which downsamples any size by 2. A dropout layer with a probability of 0.25 is included to prevent overfitting. Two fully connected layers

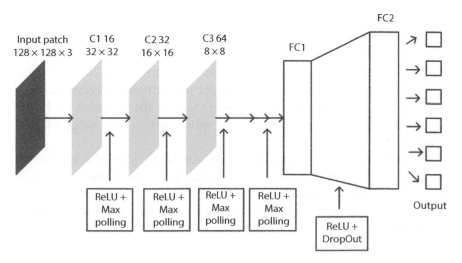

Figure 12.3 CNN model of the proposed system.

are added, where the first one is responsible for taking as input the final downside stack of feature maps. In the forward function, a pooling layer has been applied after each convolutional layer, thus the image is reduced in size to 32 × 32 then 16 × 16 and finally 8 × 8 after the last pooling layer.

The third convolutional layer produced a depth of 64 and that is how these values 16×16 from the final x, y size and 64 for the depth are obtained. That will be the number of inputs that this fully connected layer should obtain. Then, 500 outputs are produced and then fed as input into the final classification layer, which will see these as inputs and produce 36 class scores as outputs.

A sequence of convolutional and pooling layers [25–27] is added in order where input images are passed in the first convolutional layer where initially an activation function and a pooling layer are applied. The process is repeated for the second and third convolutional layer. Finally, the result obtained is flattened into a vector shape and passed into a fully connected layer proceeding to which the ReLu activation function is applied. It is noted that in between flattening and each fully connected layer a dropout layer is added to prevent over-fitting [28–32]. Thus, the result obtained is a list of 36 class scores. For classification, the dataset was divided into 36 classes from [0-9] and [A-Z] with a batch size (64), where approximately 900 images were trained per label. There were 34,153 images used in training, 7091 for validating, and 1410 for testing.

12.5   Implementation and Results

12.5.1   Performance

Model forward is used to perform the forward pass, comparing the predicted output with the actual label and measuring loss, further using cross-entropy loss to measure the model performance. Figure 12.4 and Figure 12.5 display the output character within the range of "A–J" and Output character display of numeric values within the range of "09," respectively.

The test accuracy obtained per label, for labels [0-9] the accuracy is obtained to be 100%, for [A-Z] – B is obtained to be 100%, whereas for label B it is obtained to be 97%.

Figure 12.6 shows the loss and accuracy curve for the present model of ISL prediction.

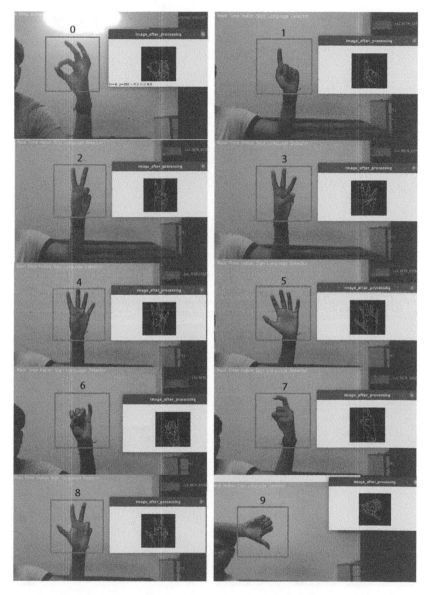

Figure 12.4 Output character display of numeric values within the range of "0–9."

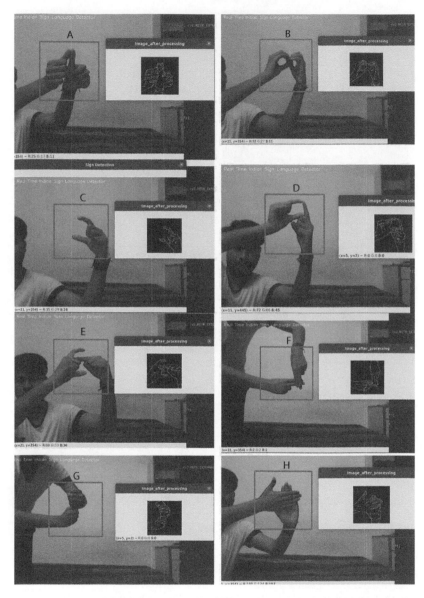

Figure 12.5 Output character display of alphabets within the range of "A–J."

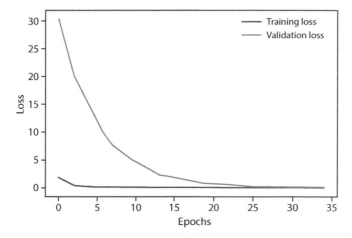

Figure 12.6 Loss-accuracy curves achieved for the ISL model.

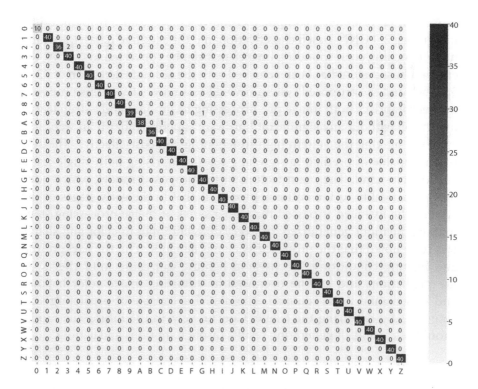

Figure 12.7 Heat map for confusion matrix for the trained model.

12.5.2   Confusion Matrix

A confusion matrix is a plot to indicate the predictions of a presented sign language recognition model on test data.

In the confusion matrix, each row signifies the predicted probability value for a mentioned actual class, whereas each column signifies the different probabilities of the classes being predicted as a specific class.

As shown in Figure 12.7, all the blue cells in the diagonal indicate the correct prediction of class labels along with a high level of confidence. The yellow color cells in each row exhibit the classes with which the model is confusing the class corresponding to that row. Thus, an overall test accuracy of around 91% is achieved. Any loss in the given model is observed to be the losses obtained due to an error in background removal or an error in edge detection. For trained models, the heat map for confusion matrix is given in Figure 12.7.

12.6   Conclusion and Future Scope

This work represented a computer vision-oriented gesture recognition by integrating 2D CNN-based classification into the acquired dataset working upon matching corresponding images that can be converted to text. At the end, a robust system has been achieved with the absolute detection of 33 of 36 labels and the proposed system's overall accuracy of 91%. This proposed model comprises both image pre-processing techniques and a deep learning model without any significant hardware support (other than webcam/camera), making the system high functioning and cost-effective. The system is capable of predicting sign languages comprising both single-hand and double-hand. The main feature of the proposed model, which adds more practicality to the entire approach, is that in the early stages of the system, the live feed is passed through a series of image processing steps making the input generalized for every different input to color and shape. It is to be noted that the proposed system is dependent on the background and lighting conditions. Inconsistent lighting can cause mild effects on the predictions and can be overawed by adjusting image color balance and background noise removal. The model is capable of predicting sign languages by capturing the frame. Conclusively, in the future, the model should be capable of generating results through motions and actions.

References

1. K. Muthukumar, K. VinothKumar, D. Hepsiba *et al.*, Performance based algorithm for DWT and DCT for ISL, Materials Today: Proceedings, https://doi.org/10.1016/j.matpr.2020.10.639.
2. Yadav, N., Thepade, S., Patil, P.H., Noval approach of classification based Indian sign language recognition using transform features. *2015 International Conference on Information Processing (ICIP)*, IEEE, 2015.
3. Wang, and Hanjie, Fast sign language recognition benefited from low rank approximation. *2015 11th IEEE International Conference and Workshops on Automatic Face and Gesture Recognition (FG)*, IEEE, vol. 1, 2015.
4. Zhang, L.-G., A vision-based sign language recognition system using tied-mixture density HMM. *Proceedings of the 6th International Conference on Multimodal Interfaces*, 2004.
5. Ekbote, J. and Joshi, M., Indian sign language recognition using ANN and SVM classifiers. *In 2017 International Conference on Innovations in Information, Embedded and Communication Systems (ICIIECS)*, IEEE, pp. 1–5, 2017, March.
6. Rekha, J., Bhattacharya, J., Majumder, S., Shape, texture and language recognition system to aid deaf-dumb people. *2011 IEEE Local Movement Hand Gesture Features for Indian Sign Language 13th International Conference on Communication Technology. IEEE, Recognition. 3rd International Conference on Trendz in*, 2011.
7. Tewari, D. and Srivastava, S.K., A visual analysis. *2014 International Conference on Medical Imaging, Mrecognition of Static Hand Gestures in Indian Sign Language Based on Health and Emerging Communication Systems (MedCom)*, IEEE, Kohonen self-organizing map algorithm. International Journal of 2014.
8. Yang, S. and Zhu, Q., Continuous Chinese sign language recognition with CNN-LSTM. *In Ninth International Conference on Digital Image Processing (ICDIP 2017)*, SPIE, Vol. 10420, pp. 83–89, 2017, July.
9. Shenoy, K., Dastane, T., Rao, V., Vyavaharkar, D., Real-time Indian sign language (ISL) recognition. *In 2018 9th International Conference on Computing, Communication and Networking Technologies (ICCCNT)*, IEEE, pp. 1–9, July 2018. link: https://arxiv.org/pdf/2108.10970.pdf
10. Indian sign language dataset. Available from: https://www.kaggle.com/vaishnaviasonawane/indian-sign language-dataset. [Accessed: 12th Feb 21].
11. Subha Rajam, P. and Balakrishnan, G., Indian sign language recognition system to aid deaf-dumb people. *2010 Second International Conference on Computing, Communication and Networking Technologies*, IEEE, 2010.
12. Sruthi, C.J. and Lijiya, A., Signet: A deep learning based Indian sign language recognition system. *2019 International Conference on Communication and Signal Processing (ICCSP)*, IEEE, 2019.

13. Wadhawan, A. and Kumar, P., Deep learning-based sign language recognition system for static signs. *Neural Comput. Appl.*, 32, 12, 7957–7968, 2020.

14. Sajanraj, T.D. and Beena, M.V., Indian sign language numeral recognition using region of interest convolutional neural network. *2018 Second International Conference on Inventive Communication and Computational Technologies (ICICCT)*, IEEE, 2018.

15. Fatmi, R., Rashad, S., Integlia, R., Comparing ANN, SVM, and HMM based machine learning methods for American sign language recognition using wearable motion sensors. *2019 IEEE 9th Annual Computing and Communication Workshop and Conference (CCWC)*, IEEE, 2019.

16. Ryan, F.D., *Deaf people in Hitler's Europe*, Gallaudet University Press, 2002. E-book: ISBN 978-1-56368-201-8. https://gupress.gallaudet.edu/bookpage/DPHEbookpage.html

17. Raghuveera, T., A depth-based Indian sign language recognition using Microsoft Knect. *Sādhanā*, 45, 1, 1–13, 2020.

18. Raheja, J.L., Mishra, A., Chaudhary, A., Indian sign language recognition using SVM. *Pattern Recognit. Image Anal.*, 26, 2, 434–441, 2016.

19. Rajam, P.S. and Balakrishnan, G., Real time Indian sign language recognition system to aid deaf-dumb people. *2011 IEEE 13th International Conference on Communication Technology*, IEEE, 2011.

20. Sehgal, S., Data analysis using principal component analysis. *2014 International Conference on Medical Imaging, m-Health and Emerging Communication Systems (MedCom)*, IEEE, 2014.

21. Rogers, D.K., Translation, validity and reliability of the British Sign Language (BSL) version of the EQ-5D-5L. *Qual. Life Res.*, 25, 7, 1825–1834, 2016.

22. Zafrulla, Z., American sign language recognition with the kinect. *Proceedings of the 13th International Conference on Multimodal Interfaces*, 2011.

23. Zeshan, U., Indo-Pakistani sign language grammar: A typological outline. *Sign Lang. Stud.*, 3, 2, 157–212, 2003. JSTOR. https://www.jstor.org/stable/26204851

24. Hietanen, K.J., Leppänen, J.M., Lehtonen, U., Perception of emotions in the hand movement quality of Finnish sign language. *J. Nonverbal Behav.*, 28, 1, 53–64, 2004.

25. Hore, S., Indian sign language recognition using optimized neural networks, in: *Information Technology and Intelligent Transportation Systems*, pp. 553–563, Springer, Cham, 2017.

26. Kong, W.W. and Ranganath, S., Sign language phoneme transcription with rule-based hand trajectory segmentation. *J. Signal Process. Syst.*, 59, 2, 211–222, 2010.

27. Krak, I.V., Barmak, O.V., Romanyshyn, S.O., The method of generalized grammar structures for text to gestures computer-aided translation. *Cybern. Syst. Anal.*, 50, 1, 116–123, 2014.

28. Luqman, H. and Mahmoud, S.A., Automatic translation of Arabic text-to-Arabic sign language. *Univers. Access Inf. Soc.*, 18, 4, 939–951, 2019.

29. Patil, S.B. and Sinha, G.R., Distinctive feature extraction for Indian Sign Language (ISL) gesture using scale invariant feature Transform (SIFT). *J. Inst. Eng. (India): Ser. B*, 98, 1, 19–26, 2017.
30. Morrissey, S. and Way, A., Manual labour: Tackling machine translation for sign languages. *Mach. Transl.*, 27, 1, 25–64, 2013.
31. Verma, K., Bhardwaj, S., Arya, R., Islam, U.L., Bhushan, M., Kumar, A., Samant, P., Latest tools for data mining and machine learning, Volume-8, Issue-9S, 2019.
32. Kholiya, P.S., Kapoor, A., Rana, M., Bhushan, M., Intelligent process automation: The future of digital transformation. *10th International Conference on System Modeling & Advancement in Research Trends (SMART)*, IEEE, pp. 185–190, 2021.

13

Recent Advances in Intelligent Transportation Systems in India: Analysis, Applications, Challenges, and Future Work

Elamurugan Balasundaram*, Cailassame Nedunchezhian, Mathiazhagan Arumugam and Vinoth Asaikannu

Department of Management Studies, Sri Manakula Vinayagar Engineering College, Puducherry, India

Abstract

The pace of urbanization in India has gathered steam and marched toward to hit a peak in the coming years. The urban region now holds one-third of the population and vows to move further. Even after initiating various policies to deal with surging transportation needs, India fails to rank at par with its peers. Failure to recognize and adopt new and emerging technologies in transportation costs India more than any other country. Intelligent transportation system (ITS) can be utilized to get rid of inefficiencies and congestion in Indian Infrastructure. It contains an expression involving information technologies, like data storage management, artificial vision, communications, control systems, sensors, etc. ITS attempts to streamline the operation of cars by managing traffic, assisting drivers with safety and other information, and providing passenger and road safety convenience applications. The future of transportation lies in the following technologies: smart infrastructure, rideshare, mobility on demand, workforce development, connected vehicles, big data and analytics, and self-driving vehicles. The major aim of the report is to learn about recent developments in the ITS and further to know its feasibility analysis, applications, challenges, and future work in India.

Keywords: Intelligent transportation system, infrastructure sector, urban infrastructure, smart infrastructure, connected vehicle

*Corresponding author: harshadhelamurugan@gmail.com

Ashok Kumar, Megha Bhushan, José A. Galindo, Lalit Garg and Yu-Chen Hu (eds.) Machine Intelligence, Big Data Analytics, and IoT in Image Processing: Practical Applications, (323–340) © 2023 Scrivener Publishing LLC

13.1   Introduction

Because every commodity produced requires transportation at all stages of production, from manufacturing to distribution, transportation is vital for any region's economic progress [1]. The economic and social development is reflected in the quality of its transportation infrastructure [2]. The rise of urban areas in developing countries like India, where over 70% of the population lives in villages, does not represent the country's overall success.

Overall, economic progress will be achieved if suitable transportation facilities are provided in both rural and urban areas [3]. Because developing countries have mixed traffic conditions, their traffic situation varies from developed ones. India's transportation network is extensive and diverse [4].

From the era of animal-drawn carriage to technologies and the development of urban transport networks, transportation has changed tremendously in recent years [5]. Roads existed between the 25th and 35th centuries B.C., according to excavations at Harappa and Mohenjo-Daro. Under British control, India's roads and transportation systems were created to facilitate commerce and administration. Transportation progress is intrinsically linked to the growth of civilization [6]. As the globe got increasingly industrialized and urbanized, different methods for transferring people and goods from one point to another were essential [6, 7]. Because of industrial growth, the rapid settlement of people in cities fosters city expansion. Because most people live distant from their places of employment, providing inexpensive and efficient transportation has become one of the most important parts of city life [8]. It tried to substitute animal power with mechanical energy. In 1881, the Tramways Company started running a public transportation system in Kolkata, India. Horses hauled the original tramcars in Kolkata. After a few years, steam engines were utilized for pulling tramcars [9]. In 1931, tramcars were phased out in favor of gasoline-powered buses. Mechanical bus road transportation has been accessible in India's leading cities since 1920. Transportation aids economic, industrial, social, and cultural advancement [10].

Most Indian cities are built densely, with high population densities and mixed land uses. The transportation sector has struggled to keep up with rising demand, putting the economy at risk [11]. As transportation capacities in rural and urban areas increase, the desire to relocate to metropolitan areas decreases, resulting in development that is more balanced. In cities, well-developed public transit is a critical component of long-term mobility. It maintains its high standards by increasing operating efficiency, standardizing rates and schedules, and enhancing accessibility and interchange facilities [12]. To achieve significant productivity and efficiency gains, the future architecture of India's transportation network must strive to encourage multimodal mobility [13].

This chapter contributes to a better understanding of the myriad difficulties transportation providers and customers face in India's transportation business. One major aim of the research would be to figure out what an intelligent transportation system (ITS) is and how it is perceived throughout the world. In addition, the researcher is interested in conducting a feasibility assessment of the ITS in India. Furthermore, the researcher is curious about the important sectors of ITS in India.

13.2   A Primer on ITS

Fairly recent technologies aim to deliver novel services concerning various traffic management and transportation systems [14]. It also enables consumers to be properly informed and use transportation networks that are more integrated, sophisticated, and safe. One purpose of the ITS is always to make things better transportation efficiency by minimizing traffic congestion [15]. This technology gives the user advance notice of seat availability, real-time running information, traffic, and other relevant information, reducing trip process, and increase consumer convenience and protection [16].

Increased accident cases, insufficient road development, and unfeasible requests to expand new roads may all be addressed with an ITS, making transportation safer, more efficient, secure, and quicker via the application of control technology [17]. Furthermore, this may assist in alleviating increased traffic congestion, which increases travel time and industrial costs, as well as reduce transportation-related environmental problems [18].

Advanced public transportation system (APTS), advanced commercial vehicle operations systems, advanced rural transportation systems, advanced traffic management systems (ATMS), and advanced vehicle control systems all require an intelligent transport system, communication systems, automated traffic prioritization signals, information processors, GPS updates, sensors, and roadside messages [19].

13.3 The ITS Stages

Traffic management center (TMC) is an integral part of the ITS [20]. The transportation authorities manage this technological system. All data is captured, processed, and analyzed in this system for traffic operation and control management, providing real-time statistics and specifics about local vehicles [21]. Automated data collecting with exact position information, data analysis to provide detailed transportation knowledge, and afterward recirculate a well-extracted features to travelers are all necessary for efficient operations and well-organized traffic control centers [22]. The four steps of ITS are data collection, transmission, analysis, and trip information. Figure 13.1 shows covers four legs of ITS and gives comprehensive data to aid us in comprehending the actual system:

Data collection and communication: Rapid, full, and accurate data collection and exchange are required for authentic surveillance, planning, and implementation in traffic management systems. A reliable data collection communication system combine's reliable data gathering hardware with a structured approach for future ITS activities. Show signs hardware devices are used, including sensor, camera, fully automated recognition, and processors which could save a massive volumes of information for later review [23].

Data cleansing, fusion, and analysis are all part of data analysis: Before delivering data to the TMC, data from the sensor and other information devices must be vetted. Inconstant data must be winnowed out, and clear data must be maintained. For more investigation, input of many sources might have to be joined or blended to evaluate and predict travel problems, the cleaned and merged traffic information will be examined. These methods for evaluating traffic patterns will be used to provide users with relevant information [24].

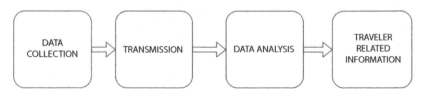

Figure 13.1 Stages of ITS.

Traveler information: The travel advisory system's capabilities are used to convey mass transit info to the people while under monitoring. Some of these possibilities include signage boards, the online, road warning broadcast, short message service companies, automatic smartphones texting, radio broadcast broadcasts, broadcast television, and other modern media technologies [25, 26]. These systems can provide real-time data on a variety of issues, including as transit times, speed, incidents, diversions, highway restrictions and detours, and work zone conditions [27].

13.4 Functions of ITS

ITS strives to help transit service consumers through enhancing dependability and comfort of sustainable mobility, and also rendering system operations and judgment better effective. As network controllers, one eventual aim of ITS would be to enhance the effectiveness of the transit system, travelers, transporters, and other consumers often in real-time [28].

The primary functionalities of ITS are depicted in Figure 13.2. Business interests and policy initiatives at the global, regional, provincial, and local levels influence public and private sectors' corporate practices, influencing ITS deployment [29].

To address typical transportation difficulties, ITS is a convenient process that prioritizes data volume, efficient decision, and increasing system flexibility. This differs from some of the more frequent approach of increasing physical ability and expanding transport network [29–31]. In instances wherein conventional tactics might not even function, like populous areas or places subjected to severe environmental rules, ITS provides options for controlling

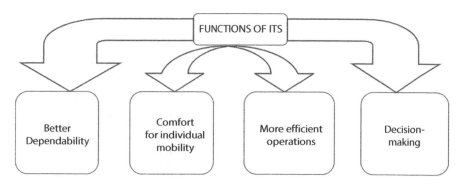

Figure 13.2 Functions of ITS.

future transportation needs. Sensors, communication, and cognitive computing are examples of tools that may have been combined through into transportation solution to enhance its efficiency, security, economy, and network operation' resilience in the event of a major disruption [32].

13.5   ITS Advantages

The capacity of ITS to give real-time traffic and trip information and a flexible method of network management is its most important characteristic [33–35]. Figure 13.3 shows major advantages of ITS which is a critical component of a long-term transportation system that:

- Satisfies the accessibility and mobility requirements of persons who live and work in a particular region or use public transportation in a safe manner.
- Provides various accessible and economical transportation solutions that make the most of existing infrastructure.
- To provide a high level of service to road users, it controls traffic and incidents.
- Allows for reducing air pollution and noise from road traffic, resulting in improved public and environmental health.

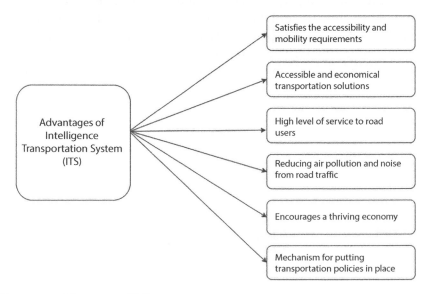

Figure 13.3 Advantages of ITS.

- Encourages a thriving economy and efficient commodities and freight transportation.
- Offers a mechanism for putting transportation policies in place, such as demand management or prioritizing public transportation.

13.6 ITS Applications

ATMS, emergency management system (EMS), advanced traveler information system (ATIS), and advanced passenger transportation management system (APTMS) are the four primary categories of ITS depending on its use in various elements of transportation management [35]. Table 13.1 shows major applications of ITS and its description.

ATIS uses the Internet, radio, television, telephones, and other technologies to help drivers and passengers make educated choices on the best routes, modes of transportation, and trip departures. This system enables

Table 13.1 Major applications of ITS.

S. no.	Major applications of ITS	Description
1	Advanced Traffic Management System	Multiple platforms, including message systems, detection and tracking, Surveillance, connectivity, and so on, should all be integrated into a single unified interface that forecasts traffic conditions for more effective planning
2	Advanced Traveler Information System	To assist motorist, uses the web, radios, TV, telephony, and other technology, passengers make educated choices on the best routes
3	Emergency Management System	Primarily designed to assist in emergencies
4	Advanced Passenger Transportation Management System	Able to relate signals controls to an integrated network for improved data collection and analysis, resulting in better traffic signal timing.

which was before & route feedback to the user, that is useful in a variety of situations [36]. Traffic information assists travelers in avoiding traffic congestion, reducing travel time, and improving traffic networks' operation. Pretrip data facilitates the utilization of highways and enables commuters to make more educated transportation decisions [35, 36].

ATMS integrates multiple systems, such as messaging systems, vehicle detection, CCTV, communication, and so on, into a single, consistent platform, which anticipates road congestion for further efficient implementation and designing actual information [37]. Expressway information systems, adaptive vehicle control, and emergency preparedness systems, among others, enable real-time reactions to changing situations [38].

The sophisticated public transportation system (APTS) is a transportation management system that uses advanced technology and information technology to maximize the productivity and effectiveness of publicly transit systems [39].

It contains a bus arrival notification system, an authentic passenger information system, and an autonomous vehicle positioning system, technologies that give buses precedence at signalized junctions, and an automatic vehicle location system [36]. Traditional light rail public transportation is being replaced by bus rapid transit (BRT) systems. Instead of using railways or subways, the BRT method incorporates vehicles that travel in a fixed path that follows the path of the said route. Pune may be the first place in India to introduce a BRT system [39, 40].

The ITS's current study topic is EMS. This mode of transportation is primarily designed to assist in emergencies [41].

Users may utilize the ATIS to get travel-related data to help with route choices, journey times, and congestion avoidance. This may be accomplished through different technologies due to its many digital signal, which require actual communication of data regarding blockages, traffic congestion, and deaths, as well as a separate route when roads are closed or repairs are being made, Through infotainment system with GPS and a website with a colored routing protocol of yellow, red, and green to illustrate freeway traffic volumes [42]. ATIS' main purpose is to bring motorists up to abreast of new traffic data and to enhance movement of vehicles. Regarding news, the web, texting, and radios are just used. These protocols have only been applied in a few Indian towns [41, 43]. The ATIS system has been deployed in Bangalore, Hyderabad, Chennai, and Delhi.

The ATMS, which helps manage heavily congested situations, are some of the most well there and commonly used divisions of ITS [44]. This method was discovered using two algorithms: data fusion for congestion analysis and control plan selection [41]. GPS is essential for gathering traffic data, such as journey duration, delay, and speed [42].

Various organizations worldwide have built ITS applications customized to provide transportation solutions for particular purposes. Congestion and demand management, as well as improved road safety and infrastructure, have become more reliant on ITS in wealthier countries [43].

13.7   ITS Across the World

Singapore has been at the forefront of intelligent traffic management for more than a decade. In recent years, there has been more investment in intelligent traffic management, with a $12 billion budget in 2018 for transportation development. Thus, high cost is attributed to linked infrastructure with sensors for traffic control and preventive maintenance [44]. On the other hand, the city is focused on connected cars, with ambitions to introduce autonomous buses as early as 2022. Singapore's information on ITS has made the place one of the lowest crowded in the world, because of its potential toward the people with authentic driving directions [45].

New York is famed for its great mass transit and is working hard to establish an ITS. At over 10,000 intersections outfitted by video surveillance, New York has already committed to linked architecture and adaptable signs. The town is also exploring with connected cars, launching a pilot program to gather and analyze data from network vehicles for various devices and services. Connected car software and hardware will be used in V2X initiatives to enhance genuine traffic and safety control [46]. Table 13.2 shows major cities in the world using ITS and its description.

London is taking its infrastructures into the future by adopting 5G technology. Sitraffic Fusion was been declared by the metropolitan area, a traffic control system that incorporates data from linked autos. This program will be an important part of London's real-time optimization system, which previously used data from integrated sensors, compatible devices, and other resources to regulate highways [47].

Mass transit and passenger mobility are important in Paris. While the city strives to improve its quasi transportation system, the government recognizes the value of automobiles [48]. Paris is replacing its entire bus fleet with electric cars to become a green city. Paris is likewise concerned with traffic safety and management, which has resulted in a 40% reduction in traffic deaths since 2010. It would spend €100 million on infrastructure

Table 13.2 Major cities using ITS.

S. no.	Major cities using ITS	Description
1	Singapore	Intelligent Traffic Management, sensors for traffic control, connected cars
2	New York	Linked infrastructure and adaptive signals, linked automobiles, V2X Projects
3	London	Sitraffic Fusion, Real Time Optimizer system, embedded cameras, linked equipment
4	Paris	Non-motorized transportation infrastructure, linked and driverless cars
5	Beijing	Connected automobile technology and integrated sensors and cameras. Extensive data analysis and artificial intelligent
6	Berlin	In-ground sensors, vehicle detection system, presence of electric vehicles, e-car sharing and E.V. fleets
7	Seoul	Use big data, collects data from sensors across the city to predict and reduce traffic jams, 5G infrastructure and linked cars
8	Barcelona	Smart mobility, actual data analysis using large amounts of information, 5G Infrastructure

adaptation to enable broad deployment of linked and driverless cars as part of its previously developed ITS [46].

China's A.V. industry is expected to be the world's biggest, and the nation recently deployed autonomous trucks to reduce the spread of the corona virus pandemic while transporting critical supplies [48]. To monitor traffic and road conditions, Beijing has used these and other connected automobile technology and integrated sensors and cameras. To keep the city moving, comprehensive data analytics and artificial intelligence are used to power smart transport monitoring systems [49].

In 2015, Berlin approved an innovative city plan, and since then, it has achieved great progress. Its most current mobility program focuses on traffic crossings with in-ground sensors. This automobile detecting device will send real-time traffic information using communication devices for optimal signalized intersections. The province likewise wants to increase the use of hybrid automobiles, launching Be Mobility, a €9 billion plan aimed at adopting e-car sharing and electric car deployments, while also increasing the metro's supercharger infrastructure [47].

In 2010, The Global Smarter & Sustainability Societies Organization was formed in Seoul to promote exporting transportation innovations and long-term growth, establishing Seoul as a Smart transit is a leading player. The metro administration is investing in knowledge to build, store, analyze, and apply big data aiming to answer metro issues. To detect and eliminate traffic congestion, the ITS gathers data from the sensors across the city [48]. In addition, the technology may be used to warn people about possible dangers and give actual alternate routes. The population is built 5G infrastructure, including linked autos, in keeping with its legacy of creativity.

Barcelona is a professional in the development of smart mobility, itself with computational center developing unique technologies for huge amount of factual data like smart urban, linked automobiles, and self-driving cars [49]. Although the city also has a comprehensive traffic management system dependent on asphalt with infrastructure sensors [50], V2X is on the way. Barcelona is spending on 5G infrastructure as well as a live lab that is pioneering the development of connected and autonomous vehicles besides efforts to realize V2X technology hub. The city has established itself as a trailblazer in urban transportation because of CLASS's achievements in sophisticated data analytics and developments in linked cars and 5G infrastructure.

13.8   India's Status of ITS

The idea of implementing ITS in India has been around for some time. Finally, the Automotive Industry Standard 140 (AIS-140), which contains ITS, was created to modify the scope of transportation in India [51]. Many cities in India, mainly metro areas, have had ITS ideas in the works for some time. However, each endeavor focused on just a few ITS components rather than adopting them in their totality [52]. While some programs were somewhat successful, most of them failed to have a meaningful effect.

Some of the fundamental contributing causes of ITS failing versions were a lack of adequate installation, regulation, and legal compliance [53].

The ARAI came up with AIS-140 as one of the successful ITS implementations. Additionally, optimizing toll collecting booths to become cashless using RFID tags comes within the ITS umbrella and has proven to be a successful endeavor [54].

13.9 Suggestions for Improving India's ITS Position

a. As India recently began to use ITS to control traffic, more comprehensive and immediate integration of advanced software and principles into conventional traffic management is needed. ITS models have already been installed in a number of areas, with the emphasis on discrete park information systems, region signal management, and enhanced tolls. On the other hand, India has a small number of fully developed ITS applications with traffic control centers. Even though various technologies have been effectively implemented in developed nations, duplicating similar systems in India faces considerable challenges.

b. In India, an ITS system must be cost-effective, easy to implement, and need minimum human contact. Because traffic cops on the ground are few and far between, technology should be tailored to the local population's dialect. Most ITS systems were installed at excessive costs paid by city governments or local traffic police in industrialized nations. In India, the expensive cost of ITS systems makes them less critical to local governments."

c. There are no defined design, structural standards, or legislation for ITS throughout the nation. ITS systems in India cannot simply be patterned after successful ITS solutions from other nations due to their unique driving behavior and surroundings.

d. India has been unwilling to adopt current technologies in ITS. Geographic, cultural, and demographic differences between the United States and other countries are apparent explanations. The country's fastest-growing automobile population occurs in poor-infrastructure urban areas. It is a self-contained process: the higher the load on the infrastructure, the more difficult it is to install the technology.

In most cases, implementing available technology will have no impact.

e. India needs a "desi" version of existing technology rather than a ready-to-use deployment. Local traffic management companies will be more significant than international firms. What worked in London may not be the ideal choice for Bangalore, and what worked in Sydney might not be the best option for Delhi.

13.10 Conclusion

The transportation business is significant to India's economy. India's total growth depends on this industry. The administration has prioritized implementing legislation to ensure the country's rapid development of excellent infrastructure. Rapid automobile development, along with expanding population, rural-to-urban migration, and economic prosperity, has put immense demand on transportation infrastructure in India's cities and towns, especially traffic control strategies. The employment of ITS seems to be a realistic solution for advanced traffic control and management, based on international experiences and best practices in nations such as the United States, Dubai, Canada, European countries, the United Kingdom, and others. ITS must overcome several physical, social, economic, and administrative challenges to succeed in India. Even though India has only just begun its adventure in ITS, there is a pressing need to extend the usage of ITS applications, first by developing an ITS-based mobility plan and later by making it mandatory in urban areas.

References

1. Tejas, R. and Devadas, V., Intelligent transportation system in India-A review. *J. Dev. Manage. Commun.*, 3, 100, 2015.
2. Boukerche, A. and Wang, J., Machine learning-based traffic prediction models for Intelligent Transportation Systems. *Comput. Networks*, 181, 107530, 2020.
3. Maitra, B. and Sadhukhan, S., Urban public transportation system in the context of climate change mitigation: Emerging issues and research needs in India, in: *Mitigating Climate Change*, pp. 75–91, Springer, Berlin Heidelberg, 2013.
4. Verma, A., Sreenivasulu, S., Dash, N., Achieving sustainable transportation system for Indian cities–problems and issues. *Curr. Sci.*, 9, 1328, 2011.

5. Kumar, A. and Anbanandam, R., Assessment of environmental and social sustainability performance of the freight transportation industry: An index-based approach. *Transp. Policy*, 13, 132, 2020.

6. Pustokhin, D.A., Pustokhina, I.V., Shankar, K., Challenges and future work directions in healthcare data management using blockchain technology, in: *Applications of Blockchain in Healthcare*, pp. 253–267, Springer, Singapore, 2021.

7. Mallik, S., Intelligent transportation system. *Int. J. Civ. Eng. Res.*, 4, 367, 2014.

8. Singh, G., Bansal, D., Sofat, S., Intelligent transportation system for developing countries-a survey. *Int. J. Comput. Appl.*, 85, 3, 2014.

9. Dubey, A., Lakhani, M., Dave, S., Patoliya, J.J., Internet of Things based adaptive traffic management system as a part of Intelligent Transportation System (ITS), in: *International Conference on Soft Computing and its Engineering Applications*, IEEE, Bengaluru, India, pp. 1–6, 2017.

10. Selvathi, D., Pavithra, P., Preethi, T., Intelligent transportation system for accident prevention and detection, in: *International Conference on Intelligent Computing and Control Systems*, IEEE, Madurai, India, pp. 442–446, 2017.

11. Bojan, T.M., Kumar, U.R., Bojan, V.M., An Internet of Things based intelligent transportation system, in: *International Conference on Vehicular Electronics and Safety*, IEEE, Hyderabad, India, pp. 174–179, 2014.

12. Illahi, U. and Mir, M.S., Maintaining efficient logistics and supply chain management operations during and after coronavirus (COVID-19) pandemic: Learning from the past experiences. *Environ. Dev. Sustain.*, 23, 11157, 2021.

13. Khadhir, A., Kumar, B.A., Vanajakshi, L.D., Analysis of global positioning system based bus travel time data and its use for advanced public transportation system applications. *J. Intell. Transp. Syst.*, 25, 58, 2021.

14. Gohar, M., Muzammal, M., Rahman, A.U., SMART, T.S.S., Defining transportation system behavior using big data analytics in smart cities. *Sustainable Cities Soc.*, 41, 114, 2018.

15. Veres, M. and Moussa, M., Deep learning for intelligent transportation systems: A survey of emerging trends. *Trans. Intell. Transp. Syst.*, 21, 3152, 2019.

16. Cristobal-Salas, A., Tchernykh, A., Nesmachnow, S., García-Morales, C.Y., Santiago-Vicente, B., Herrera-Vargas, J.E., Solís-Maldonado, C., Luna-Sánchez, R.A., ETL processing in business intelligent projects for public transportation systems, in: *International Conference on Supercomputing*, Springer, Cham, pp. 42–50, 2019.

17. Gharehbaghi, K. and Farnes, K., Process automation in intelligent transportation systems (ITS). *Int. J. Mach. Learn. Comput.*, 8, 294, 2018.

18. Yang, W., Wang, X., Song, X., Yang, Y., Patnaik, S., Design of intelligent transportation system supported by new generation wireless communication technology, in: *Intelligent Systems: Concepts, Methodologies, Tools, and Applications*, pp. 715–732, IGI Global, China, 2018.

19. Mazinan, A.H. and Sarikhani, M., Providing an efficient intelligent transportation system through detection, tracking and recognition of the region

of interest in traffic signs by using non-linear SVM classifiers in line with histogram oriented gradient and Kalman filter approach. *Sadhana*, 39, 1, 27, 2014.

20. Li, B., Analysis and evaluation of application technology of smart transportation system, in: *International Symposium on Traffic Transportation and Civil Architecture*, IEEE, Suzhou, China, pp. 65–69, 2021.

21. Liu, Y., Big data technology and its analysis of application in urban intelligent transportation system, in: *International Conference on Intelligent Transportation, Big Data & Smart City*, IEEE, Xiamen, China, pp. 17–19, 2018.

22. Malygin, I., Komashinsky, V., Tsyganov, V.V., International experience and multimodal intelligent transportation system of Russia, in: *International Conference Management of Large-Scale System Development*, IEEE, Russia, Moscow, Russia, pp. 1–5, 2017.

23. Mittal, P. and Singh, Y., Development of intelligent transportation system for improving average moving and waiting time with artificial Intelligent. *Indian J. Sci. Technol.*, 9, 1, 2016.

24. Chowdhury, M. and Sadek, A.W., Advantages and limitations of artificial Intelligent. *Artificial Intelligent App. to critical transportation*, vol. 6, p. 360, 2012.

25. Deng, Y., Application of intelligent transportation system based on 5G, in: *International Conference on Human-Machine Interaction, Association for computing Machinery*, Guangzhou, China, pp. 35–39, 2021.

26. Kuo, C.C., Lin, J.N., Wu, S.H., Multi-system integration scheme for intelligent transportation system applications. *Int. J. Wirel. Netw. Broadband Technol.*, 4, 21, 2014.

27. Nagar, D.M. and Baghela, V., Efficient emergency message DHC broadcasting in vehicular ad hoc networks, in: *Advances in Information Communication Technology and Computing*, pp. 57–63, Springer, Singapore, 2021.

28. Kapetanakis, K., Zampoglou, M., Malamos, A.G., An MPEG-DASH methodology for QoE-Aware Web3D streaming. *Int. J. Wirel. Netw. Broadband Technol.*, 4, 31, 2014.

29. Rajput, S.A., Intelligent urbanism guiding the smart city region development: Case study of bhopal, in: *Smart Cities—Opportunities and Challenges*, pp. 423–442, Springer, Singapore, 2020.

30. Parulekar, G.D., Desai, D.B., Gupta, A.K., Vehicle detect and monitor techniques for intelligent transportation—A survey. *IOSR J. Mech. Civ. Eng.*, 13, 57, 2015.

31. Zhu, L., Yu, F.R., Wang, Y., Ning, B., Tang, T., Big data analytics in intelligent transportation systems: A survey, in: *IEEE Transactions on Intelligent Transportation Systems*, pp. 383–398, Taipei, 2018.

32. Sun, N., Intelligent transportation system planning in the age of artificial Intelligent, in: *International Conference on Environmental and Engineering Management, E3S Web Conf*, Beijing, China, pp. 1036–1044, 2021.

33. Vasudevan, M., Townsend, H. *et al.*, *Identifying real-world transportation applications using artificial intelligent (AI)-real-world AI scenarios in transportation for possible deployment*, pp. 6–8,Washington DC, 2020.

34. Dongbin, Z., Yujie, D. *et al.*, Computational Intelligent in urban traffic signal control: A survey, in: *Transactions on Systems, Man, and Cybernetics*, pp. 485–494, IEEE, Malaysia, 2011.

35. Naveen, K.H., *Artificial intelligent: Learning outcomes of classroom research*, pp. 82–86, Madurai, 2021.

36. Ibáñez, J.G., Zeadally, S., Castillo, J.C., Sensor technologies for intelligent transportation systems. *Sensors*, 4, 1212, 2018.

37. Raul, M., Discovering mobility patterns on bicycle-based public transportation system by using probabilistic topic models, in: *Ambient Intelligent-Software and Applications*, pp. 145–153, Springer, Salamanca, Spain, 2012.

38. Luc, S. and Mantaras, R.L., The Barcelona declaration for the proper development and usage of artificial intelligent in Europe. *AI Commun.*, 6, 485, 2018.

39. Agarwal, P.K., Gurjar, J., Agarwal, A.K., Birla, R., Application of artificial Intelligent for development of intelligent transport system in smart cities. *Int. J. Transp. Eng. Traffic Syst.*, 2, 20, 2015.

40. Shit, R.C., Crowd intelligent for sustainable futuristic intelligent transportation system: A review, in: *IET Intelligent Transport Systems*, pp. 480–494, IEEE, Hyderabad, India, 2020.

41. Chavhan, S., Gupta, D., Chandana, B.N., Chidambaram, R.K., Khanna, A., Rodrigues, J.J., A novel emergent Intelligent technique for public transport vehicle allocation problem in a dynamic transportation system, in: *Transactions on Intelligent Transportation System*, pp. 5389–5402, IEEE, Chennai, India, 2020.

42. Puri, V., Le, C.V., Kumar, R., Jagdev, S.S., Fruitful synergy model of artificial intelligent and Internet of Thing for smart transportation system. *Int. J. Hyperconnect. Internet Things*, 4, 43, 2020.

43. Mishra, P., Shampritha, C., SanthoshKumar, P.S., Quadri, W., Intelligent transportation system based on IoT. *Intelligent*, 6, 6, 2019.

44. Sharma, S., Nanda, M., Goel, R., Jain, A., Bhushan, M., Kumar, A., Smart cities using Internet of Things: Recent trends and techniques. *Int. J. Innov. Technol. Exploring Eng.*, 8, 24, 2019.

45. Mangla, M., Kumar, A., Mehta, V., Bhushan, M., Mohanty, S.N., *Real-life applications of the Internet of Things: Challenges, applications, and advances*, Apple Academic Press, New York, 1st Edition, pages 536, ISBN: 9781003277460, 2022. https://doi.org/10.1201/9781003277460 2022.

46. Kapur, P., Chavhan, S. *et al.*, Industry 4.0: Hyperloop transportation system in India, in: *Computational Intelligent for Sustainable Transportation and Mobility*, pp. 113–125, Bentham Science, Sharjah, UAE, 2021.

47. Jing, B., Yu, X., Du, W., Design of intelligent public transportation system based on ZigBee technology. *Int. J. Performability Eng.*, 14, 483, 2018.

48. Chavhan, S., Gupta, D., Chandana, B.N., Chidambaram, R.K., Khanna, A., Rodrigues, J.J., Cognitive and ontological modeling for decision support in the tasks of the urban transportation system development management, in: *International Conference on Information, Intelligent Systems and Applications*, IEEE, Corfu, Greece, pp. 1–5, 2015.

49. Heaslip, K.K., Khalilikhah, M., Fuentes, A., Intelligent transportation system security: Hacked message signs. *SAE Int. J. Transp. Cybersecur. Priv.*, 1, 75, 2018.

50. Samant, P., Bhushan, M., Kumar, A., Arya, R., Tiwari, S., Bansal, S., Condition monitoring of machinery: A case study, in: *International Conference on Signal Processing, Computing and Control*, IEEE, Solan, India, pp. 501–505, 2021.

51. Mohammed, E., Glaser, S., Bond, A., A framework for testing independence between lane change and cooperative intelligent transportation system. *PLoS One*, 15, 2, 2020.

52. Goel, R., Jain, A., Verma, K., Bhushan, M., Kumar, A., Negi, A., Mushrooming trends and technologies to aid visually impaired people, in: *International Conference on Emerging Trends in Information Technology and Engineering*, IEEE, Vellore, India, pp. 1–5, 2020.

53. Subramanyam, M. and Kumar, A., RLTS: Recommendation for local transportation system using ambient intelligent, in: *International Conference on Emerging Research in Electronics, Computer Science and Technology*, IEEE, Mandya, India, pp. 65–69, 2015.

54. Priyanka, E.B., Thangavel, S., Madhuvishal, V., Tharun, S., Raagul, K.V., Krishnan, S., Application of integrated IoT framework to water pipeline transportation system in smart cities, in: *Intelligent in Big Data Technologies— Beyond the Hype*, pp. 571–579, Springer, Singapore, 2021.

14

Evolutionary Approaches in Navigation Systems for Road Transportation System

Noopur Tyagi, Jaiteg Singh and Saravjeet Singh*

Chitkara University Institute of Engineering and Technology, Punjab, India

Abstract

In modern navigation systems, integration of navigation systems and the Global Positioning System has been successfully used in practice. To find the position in terms of longitude and latitude, different methods are being used. These navigation systems consist of Global Positioning System receivers, processing devices, algorithms, and spatial databases. Differential evolution is a technique that optimizes a problem repetitively trying to enhance a solution concerning a given measure of quality. OpenStreetMap is a nonpropriety spatial database and very frequently used for research and development in navigation systems. OpenStreetMap has many quality issues as a volunteer spatial database and to identify and correct these quality issues, many algorithms have been developed. To process and use Global Positioning System output for localization in navigation systems, the research community developed many different methods, such as navigation based on a bio-inspired algorithm. The hybrid genetic algorithm combines genetic algorithm and Tabu algorithm based on different mathematical models and techniques to solve dynamic multiobjective problems. This chapter provides a survey of evolutionary techniques used in navigation to create opportunities for analysts and researchers seeking to understand the broad pattern of different algorithms used in the navigation system.

Keywords: Global positioning system, differential evolution, navigation, genetic algorithm, evolutionary algorithm, metaheuristic

Corresponding author: saravjeet.singh@chitkara.edu.in

Ashok Kumar, Megha Bhushan, José A. Galindo, Lalit Garg and Yu-Chen Hu (eds.) Machine Intelligence, Big Data Analytics, and IoT in Image Processing: Practical Applications, (341–368) © 2023 Scrivener Publishing LLC

14.1   Introduction

Over the years, people have used many traditional techniques to navigate from across the globe. They relied on stars, compasses, maps to travel between different locations; it helped to prevent people from getting lost. Navigation is the combination of science and art to determine the position of different vehicles and guiding them to a specific destination [1]. The navigation is used in every aspect of life without giving a second thought. Everyone has experienced navigation—finding the way in a city or going from a single point to a different point. In these cases, markers, landmarks, hints, or technology are used to guide directions gradually. From thousands of years, solar physics to geology—science has a concrete approach for navigation techniques. It determines the position, distance by avoiding collisions, conserving fuel, and meeting schedules. Longitude and latitude are essential in recognizing the craft of navigation [2]. Latitude and longitude are imaginary lines on the Earth's surface, which are used to determine the location of a particular point or location. Latitude forms the circle around the Earth, and longitude runs from North Pole to South Pole, and it lies in between 0 and 360 degrees. The Earth is divided in two equal halves known as the equator.

The earliest navigation methods involved observing the constellation to mark their position by using stars, moon, horizons, and sun then calculate their position out in the open ocean known as celestial navigation [3]. Today, some space agencies, like NASA are using celestial navigation for many of their missions outside the Earth's atmosphere. Compass is another familiar tool for navigation on land, at sea, and in the air. It works by responding and detecting the Earth's natural magnetic field. In the last century, radar, long-range navigations, the gyroscopic compass, and Global Positioning System (GPS) aided progress quickly [4].

Modern years have seen a new approach of gathering topographical information and data via the people. OpenStreetMap (OSM) is the best instance of this approach [5]. Over the past few years, more individuals consider it as a thinkable alternative to certified and commercial information. Freely available geographic information and the availability of richer data induced the well-built interest of people, analysts, and researchers. Traditionally, government agencies, mapping agencies, and related agencies were the only source of spatial data. High fees, copyrights, licenses, and some other restrictions were imposed if a user wants to seek any information. The Internet and social media have changed everything; an OSM project is one of them. The agenda was simple to build and maintain spatial data for the end-user by the end-users. Initially, the focus was on streets and maps; now it contains a very rich variety of content being mapped by thousands of volunteers from all over the world [6].

14.1.1   Navigation System

Navigation is a method to determine the location of an object by calculating the speed, time, position, and direction by measuring various changes in the state (such as acceleration and velocity). Today's navigation system uses GPS, satellites, and other devices to pinpoint an object's location [7]. To determine positional coordinates, GPS uses radio waves emitted by a constellation of orbiting satellites. Figure 14.1 describes the block diagram of the global navigation satellite system.

Major Components of Navigation System are as below:

- **Standard positioning services (SPS):** All GPS users use L1, L2, and L5 frequencies, timing and positioning is dependent on SPS. For military purposes, it is mandatory to have highly accurate and precise positioning for that Precise Positioning

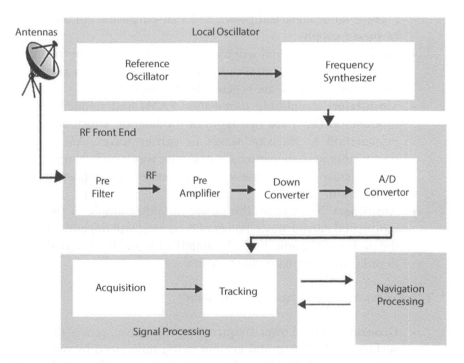

Figure 14.1 Block diagram of global navigation satellite system.

Service broadcasts at the GPS L1 and L2. It is necessary to encrypt navigation data because both frequencies contain a precision (P/Y) code ranging signal for authorized users; these signals can be transmitted through antennas and other devices.

- **Radio waves:** It is an electric current that is responsible for transmitting and receiving electromagnetic waves. By the use of an antenna transmitter emits this field outward receiver receives it and converts it into the desirable radio frequency. There are different types of electromagnetic radiation that are generated artificially to transmit signals to fixed and mobile communication, broadband, radar, navigation.

- **Antennas:** Antennas receive and transmit the different types of radio waves. These are available in different shapes and sizes

- **Radio transmitters:** Several elements combine together to produce radio waves. These radio waves hold useful information in different forms of data such as audio, video, or digital.

- **Power supply:** To operate the transmitter, it is necessary to provide some electrical power and that purpose is fulfilled by power supply.

- **Oscillator:** Transmitter will transmit only at a particular frequency; alternating current and a sine wave which is also known as a carrier wave is generated by oscillator.

- **Modulator:** Amplitude modulation (AM) and frequency modulation (FM) are two types of modulation which add information to the sine waves or carrier waves. AM is responsible to change intensity of carrier waves and changes in frequency of the carrier wave is dependent on frequency modulation.

- **Amplifier:** To increase the power amplifier amplifies the modulated carrier wave. Power of broadcast will depend on the power of the amplifier. The amplified signal is converted to radio waves by using the antenna.

Types of waves used in navigation system:

- **Ground waves:** Lower-frequency waves travel across the surface of the Earth and are not being influenced by outside factors [8]. The lower frequency will assure the signals will travel as far as possible. Ground waves travel does not change its route easily, so it is a reliable wave for navigation. The ground wave frequency ranges from 100 Hz to approximately 1 MHz.

- **Sky waves:** High-frequency waves are refracted by the ionosphere and transmit signals, the range of this frequency lies in between 1 and 30 MHz due to high frequency [9]. It is fun and reliable to use sky waves for navigation but gradually high frequency is substituted by more authentic satellite communication.

- **Space waves:** Very high frequency. Passing through the ionosphere straightly radio waves of 1.5 MHz and above are considered space waves. Most navigation systems use those signals that originated as space waves. The ionosphere creates some propagation error, so to minimize those errors, a navigation signal is used before it reaches the ionosphere.

- **Nondirectional radio beacons:** It marks or locates the Instrumental Landing System (ILS) approach [10]. It is one of the ancient aviation navigation systems that broadcast the signals and emits a constant signal in every direction also known as an Omni-directional beacon.

Global Positioning System

GPS is a satellite-centered navigation system that comes up with location, velocity, and time synchronization. GPS helps to identify the current location and find the direction between two positions. These satellites move in a particular orbit and to find the location, signals from at least 3 satellites are required. By using these three satellites, an accurate location of a geographic entity can be obtained. A fourth satellite is required to measure the time. Figure 14.2 shows a sample scenario to find the location using the signals from four different satellites. Trilateration is a mathematical process used by GPS receivers, it requires four distances instead of three to form a unique solution by involving spheres [11].

Theory of Relativity

Einstein's theory of special relativity is popular to predict that time changes related to speed. If there is no accuracy with time, then there would be some lapse in GPS positioning. The satellite sends updated information to the entire GPS receiver based on the location of four satellites above it

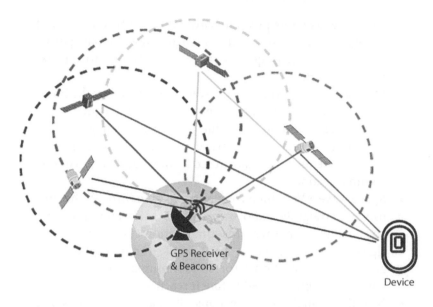

Figure 14.2 Trilateration of GPS.

and the distance between those satellites and receivers by analyzing radio signals transmitted from GPS satellites. The atomic clock is integrated into both the system receiver and transmitter. Several satellites will decide the accuracy of trilateration. The transmitter transmits the signal at time T2, and the receiver receives the signal at time T1, so distance can be calculated using equation 14.1.

$$(T2 - T1) * C \dots \dots .. \qquad (14.1)$$

where C is the speed of light.

Instead of using an atomic clock, smartphones have a crystal clock in them. By solving some coordinates, the exact location (x, y, z, T offset) can be retrieved. T offset can be calculated using equation 14.2. T offset will remain the same for every satellite.

$$T \textit{ offset} = \text{“Atomic Clock – Crystal Clock”} \dots \dots \qquad (14.2)$$

The GPS system consists of three segments as shown in Figure 14.3. Space segment consists of 24 satellites and communication links, the control segment consists of several monitoring stations and master control stations and the user segment consists of GPS receivers that are located

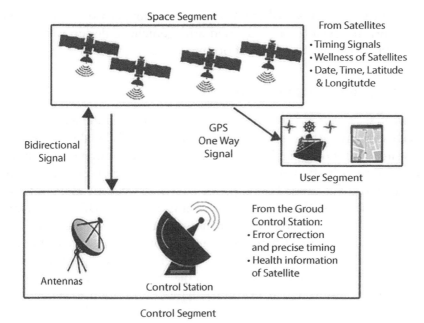

Figure 14.3 Segment of GPS.

in various devices such as ships, cars and various small devices like cell phones.

14.1.2   Genetic Algorithm

John Holland coined the concept of a meta-heuristic algorithm that finds the best solution called a Genetic Algorithm (GA) which is founded on the principle of evolution and biogenetics inspired by Darwin's theory that mimics some aspects of bio-inspired evolution [12]. By using genetic inheritance GA models the natural evolution which emulates the mechanism of evolution in solving optimization problems [13]. The basic element of GA is the chromosome which contains a parameter called genes. A string of all genes of one individual is known as a chromosome which is coded in the binary form [14, 15]. Several techniques using operators are implemented to perform the algorithm and find the main evaluating factor that is the fitness of an individual. The objective function defines how the problem is solved. The way of implementation decides the result of the randomly generated population; if proper techniques are implemented in GA then the performance of GA fitness value will be raised. Strength towards the problem can be evaluated with the fitness value of chromosomes [16].

By using the mechanism of initialization, selection, crossover, and mutation algorithm explores a predefined arrangement of design variables in the search space with the focus of optimization.

14.1.3　Differential Evolution

Differential Evolution (DE) algorithm is a novel Evolutionary Algorithm (EA) based on GAs first introduced in 1997 [17]. Traditionally DE explores the search with no particular information about the search direction that results in a stochastic behavior that aims to discover an optimal answer over a continuous range [18, 19]. In achieving stochastic global optimization DE strategies have a noteworthy impression on DE performance. Outcome involves randomness and has uncertainty. For handling optimization problems meta-heuristic search algorithms are one of the best approaches. If there is no satisfactory problem-specific algorithm then it can be generally applied. These days, in the area of research and technology different bio-inspired algorithms [20] for optimization, are in demand. Evolutionary computation algorithms refer to the whole family of evolutionary optimization algorithms. GA [14], genetic programming, DE, the evolution strategy, and evolutionary programming can be mentioned as the main algorithms in the evolutionary computation domain. Both the GA and DE algorithms use the same operator: selection, crossover, and mutation. To construct the better solution GAs rely on crossover whereas DE relies on mutation and other two components; those are chosen strategies and control parameters. In fact, the mutation is the main component that differentiates the further population algorithm from DE. DE is designed to work for continuous objective functions with multi-dimensional valued candidate solutions and does not make use of gradient information in the search whereas GA uses a sequence of bits. Sensitivity to control parameters and other issues cause stagnation and it becomes inefficient to solve a particular problem in a precise domain space. DE with modification improves a population's proficiency for developing new suitable offspring based on the recent distribution with the area space. Exploration and exploitation are two vital processes that drive the evolution of the differential population. Exploration explores the new solution in new space whereas exploitation works on the selection process for the fitness of new offspring from the refinement of existing solutions [21]. Selecting a base solution and applying some strategies to add mutation to create new candidate solutions and other random solutions from the population from which the amount of mutation is calculated is called difference vector, then DE produces new parameter vectors by calculating the alteration between two populations vectors. If the offspring has a better objective function then it is replaced by the base solution. Each parent created a single offspring and in the current population if the offspring is healthier than the parent then it replaces the parent.

14.2   Related Studies

Throughout the years, researchers have suggested numerous types of routing algorithms using genetic information with DE algorithms. The basic objective of this section is to look at the research activities in the field of navigation, GAs, and DE.

GA based route planning algorithm which has the ability to tackle the problem when there is more than one optimal solution or path [18]. More than 1000 nodes in a particular area were tested by using this GA-based route planning algorithm for emerging an intelligent route planning structure. Using DE and GA to resolve the vehicle routing problem with Simultaneous Delivery and Pick-up and Time Windows [22]. By setting parameters it can be transformed into a classical vehicle routing problem. Combination of DE theory and GA with a Hybrid Optimization Algorithm (HOA) was suggested. Genetic operators and self-adapting differential evolution crossover operators were utilized to circumvent premature convergence, according to the discoveries.

The problem of selection of a route to a given source and destination on the real map under a static environment. For selecting the best next neighbor GA was proposed with variable length chromosomes, a new suggested selection method and tournament method are used and compared. To study the behavior of several crossovers and different mutation operators on route navigation systems are used and tested. It showed that calculating an optimal route is more efficient by using the suggested route navigation system. A routing algorithm built on DE to enrich the LEACH routing protocol (arranges the collection in such a way that the energy is distributed evenly among all of the sensor nodes in the network) [23]. To improve the multi-objective selection of cluster head algorithms uses a feature of DE and prevents blind nodes. To discover the routes in best routing algorithms based on the shortest path evaluation of each path are done. If the path becomes blocked or overloaded, further optimization factors will be studied to provide a better solution [24]. A GA for providing better alternative paths instead of overloaded paths was forwarded. Solution improvement can be obtained by searching facilities provided by crossover and mutation and increasing the speed of convergence [25].

A GA approach to discover the shortest path for driving time in a route guiding system [26]. Influential factors of traffic flow should be considered inefficient route guidance. To achieve optimum solution mutation operators provide the best contribution. It's hard to determine the exact optimal solution in real time using deterministic approaches like the Dijkstra algorithm on navigation systems with restricted processing speed. GA provides approximate solutions in real-time on vast networks.

The DE method was revised, and the concept of jitter factors was introduced, allowing for a better balancing between exploration and exploitation through control settings [27]. Numerous benchmark functions in solving dynamic route planning problems were used to test the algorithm's practicality and efficiency.

Diversified vicinity process with DE is an improvised search operator [28]. By varying the number of sensors, the algorithm has been investigated on various situations of wireless sensor networks. A dynamic control-based mechanism and clustering-based GA with polygamy. Polygamy crossover and dynamic population were discussed by using chromosomes [29]. GA is applied to the route optimization problem. This technique allows quick convergence to best solutions and diversity in the population.

Using a combination of Open Shortest Path First (OSPF) and GA, a method for computing the shortest path for a network to improve the router's performance was introduced [30]. To find an optimal solution, proper setting of the GA gives better results than OSPF. A vector generation scheme that is modified by a new alternative of DE for solving the stagnation problem. Comparing Continuous Optimizers (COCO) framework composed of 24 different benchmark functions was tested [31]. Increasing the convergence rate and removing the inactive problem gives the best result for unimodal and multimodal problems.

A data management-based technique for efficiently computing the shortest path in real-world road maps [32]. The tilemap partitioning included real implementation of a full solution stack on real road maps. For mobile ad hoc networks, a DE algorithm-based ad hoc on-demand multipath distance vector protocol was proposed [33]. In the route recovery process to find the optimal path from diverse paths between starting point and ending point, this algorithm provides backup to avoid any link failure between nodes.

To reduce the processing load on receivers, a satellite selection technique based on the Gibbs sampler was developed [34]. The investigation of a novel hybrid DE algorithm and an integer linear programming formulation with a genetic operator and a fuzzy logic controller for deciphering the multi-trip vehicle routing issue with backhauls and a heterogeneous fleet [35]. To demonstrate its practical viability experiments were designed.

14.2.1 Related Studies of Evolutionary Algorithms

In this chapter, 80 papers regarding different approaches used in navigation systems using The GAs have been examined. To handle optimization or routing difficulties, certain new variants of GA and DE algorithm approaches with local search strategies have been made. Developers have utilized numerous methodologies in a wide range of technological sectors because to the rapid emergence of DE as a modest and strong optimizer, machine intelligence, and cybernetics. The usage of GAs for Dynamic Shortest Path Routing Problems (DSPRP) was investigated [36]. On the dynamic scale, it is improper due to high computation complexity, although bellman-ford and Dijkstra perform well in fixed infrastructure. The shortest path problem is solved using MANET to construct a special ized GA [37]. The study of mobile robot navigation is being carried out in order to better understand path planning strategies and identify research gaps in diverse environmental situations. A step-by-step investigation of traditional and reactive methodologies is being carried out [38]. Different approaches like fuzzy logic, roadmap approach, GA, neural network, and many more are considered for the study. It closes with graphs and tables comparing the frequency of different navigation tactics. The implementation of a GA using python is quick and easy without any need for extra libraries [39]. A paper was proposed to deal with the optimal navigation co scheduling for two dams in the Three Gorges project using a hybrid simulated annealing algorithm. While some experiments showed that the algorithms are feasible, the algorithm used in the paper has not been used in a practical navigation scheduling system due to its complexity [40]. To boost population variety and strengthen local search capabilities, a proximity crossover strategy and greedy backward mutation technique were created [41]. DE optimization combined with TRANSYT traffic simulation software was proposed in a study on the traffic congestion minimization model [42]. To handle traffic assignment problems, a path flow estimator is utilized as a stochastic user equilibrium evaluation. The same result was emphasized in a new edition of GA with local search approach that was created to handle TSP problems by using Android OS integrated with Google API [43]. Python is an interpreted language that supports multiple programming paradigms like object-oriented programming, functional and procedural programming. Similarly, TSP is one of the optimization approaches to find the shortest path routing and for local search techniques hill-climbing methods have been used [44]. Android OS is used for implementation because of its accessibility around the world. To get the geographical location Google API is integrated with android OS. To compare the effectiveness of a Hybrid Genetic Algorithm (HGA) with a simple GA experiment is performed on different locations of several cities [45]. The result showed that HGA performs better than simple GA.

14.3   Navigation Based on Evolutionary Algorithm

DE is an EA [17] and emerging very rapidly among other EAs. In the Road network, there are several routing planning algorithms applied as displayed in Figure 14.4. To find the lowest cost path Dijkstra algorithm is worthwhile to consider because it terminates when the destination node is labeled. A variant of Dijkstra A* uses heuristic search and reduces the computational time. Tabu search is a local-based Meta-heuristic search to discover the best possible key with a run through several iterations. Genetic Algorithm is a population-based metaheuristic, applied to different routing problems. A GA is a search procedure based on principles derived from dynamic natural population genetics. DE is a population-based optimization algorithm that depends on various parameters as crossover and mutation exist in the GA as presented in Figure 14.5. The sort of genetic operators used in a GA determines its performance: Initialization, Selection, Crossover, Mutation, and Replacement. There is no guarantee that any optimization method will generate the best optimal solution, but all evolutionary algorithms, including GAs, can find a near-optimal solution [13]. Techniques

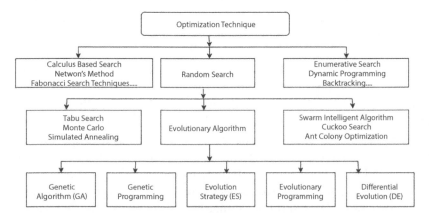

Figure 14.4 Taxonomy of evolutionary algorithms.

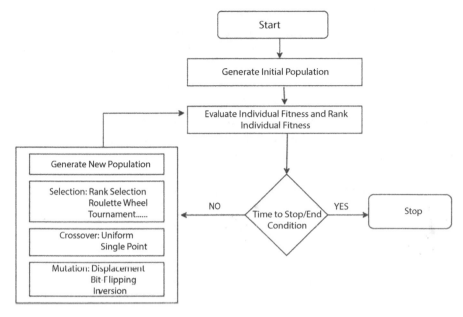

Figure 14.5 Flow of differential evolution and genetic algorithm.

are presented in the form of pseudo-code. Any programming language can easily implement pseudo code. To compute the shortest path in the road, routing modules such as Dijkstra's algorithm and bidirectional search are used. To discover the shortest path from source(s) to destination (t) it is necessary to skip the unimportant nodes from the query and shortcut paths are added to preserve the shortest path from all remaining nodes.

14.3.1   Operators and Terms Used in Evolutionary Algorithms

DE is divided into two stages: startup and evolution. The population is randomly initialized in the first phase, and the evolution-generated population passes through several stages such as selection, crossover, and mutation in the second phase.

Encoding Operators

For most of the computation problem, given information has to be encrypted in the form of bit strings and differentiated to the problem domain accordingly. It should have complete knowledge of the path from the start point to the destination point. Encoding schemes are Binary in which each bit is represented as a string of 0 or 1 to represent the characteristics of the solution as indicated by Table 14.1. Suppose if there is any path from the source point to destination 0 bit else 1 bit and it is easy to implement [46–48]. Octal Encoding scheme, chromosomes are represented in the form of octal numbers ranging from 0 to 7 [49]. Similarly (0–9, A–F) are represented in the hexadecimal encoding scheme. In ordering problems, a permutation encoding scheme is used as described in Table 14.2. When sophisticated values, such as real numbers, are utilized and binary encoding is insufficient, a value encoding method is used as represented in Table 14.3. In the tree, the encoding scheme operator can easily be applied as shown in Table 14.4.

Table 14.1 Binary encoding.

Chromosome i	101010010011010101
Chromosome ii	110010100100101010

Table 14.2 Permutation encoding.

Chromosome i	2 4 6 1 3 5 8 7 9
Chromosome ii	9 3 1 4 2 7 5 6 8

Table 14.3 Value encoding.

Chromosome i	1.2333 4.3435 5.6774
Chromosome ii	ACEFDBAGBCADE
Chromosome iii	(forward)(back)(left)(right)

Table 14.4 Tree encoding.

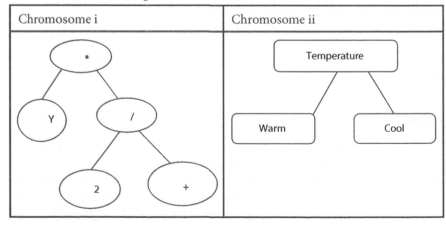

Selection Operators

Whether the particular chromosome (route) will participate and generate the fittest individual or not is determined by the selection operator some-times known as the reproduction operator [50]. Selection pressure decided the convergence rate of DE and GA and two or more candidates for cross-over [51]. Tournament and Roulette wheel selection are some of the selection techniques as presented in Figure 14.6 and Figure 14.7, respectively. For crossover, highly valued individuals have higher chances to replace the existing individual with the fittest generated individual [52]. All possible strings are mapped onto the wheel; a portion on the wheel is allocated by the fitness value in Roulette wheel selection [53]. To get the fittest result, this wheel is then rotated and a randomly generated value will be respon-sible for the generation of a new value. Rank selection uses rank instead of fitness value and it is a modified form of roulette wheel selection tech-nique. The tournament selection technique preserves diversity; no sorting is required for parallel implementation [54].

Crossover Operators

To produce offspring, the genetic information of two or more parents is joined [55]. For the pool of next-generation two or more parents may or may not be replaced in the existing population. Single point, two-point, intermediate, heuristic, arithmetic, shuffle, uniform, partially matched are well-known crossover operators [56]. Offspring from two dominant par-ents chosen by the selection operator will most probably have dominant traits. Single point crossover is shown below in Table 14.5.

Mutation

Eventually, all the individuals in the pool will be the same without muta-tion [57]. Flipping one gene of a candidate diversifies the actual change in

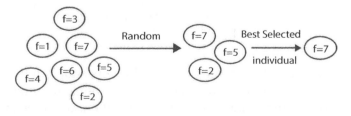

Figure 14.6 Tournament selection.

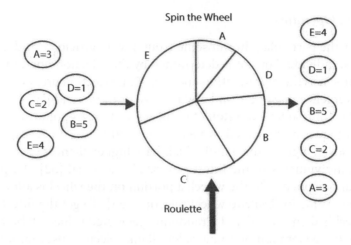

Figure 14.7 Roulette wheel selection.

Table 14.5 Single point crossover (| is the crossover point).

Chromosome 1	1101 | 11000011110
Chromosome 2	1101 | 00100110110
Offspring 1	1101 | 00100110110
Offspring 2	1101 | 11000011110

Table 14.6 Mutation.

Current Descendent i	1001111000011111
Current Descendent ii	1101100100010100
Mutated Descendent i	1000111000011111
Mutated Descendent ii	1101101100010100

the exchange of genetic information. This bias operator is influenced by selection and crossover operators. Displacement mutation, simple inversion mutation, scramble mutation, uniform mutation, and reversing mutation are few mutation techniques. Table 14.6 shows the mutation.

14.3.2   Operator and Terms Used in Evolutionary Algorithm

These DE operations have similar terms and terminologies as other evolution algorithms. Following are the prominent operations used in both DE and Gas. Figure 14.8 presents the cast of characters for genetic algorithm and differential evolution. Comparison between evolutionary algorithm and mathematical programming is indicated in Table 14.7.

Agents: Basic infrastructure or building block of any GA is known as a gene, a set of genes decides the allele set responsible for chromosomes. It may bring a possible solution without being the solution.

Chromosomes: It is subdivided into genes; the number of its genes decides the length of the chromosomes [58]. A possible solution in the genetic algorithm is represented by chromosomes; chromosomes can have variable length and must define one unique solution.

Population: It consists of copies of all the possible solutions of all the genes present in the population generation and can be described as a set of chromosomes. The first generation of the population pool is mostly randomly generated.

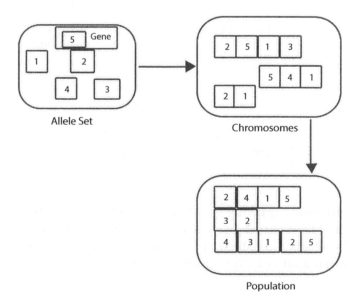

Figure 14.8 Cast of characters for genetic algorithm and differential evolution.

Table 14.7 Comparison between evolutionary algorithm and mathematical programming.

| Terminologies | Evolutionary algorithm | | Mathematical programming |
	Genetic algorithms	Differential algorithm	
Allele	Value of genes, feature value	Value of agent	-
Genes	Building block, bit, feature, character	Part of solution	Coded particular variable
Chromosomes	One solution to a given problem, String	Solution	Coded vector of control variables
Phenotype	The population in real world solution space	Decoded solution	Decoding function
Population	Subset of all possible encode solutions	Subset of all possible encode solution	Set of vectors of control variable
Fitness Function	Evaluation Function	Evaluation Function	Normalized objective function at iteration t
Encoding and Decoding	Process of transforming from phenotype to genotype	Process of initialization of population to get best particle	Decoding function

Fitness Function: By taking the candidate solution to a problem as input and deciding how near a given answer is to the optimal solution of the preferred problem, this function, also known as the evaluation function, generates an output as a fittest result of any problem [59].

Algorithm 1: Outline of DE
Begin

Step 1: *Initialize or generate a population of n chromosomes to determine the objective function and assume that this could be the probable optimal solution.*
Step 2: *Fitness Function: Fitness function f(x) of each chromosome is calculated and applied to each generation.*
Step 3: *Next pool of population: To obtained the desired candidate of next-generation perform the following steps until n chromosomes are obtained*
  i) Selection: Select any best fit parents chromosomes from the population
  ii) Crossover: To get the offspring to apply crossover technique on the selected parents
  iii) Mutation: at desired Position offspring are mutated as per their mutation rate.
Step 4: *Test whether the child is fit for the pool of next- generation or not. Place the offspring in the next generation if it is of desired size and better than already existing chromosomes.*
Step 5: *Termination condition: Stop if the desired condition is satisfied*
Step 6: *Repeat : Go back to fitness function*

End

14.4   Meta-Heuristic Algorithms for Navigation

DE is a meta-heuristic search technique based on population that optimizes a problem by iteratively refining a potential solution through an evolutionary process [60]. To assist the positioning requirement of advanced transport telemetric (ATT) services development of a robust and reliable navigation system is achieved through the high integrity of map-matching algorithms [61]. From positioning technologies, either a probabilistic or a conventional topology analysis map matching algorithms has been generated for vehicle navigation. Input comes from GPS; improved topological and fuzzy logic map matching is high integrity map matching algorithms in terms of link identification and location determination. The status of both positioning and spatial road network has been upgraded due to the application of progressive techniques like Kalman filter [62], belief theory, fuzzy logic, and Dead reckoning. High precision data from the intelligent transport system (ITS) are rarely available. Hidden Markov Model (HMM) is an online map-matching algorithm that is robust to sparseness and disturbance [63]. In complex areas, high required navigation performance is difficult to handle because these algorithms and methods are based on different mathematical models and techniques. In road networks, different route planning algorithms are applied. Some of them are briefly described below:

Dijkstra algorithm: It is the way to discover the shortest path or lowest cost from the source to the destination node in an area [64, 65]. Dijkstra is one of the finest techniques for creating a least weight path tree with the origin, with a computation complexity of O(n2), where n is the number of nodes in the network. If the destination node is found, the algorithm is terminated, which also means the least-cost path is found.

Tabu Search: A local search starts with an initial, feasible solution generated with several iterations [66]. During iterations, the finest solution in the neighborhood is replaced with the existing solution even if the solution cost is increased. In tabu search, a short memory stores attributes of the recently visited solution. It stops after a fixed number of cycles without any improvement in the finest solution. Combination of arithmetic coding differential evolution with Tabu solves difficult problems very reliably [67]. The idea of combining DE with Tabu to handle numerous problems, such as job shop scheduling challenges, is competitive with other state-of-the-art methodologies [68].

Travel Salesman Problem (TSP) is one of the most well-studied examples of combinatorial optimization and best NP-hard problems for determining the shortest path to numerous target nodes without passing the same node twice and returning to the first node [69, 70]. Though it is unchallenging to declare but hard to solve, the main problem of TSP is to traverse all the nodes with minimum distance and return to the starting node. Required computation will increase exponentially as the problem increases when completed in the exact time. It has a variety of applications, including vehicle routing, scheduling, and planning, and it could be used to handle middle-sized TSP hybrid DE problems [71]. Illustration of TSP is displayed in Figure 14.9.

TSP is denoted by a graph that is both complete and weighted. Enumerating (n-1)!/2, where n specifies the number of nodes, and then selecting the shortest route can be used to solve Graph = (Vertices, Edges).

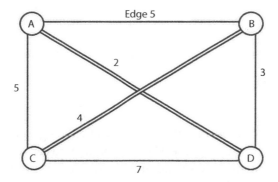

Figure 14.9 Illustration of TSP.

Simulated Annealing is a stochastic search strategy for nonlinear objective functions, which implies it uses randomization and iteratively updates one candidate solution [72]. Like the stochastic hill-climbing local search algorithm, it examines a very small part of the search space and updates the local search engine [73]. It may accept the worst points for the processing at high starting temperature inspired by annealing in metallurgy where the temperature of metal raised high so quickly and then cooled slowly. Multiobjective DE with simulated annealing was proposed for multiobjective optimization problem [74].

Hill climbing: It is a single-objective optimization process in which the currently known best individual generates the best offspring through iteration [75]. If DE and GA run multiple times, they will merge at different best chromosomes each time. To identify the best solutions in a region, GA is usually combined with a local search mechanism, resulting in a hybrid algorithm. Multidimensional problems that are not solved by traditional methods, DE is a powerful met heuristic method and hybridization of DE, and Hill Climbing is used to improve DE's performance [76].

Algorithm 2: Hybrid Differential Algorithm
Begin
Step 1: Initial population: Generate initial population and by using path representation encode the problem. *Step 2: Evaluation: By using the fitness function evaluate the current population.* *Step 3: Operators: Selection: By using tournament selection select two parents Cross two chromosomes chosen by order crossover and mutate the offspring using exchange mutation.*

Step 4: *Local Search: to explore a new best individual, hill-climbing method is used.*

Step 5: *Closing condition: If this end condition is attained and stop the iterations.*

End

It is a hard task to increase the efficiency of existing algorithms in route planning and determining the shortest path between two points [77]. There is some information for the input to enhance the efficiency of route planning algorithms. Road information like congestion conditions, traffic conditions, the surface of roads and incidents, etc. can affect the performance of the routing algorithm [78]. Destination information such as what is the purpose of travel and other information like vehicle-related conditions, road condition, fuel, etc.

14.4.1   Drawbacks of DE

- The "no free lunch" theorem is one of the weaknesses of DE, as all other evolutionary search techniques do. DE is slow at the exploitation of the solution though it has good global exploitation ability [78].
- The DE family of algorithms has still not made a significant development, whereas GAs, ES, and EP have made considerable progress in theoretical understanding [19].
- It faces difficulty on nonlinearly separable functions, it must rely on its differential mutation procedure [79].
- DE applications are also limited for multiobjective optimization problems [79].

14.5   Conclusion

Differential evolution algorithm based on meta-heuristic algorithm is used to get the optimal solution of a given problem. Treatment of hurdles in binary tree schemes is not ideal, and the calculation efficiency is also low. An algorithm built on an improved ant colony algorithm is time-consuming, the computational cost is too high and stagnation phenomena occur easily. Simulated annealing-based multioptimizer cannot find the perfect solution and the division is too coarse. The best optimal result can be obtained by introducing the best particle into the initial production.

It is further opined that other local search techniques may also be mixed in different steps of DE and can be tested on several different benchmark problems. The Tabu search- Meta-heuristic search is compared with other random initialization. The result shows that the tabu initialization approach effectively generates the more optimal solution and enhances the convergence speed towards optima. Apart from heuristic, several meta-heuristic algorithms are used by many researchers to solve the routing in navigation. To analyze the data from satellites, the Los Alamos National Laboratory has used it to create realistic special effects. The inclusion of elements will make the problem more difficult, thus finding a better method to optimize is an important point to research in the future. The chapter's content suggests that DE will continue to be a lively and busy multidisciplinary field in the future.

References

1. Hunt, M.M., Marquet, W.M., Moller, D.A., Peal, K.R., Smith, W.K., R.C.S., Acoustic navigation systems. *Nav. Eng. J.*, 76, 6, 937–940, 1964.
2. Bagali, M.U., Reddy, N.K., Dias, R., Thangadurai, N., The positioning and navigation system on latitude and longitude map using IRNSS user receiver. *Proc. 2016 Int. Conf. Adv. Commun. Control Comput. Technol. ICACCCT 2016*, pp. 122–127, 2017, (978).
3. Van Allen, J.A., Basic principles of celestial navigation. *Am. J. Phys.*, 72, 11, 1418–1424, 2004.
4. Grewal, M.S., Weill, L.R., Andrews, A.P., Signal characteristics and information extraction, Volume: 2nd Edition, John Wiley & Sons, Hoboken, 2007.
5. Haklay, M. and Weber, P., OpenStreet map: User-generated street maps. *IEEE Pervasive Comput.*, 7, 4, 12–18, 2008.
6. Jokar Arsanjani, J., Zipf, A., Mooney, P., Helbich, M., An introduction to OpenStreetMap in geographic information science: Experiences, research, and applications. In OpenStreetMap in GIScience, 1-15, Springer, Cham, 2015.
7. Huang, J.Y., Tsai, C.H., Huang, S.T., The next generation of GPS navigation systems. *Commun. ACM*, 55, 3, 84–93, 2012.
8. Norton, K.A., The calculation of ground-wave field intensity over a finitely conducting spherical earth. *Proc. IRE*, 29, 12, 623–639, 1941.
9. Johler, J.R., Sky wave propagation at low frequencies, in: *Encycl. RF Microw. Eng.*, 2005.
10. Koirala, S., Improving safe approach and landing system in TIA, 2018.
11. Gray, L., *GPS in schools-how does GPS works?*, pp. 1–7, High-Tech Science Ser., University of Tasmania, 2014.
12. Mijwel, M.M., Genetic algorithm optimization by natural selection. *Comput. Sci. Sci.*, 1, 1, 1–6, 2016.

13. Arora, R.K., 1-D optimization algorithms. *Optimization*, 51–70, Chapman and Hall/CRC, 2015.
14. Goldberg, D.E. and Holland, J.H., Guest editorial genetic algorithms and machine learning. *Mach. Learn.*, 3, 95–99, Pearson Education India, 1988.
15. Verma, K., Bhardwaj, S., Arya, R., Ul Islam, M.S., Bhushan, M., Kumar, A., Samant, P., Latest tools for data mining and machine learning. *Int. J. Innov. Technol. Exploring Eng.*, 8, 9, Special Issue, 18–23, 2019. https://doi.org/10.35940/ijitee.I1003.0789S19
16. Kumar Bhattacharjya, R. and Holland, J.H., Kalyanmoy Deb, An introduction to genetic algorithms. *Sci. Am. J.*, 24, 1–90, November, 1992.
17. Storn, R. and Price, K., Differential evolution – A simple and efficient heuristic for global optimization over continuous spaces. *J. Glob. Optim. 1997*, 114, 11, 4, 341–359, 1997.
18. Mezura-Montes, E., Velázquez-Reyes, J., Coello Coello, C.A., A comparative study of differential evolution variants for global optimization. *GECCO 2006 - Genet. Evol. Comput. Conf.*, 1, 485–492, 2006.
19. Das, S. and Suganthan, P.N., Differential evolution: A survey of the state-of-the-art. *IEEE Trans. Evol. Comput.*, 15, 1, 4–31, 2011.
20. Selvaraj, C., Siva Kumar, R., Karnan, M., A survey on application of bio-inspired algorithms. *Int. J. Comput. Sci. Inf. Technol.*, 5, 1, 366–370, 2014.
21. Crepinsek, M., Liu, S.H., Mernik, M., Exploration and exploitation in evolutionary algorithms: A survey. *ACM Comput. Surv.*, 45, 3, 1–33, 2013.
22. Erbao, C., Mingyong, L., Kai, N., A differential evolution & genetic algorithm for vehicle routing problem with simultaneous delivery and pick-up and time windows. *IFAC Proceedings*, 41, 2, 10576–10581, 2008.
23. Li, X., Xu, L., Wang, H., Song, J., Yang, S.X., A differential evolution-based routing algorithm for environmental monitoring wireless sensor networks. *Sensors*, 10, 6, 5425–5442, 2010.
24. Journal, D., Routing using genetic algorithm for large networks. *Diyala J. Eng. Sci.*, 03, 02, 53–70, 2010.
25. Fadil, Y.A., Routing using genetic algorithm for large networks. *J. Eng. Sci.*, 03, 02, 53–70, 2010.
26. Karaş, İ.R., Yaman, B., Atila, U., Rakip Karas, I., Gologlu, C., Orak, I.M., Design of a route guidance system with shortest driving time based on genetic algorithm, *ACACOS' 11: Proceedings of the 10th WSEAS international conference on Applied computer and applied computational science, Researchgate. net*, 61–66, 2011.
27. Zhao, Y.X., Li, W., Feng, S., Ochieng, W.Y., Schuster, W., An improved differential evolution algorithm for maritime collision avoidance route planning. *Abstr. Appl. Anal.*, 2014, 2014.
28. Sumithra, S. and Victoire, T.A.A., Erratum: Differential evolution algorithm with diversified vicinity operator for optimal routing and clustering of energy efficient wireless sensor networks. *Sci. World J.*, 2016, 2016.

29. Aibinu, A.M., Bello Salau, H., Rahman, N.A., Nwohu, M.N., Akachukwu, C.M., A novel clustering based genetic algorithm for route optimization. *Eng. Sci. Technol. Int. J.*, 19, 4, 2022–2034, 2016.

30. Moza, M. and Kumar, S., Routing in networks using genetic algorithm. *Bull. Electr. Eng. Inform.*, 6, 1, 88–98, 2017.

31. Shukla, R., Hazela, B., Shukla, S., Prakash, R., Mishra, K.K., Variant of differential evolution algorithm. *Adv. Intell. Syst. Comput.*, 553, 601–608, 2017.

32. Camero, A., Arellano-Verdejo, J., Alba, E., Road map partitioning for routing by using a micro steady state evolutionary algorithm. *Eng. Appl. Artif. Intell.*, 71, February, 155–165, 2018.

33. Sharma, A. and Sinha, M., A differential evolution-based routing algorithm for multi-path environment in mobile ad hoc network. *Int. J. Hybrid Intell.*, 1, 1, 23, 2019.

34. Xia, N., Zhi, Q., He, M., Hong, Y., Du, H., A navigation satellite selection algorithm for optimized positioning based on Gibbs sampler. *Int. J. Distrib. Sens. Netw.*, 16, 6, 2020.

35. Sethanan, K. and Jamrus, T., Hybrid differential evolution algorithm and genetic operator for multi-trip vehicle routing problem with backhauls and heterogeneous fleet in the beverage logistics industry. *Comput. Ind. Eng.*, 146, 106571, May 2020.

36. Sabar, N.R., Song, A., Tari, Z., Yi, X., Zomaya, A., A memetic algorithm for dynamic shortest path routing on mobile ad-hoc networks. *Proc. Int. Conf. Parallel Distrib. Syst. - ICPADS*, 60–67, January 2016.

37. El-Sayed, H.H., Shortest paths routing problem in MANETs. *Appl. Math. Inf. Sci.*, 10, 5, 1885–1891, 2016.

38. Patle, B.K., Babu L, G., Pandey, A., Parhi, D.R.K., Jagadeesh, A., A review: On path planning strategies for navigation of mobile robot. *Def. Technol.*, 15, 4, 582–606, 2019.

39. Lee, W. and Kim, H.Y., Genetic algorithm implementation in Python. *Proc. - Fourth Annu. ACIS Int. Conf. Comput. Inf. Sci. ICIS 2005*, 2005, 8–12, 2005.

40. Zhang, X., Yuan, X., Yuan, Y., Improved hybrid simulated annealing algorithm for navigation scheduling for the two dams of the Three Gorges Project. *Comput. Math. Appl.*, 56, 1, 151–159, 2008.

41. Yu, Z., Ni, M., Wang, Z., Zhang, Y., Dynamic route guidance using improved genetic algorithms. *Math. Probl. Eng.*, 2013, 2013.

42. Ceylan, H. and Bell, M.G.H., Traffic signal timing optimisation based on genetic algorithm approach, including drivers' routing. *Transp. Res. Part B Methodol.*, 38, 4, 329–342, 2004.

43. Narwadi, T. and Subiyanto, An application of traveling salesman problem using the improved genetic algorithm on android Google maps. *AIP Conf. Proc.*, 1818, 2017.

44. Fronita, M., Gernowo, R., Gunawan, V., Comparison of genetic algorithm and hill climbing for shortest path optimization mapping. *E3S Web Conf.*, 31, 1–5, 2018.
45. Wan, W. and Birch, J.B., An improved hybrid genetic algorithm with a new local search procedure. *J. Appl. Math.*, 2013, 14–19, 2013.
46. Lee, J.Y., Kim, M.S., Kim, C.T., Lee, J.J., Study on encoding schemes in compact genetic algorithm for the continuous numerical problems. *Proc. SICE Annu. Conf.*, 2694–2699, 2007.
47. Shop, J., A new encoding scheme. December, 4395–4400, 1996.
48. Tuson, A., *Soft computing in industrial applications*, 2007.
49. Kumar, A., Properties for selection of encoding scheme, Int. J. Adv. Res. Comput. Commun. Eng., 2, 3, 1–7.
50. Katoch, S., Chauhan, S.S., Kumar, V., A review on genetic algorithm: Past, present, and future. *Multimed. Tools Appl.*, 80, 5, 8091–8126, 2021.
51. Fajfar, I., Puhan, J., Tomažič, S., Burmen, Á., On selection in differential evolution. *Elektroteh. Vestn./Electrotech. Rev.*, 78, 5, 275–280, 2011.
52. Singh, S., Singh, J., Sehra, S.S., Genetic-inspired map matching algorithm for real-time GPS trajectories. *Arab. J. Sci. Eng. 2019*, 45, 4, 2587–2603, 2019.
53. Qian, W., Chai, J., Xu, Z., Zhang, Z., Differential evolution algorithm with multiple mutation strategies based on roulette wheel selection. *Appl. Intell.*, 48, 10, 3612–3629, 2018.
54. Prayudani, S., Hizriadi, A., Nababan, E.B., Suwilo, S., Analysis effect of tournament selection on genetic algorithm performance in traveling salesman problem (TSP). *J. Phys. Conf. Ser.*, 1566, 1, 0–7, 2020.
55. Soon, G.K., Guan, T.T., On, C.K., Alfred, R., Anthony, P., A comparison on the performance of crossover techniques in video game. *Proc. - 2013 IEEE Int. Conf. Control Syst. Comput. Eng. ICCSCE 2013*, 493–498, 2013.
56. A.J., U. and P.D., S., Crossover operators in genetic algorithms: A review. *ICTACT J. Soft Comput.*, 06, 01, 1083–1092, 2015.
57. Hassanat, A., Almohammadi, K., Alkafaween, E., Abunawas, E., Hammouri, A., Prasath, V.B.S., Choosing mutation and crossover ratios for genetic algorithms-a review with a new dynamic approach. *Inf.*, 10, 12, 390, 2019.
58. Michalewicz, Z., Evolutionary algorithms for constrained parameter optimization problems. *Evol. Comput.*, 4, 1, 1–32, 1996.
59. Lima, J.A., Gracias, N., Pereira, H., Rosa, A., Fitness function design for genetic algorithms in cost evaluation based problems. *Proc. IEEE Conf. Evol. Comput*, 207–212, January 1996.
60. Georgioudakis, M. and Plevris, V., A comparative study of differential evolution variants in constrained structural optimization. *Front. Built Environ.*, 6, 102, 2020.
61. Quddus, M.A., Noland, R.B., Ochieng, W.Y., A high accuracy fuzzy logic based map matching algorithm for road transport. *J. Intell. Transp. Syst. Technol. Plann. Oper.*, 10, 3, 103–115, 2006.

62. Kim, Y. and Bang, H., Introduction to kalman filter and its applications, in: *Introd. Implementations Kalman Filter*, 1–16, 2019.
63. Goh, C.Y., Dauwels, J., Mitrovic, N., Asif, M.T., Oran, A., Jaillet, P., Goh_ Online map matching based HMM for real time traffic sensing applications. *In 2012 15th International Conference on Intelligent Transportation IEEE Systems*, 117543, 776–781, 2012.
64. Madkour, A., Aref, W.G., Rehman, F.U., Rahman, M.A., Basalamah, S., A survey of shortest-path algorithms, 1–26, 2017.
65. Guerreiro, P., Jesus, M., Márquez, A., Differential evolution in shortest path problems. Avances en Matemática Discreta en. Andalucıa, 2, 1–6.
66. Glover, F., Kelly, J.P., Laguna, M., Genetic algorithms and tabu search: Hybrids for optimization. *Comput. Oper. Res.*, 22, 1, 111–134, 1995.
67. Watchanupaporn, O., Suwannik, W., Chongstitvatana, P., Arithmetic coding differential evolution with Tabu search. *2014 Int. Comput. Sci. Eng. Conf. ICSEC 2014*, pp. 174–179, 2014.
68. Ponsich, A. and Coello Coello, C.A., A hybrid differential evolution—Tabu search algorithm for the solution of job-shop scheduling problems. *Appl. Soft Comput.*, 13, 1, 462–474, 2013.
69. Sharma, S. and Gupta, K., Solving the traveling salesmen problem through genetic algorithm with new variation order crossover, 274–276, 2012.
70. Liao, Y.F., Yau, D.H., Chen, C.L., Evolutionary algorithm to traveling salesman problems. *Comput. Math. Appl.*, 64, 5, 788–797, 2012.
71. Wang, X. and Xu, G., Hybrid differential evolution algorithm for traveling salesman problem. *Proc. Eng.*, 15, 2716–2720, 2011.
72. Dimitirs, B. and John, S., Simulated annealing.pdf. *Stat. Sci.*, 8, 1, 10–15, 1993.
73. Solving, P., Problem solving and search in artificial intelligence. *INFORMS J. Comput.*, 1, 1, 1–5, 2005.
74. Chen, B., Zeng, W., Lin, Y., Zhong, Q., An enhanced differential evolution based algorithm with simulated annealing for solving multiobjective optimization problems. *J. Appl. Math.*, 2014, 2014.
75. Chowdhary, K.R., Fundamentals of artificial intelligence. *Fundam. Artif. Intell.*, fig 1–716, 2020.
76. Hernández, S., Artificial, G.L.-... de I., and 2013, undefined, Hybridization of differential evolution using hill climbing to solve constrained optimization problems. *Inteligencia Artificial*, 16, 52, 3–15, 2013, redalyc.org.
77. Mangla, M., Kumar, A., Mehta, V., Bhushan, M., Mohanty, S., Real-life applications of the Internet of Things: Challenges, applications, and advances, Apple Academic Press, New York, 1st Edition, pages 536, ISBN: 9781003277460, 2022. https://doi.org/10.1201/9781003277460.
78. Lin, J., Yu, W., Yang, X., Yang, Q., Fu, X., Zhao, W., A real-time en-route route guidance decision scheme for transportation-based cyberphysical systems. *IEEE Trans. Veh. Technol.*, 66, 3, 2551–2566, 2017.

79. Bilal, Pant, M., Zaheer, H., Garcia-Hernandez, L., Abraham, A., Differential evolution: A review of more than two decades of research. *Eng. Appl. Artif. Intell.*, 90, 103479, 2020.

IoT-Based Smart Parking System for Indian Smart Cities

E. Fantin Irudaya Raj[1]*, M. Appadurai[2], M. Chithamabara Thanu[2]
and E. Francy Irudaya Rani[3]

[1]Department of Electrical and Electronics Engineering, Dr. Sivanthi Aditanar
College of Engineering, Tamilnadu, India
[2]Department of Mechanical Engineering, Dr. Sivanthi Aditanar College of
Engineering, Tamilnadu, India
[3]Department of Electronics and Communication Engineering, Francis Xavier
Engineering College, Tamilnadu, India

Abstract

The concept of smart cities has recently received a lot of attention. Most governments throughout the world are undertaking various initiatives to construct smart cities. The Indian Government has also announced the "National Smart Cities Mission" for urban retrofitting, revitalization, and creating smart cities across India. The fundamental goal of this mission is to utilize the Internet of Things (IoT) technology to build new cities or to convert existing cities and make them more sustainable and citizen-friendly. Vehicle parking will become a major issue in Indian cities. It is mainly due to poor planning and land scarcity, and it has led to a slew of other issues, such as traffic congestion, pollution, economic losses, and so on. IoT makes extensive use of sensors and transducers to collect real-time data from the physical world. Problems, such as vehicle parking facilities, traffic congestion, traffic management, and road safety, are being addressed by IoT. The current chapter examines the IoT-based vehicle parking system in Indian cities. Additionally, it discusses vehicle parking and its basic requirements, various technologies incorporated in modern parking systems, different sensors utilized in parking facilities, and the advantages of IoT-based vehicle parking systems in detail.

Keywords: Smart cities, Internet of Things (IoT), sensors, smart vehicle parking system, traffic management system

Corresponding author: fantinraj@gmail.com

Ashok Kumar, Megha Bhushan, José A. Galindo, Lalit Garg and Yu-Chen Hu (eds.) Machine Intelligence, Big Data Analytics, and IoT in Image Processing: Practical Applications, (369–398) © 2023 Scrivener Publishing LLC

15.1   Introduction

After the introduction of the motorized vehicle toward the end of the 19th century, the use of these vehicles gradually grew until the mid-20th century. However, in the 21st century, it has expanded tremendously. This increase is primarily seen in highly developed cities around the world, and it has begun to cause issues with vehicle parking, traffic congestion, and environmental pollution. Vehicle parking challenges have emerged primarily as a result of a scarcity of land in the city as a result of poor city design, as well as a growing population density [1]. In metropolitan regions, such as modern smart cities, this problem will be much worse than in rural ones. It is primarily due to an increase in people relocating from rural areas to urban areas, such as metropolitan areas, in search of work and to meet their financial necessities. Following the industrial revolution, the rate at which people migrated from rural to urban environments increased exponentially [2].

People in cities will travel from one location to another within the city for a variety of reasons, including work trips, educational trips, recreational trips, and business trips. Because of the increased movement of vehicles from one location to another inside the city, there is a large demand for parking, which, if not provided, produces heavy traffic congestion owing to vehicles occupying road space for parking. As a result, planners were forced to include parking structures in the construction of cities, recreational areas, and industrial areas in order to ensure a smooth flow of vehicles to prevent traffic congestion, accidents, and pollution, and to increase economic activity, among other things [3]. The planners allocate open land and roadside space for parking in the traditional method of parking, but this is not an effective solution to meet the parking demand due to the over vehicle population saturation of cities. As a result, they have turned to advanced methods to manage the parking demand, such as metered parking, multilevel parking, automated parking, and so on. However, the problem of satisfying the parking demand has not been remedied [4].

To achieve sustainability, most cities in both developed and developing countries have implemented a new idea of city development known as smart cities. They hope to reduce traffic congestion, pollution, delays, travel time, travel costs, and accidents by doing so. Another major element that they have considered is how to manage the demand for parking. The integrated transportation system, exclusive bus lanes, and improved mass transportation are all part of the smart city plan. Specifically, the smart parking system with the Internet of Things (IoT) technology, which is more efficient and effective, will be used to meet the parking demand [5, 6]. It is used all across the world using the following components: sensors, geographic information system circuits, and mobile phones with Internet access, among others. The Indian government has also taken initiatives for urban remodeling, revitalization, and the development of smart cities across the country. The mission's main purpose is to employ information and communication technologies (ICT) to develop new cities or to turn existing cities into more sustainable and citizen-friendly environments [7]. The mission is primarily focused on transforming existing cities in India into smart cities. The present chapter provides a review and further explores IoT-based smart parking systems for Indian smart cities in detail.

15.2   Indian Smart Cities Mission

India is one of many developing countries seeing a rapid transition from rural to urban areas. The larger decadal rise of the urban vs. rural population reflects this shift. India's urban population grew from roughly 27.8% in 2001 to 31.2% in 2011 and is expected to expand to 40% by 2030 and more than 50% by 2050 [8]. Cities are experiencing infrastructure management and service delivery issues as their populations rise [9]. One technique being used to successfully and efficiently deal with these challenges is the development of smart cities. The Smart Cities Mission of India is a 5-year national initiative led by the Ministry of Urban Development (MoUD), the Government of India to provide the groundwork for 100 smart cities across the country [10]. This mission does not, however, identify the specific traits that must be present in a "smart" city. Instead, they are concentrating on transforming current cities into smart cities [11]. Huge sums of money have been spent, and new projects and infrastructure have been established for this aim. The various components of a smart city are illustrated in Figure 15.1.

According to data from the Indian government, the automotive industry is a significant contributor to the country's GDP. Its current percentage of GDP is around 6.7%, and by 2026, it is expected to reach around 12%. It would be one of the most important sources of employment. With about 47 lakh kilometers of road, India possesses one of the world's greatest road networks. Roads carry over 65% of total freight and 80% of passenger traffic, but the state of road infrastructure is a source of worry. Other issues include traffic congestion, air pollution from vehicles running on combustible fuel, insufficient public transportation, underutilization of water

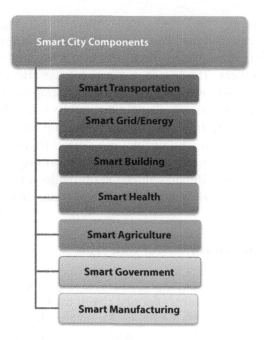

Figure 15.1 Smart city components.

transportation, and a lack of technological innovation in transportation systems [12].

Intelligent transportation system (ITS) is a collection of ICT initiatives used to efficiently manage transportation in a smart city setting. One of the key areas of the ITS is real-time parking management and multilevel parking. Real-time parking management systems also display information about available parking lots on an electronic signboard that is visible to the public. This amenity is beneficial to both parking lot employees and end-users. With minimal land utilization, rapid entry and departure, and several sensors and safety systems, multilevel parking offers cheap operating and maintenance costs [13].

The usage of IoT-based sensors and devices is a recent advancement in automotive technology. The vehicle's chassis and driveline are also integrated with cameras and sensors that aid during parking. The technology improves safety, efficiency, and comfort in parking and driving, making it a beneficial tool for inexperienced drivers and learners. It is also helpful in many ways for the citizen in modern smart city setup [14].

15.3   Vehicle Parking and Its Requirements in a Smart City Configuration

Stopping the car and halting in a safer location for a short period of time for the following purposes: loading and unloading of products, shopping, taking a short break, and repairing a broken-down vehicle are all examples of parking. And for a long period of time for the following purposes: park and work, leisure activity, park and mass transportation, and residential parking. Parking is one of the most important components of a successful transportation system, as parking spaces should be available at every destination, allowing for the convenience of driving and thus increasing overall accessibility [15].

Nowadays, parking has become a social issue due to rising construction costs, building costs, pollution, traffic congestion, delays, and accidents, among many other things. All economic classes of people should be able to access all available travel options, such as mass transportation, monorail, metro rail, and airports, and they should have accurate

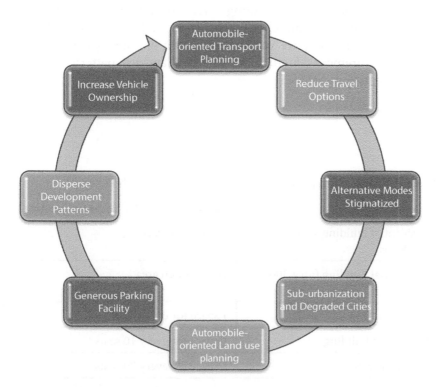

Figure 15.2 Parking planning cycle.

information on parking facilities, such as the cost of parking various vehicles, aesthetics accessibility, user information, and vehicle security [16]. It should be planned in such a way that it can accept uncertainties, such as peak demand, while yet allowing different destinations to be served. Because of fast infrastructure development, industrialization, and the migration of people from rural areas to cities for jobs, education, and business, high mobility results in increased parking demand, requiring a paradigm shift in parking planning. In the past, a parking difficulty meant a lack of space to park the vehicle; presently, it means a lack of user information, space saturation, and a high cost [17]. Figure 15.2 shows how parking is planned based on estimating current and projected demand.

In order to achieve the comprehensive and integrated strategy by offering optimal parking facilities, the planner should carefully deal with sorting out the parking problem by considering the following parameters, such as when and where the parking deficit occurs, as well as the category of problem. Parking management should be more cost-effective, as parking has an indirect impact on other costs, such as taxes, rents, and the cost of retail products, among several others. According to Indian Road Congress (IRC) guidelines created as stated in Table 15.1, different types of buildings require varying minimum parking spaces. However, parking requirements fluctuate for different land-use patterns according to IRC regulations.

Table 15.1 Minimum parking space for different types of buildings.

Type of building	Parking space (IRC recommendation)
Residential building area less than 300 m²	Communal parking
Residential building (area 300 to 500 m²)	1/4th of open area
Office building	1 space for every 70 m²
Restaurant building	1 space for every 10 seats
Cinema theatre	1 space for every 20 seats

15.4   Technologies Incorporated in a Vehicle Parking System in Smart Cities

The number of automobiles on the road is growing at a similar rate to the rapid growth in population in major cities. People prefer private mobility over public transit, which has resulted in a rise in the number of vehicles on the road. As a result, obtaining an empty parking spot becomes increasingly difficult, causing a variety of issues such as higher energy use, time wastage, and gasoline wastage. Various methods have been implemented in an attempt to alleviate traffic congestion. Although there are various things that should be considered, one of them is to implement a smart parking system. Advanced technology and research from numerous academic disciplines are used to produce smart parking systems, which are mostly established in many developed countries. It is intended that deploying the technology in the parking area will eliminate the aforementioned issues that users in the vehicle park are experiencing [18].

The carpark system is advantageous to parking operators, parking clients, and environmental observers. Car park operators can use the information gathered through the implementation of a smart parking system to forecast future parking patterns. Pricing strategies can also be manipulated based on the data collected in order to boost the company's profit. Pollution levels can be reduced in terms of environmental preservation by reducing vehicle emissions in the air. This can be attributed to the reduction in road transportation. Because fuel consumption is directly proportional to vehicle miles traveled, it will be significantly reduced [19].

Users can also profit from a smart parking system because parking spaces can be fully utilized with a safer, more efficient system in place. Because of the information provided by the smart parking system, the system is more efficient because the vehicle travel time and search time are greatly decreased. With the information supplied, drivers can easily avoid fully populated car parks and discover vacant parking places elsewhere. As more vehicles are incorporated into car parks, the number of vehicles parked illegally along the roadside, which causes traffic congestion, decreases [20]. The most essential benefit is that traffic congestion is decreased. All of this would eventually contribute to public convenience. The smart parking system is further categorized into five different categories. They are a) automated parking, b) E-parking, c) smart payment system, d) transmit-based information system, and e) parking guidance and information system [21]. There are various modern technologies incorporated into vehicle parking systems and make them modernized and effective. The various important works relevant to the modern vehicle parking system are listed in Table 15.2.

There are numerous works listed in the literature. A few of the important ones are getting discussed in this section. Among these technologies, the IoT-based smart parking system is more reliable and cost-effective. At the present time, there are numerous developments in IoT-based sensors and devices. The cost of the entire setup also becomes so less. All these things, finally make IoT-based technology more advantageous in using smart parking systems in Indian smart cities. The forthcoming section explains the IoT-based modern smart parking system, which is most suited for Indian smart cities.

Table 15.2 Notable works in the literature regarding modernized vehicle parking system.

Technology adopted	Description	Merits	Demerits
Parking fee collection and automatic parking management system based on number plate recognition [22].	As a sensor, a wide-angle camera is being used to record or identify unoccupied parking spaces.	Proper utilization of parking space.	The use of a camera degrades the system's reliability and increases its cost.
The IoT is being used to develop an automatic smart parking system (IoT) [23].	The system creates a simple smart auto parking system that is both cost-effective and helps to reduce carbon dioxide emissions, making this system environmentally friendly.	Cost-effective and environmentally friendly.	Not resolving the congestion issues.

(Continued)

Table 15.2 Notable works in the literature regarding modernized vehicle parking system. (*Continued*)

Technology adopted	Description	Merits	Demerits
IoT-based Smart City Architecture and Applications [24].	The author focuses on customized communication services using Satellite Communication, Zig-Bee, WI-Max, Wi-Fi, and other technologies to transform a city into a smart city. For example, we may merge the health and transportation domains by creating a sensor that can readily assess all health-related parameters of the driver while driving, such as blood pressure and pulse rate, and deliver real-time health information to the driver.	Communication becomes easier, and the environment is safer.	A large number of sensors are used. In comparison to another system, it took a bit longer.
Facilitating the Development of Reliable and Secure IoT-based Smart City Applications [25].	To lower total traffic in the downtown region, IoT is employed as a real-time traffic monitor.	Keep an eye on the traffic in real-time. Assists in reducing overall traffic.	Need IoT-based sensors placed everywhere.

(*Continued*)

Table 15.2 Notable works in the literature regarding modernized vehicle parking system. (*Continued*)

Technology adopted	Description	Merits	Demerits
The application, architecture, standardization, and research issues of visible light communication [26].	The one-of-a-kind computer code was created to interface with a variety of roadside equipment. Visible light communication is also employed for vehicle-to-vehicle communication.	Machine-to-machine communication is excellent.	Implementation will take a long time.
Intelligent Transportation System based on the Cloud database [27].	IoT technologies and cloud computing were used to enable the multilayered vehicle data cloud platform.	Support a variety of technologies.	The issue with data storage. It is difficult to keep track of a vast volume of data.
A state-driven autonomous passing through method for driverless vehicles at crossings [28].	The reservation-oriented centralized scheduling algorithm is adopted. This guarantees that the high volume of requests is handled appropriately. This algorithm is tested on automobiles with a high priority.	Quicker response after the implementation	Implementation is a little bit complex. Difficulty in understanding.

(*Continued*)

Table 15.2 Notable works in the literature regarding modernized vehicle parking system. (*Continued*)

Technology adopted	Description	Merits	Demerits
Analyses of cooperative planning algorithms at intersections [29].	The authors created a fully automated car with significant distinctions from what is available today. This article also discusses how the reservation method might be improved.	On the basis of the reservation algorithm, it is implemented. Some sections are not getting concentrated.	Practically not implemented.
Using an integrated simulator, evaluating the mobility and environmental benefits of reservation-based intelligent intersections [30].	This intersection control system is built on a reservation system to make use of the exceptional connectivity that gives dynamism to connected transports.	Reservations can be made via the internet. Transport connectivity dynamism.	The cost of making a reservation is quite high. Congestion on the network.
Control and coordination of linked and autonomous cars at urban traffic crossings [31].	A decentralized, effective system for decreasing fuel consumption during any expedition or stoppage. This algorithm provides a congestion-control method and aids in the reduction of trip time.	The framework that is effective. Solve the traffic problem. Travel time is cut down.	Implementation will take a long time. The price is very high.

Table 15.2 Notable works in the literature regarding modernized vehicle parking system. (*Continued*)

Technology adopted	Description	Merits	Demerits
Innovative Technology for Intelligent Roads with the use of IoT devices [32].	The RASPBERRY PI 2 is utilized in this research to create a smart high-definition image. Ultrasonic sensors are utilized to measure vehicle strength and convey the signal to the RASPBERRY PI 2, which manages the traffic problem.	Detect the vehicles automatically. Controls traffic congestion issues.	Costlier than the rest of the methods.
Intelligent Transportation System Powered by the IoT [33].	The IoT-based transportation system is being created with a smart city perspective in mind.	Use for real-time traffic and provide immediate results.	The issue of data storage will be the main concern.
Smart Traffic Management system [34].	RFID is a novel technology that is being utilized to save installation costs and time.	Reduce the time and cost of installation.	Global Positioning System (GPS) not utilized.
Integrating the IoT with Agent Technology to Create an Intelligent Traffic Information System [35].	RFID, WSN, object ad-hoc network and information systems are used to track and control traffic items automatically over a network	This system is more reliable because it has a tracking feature and uses WSN.	Congestions in the network.

(*Continued*)

Table 15.2 Notable works in the literature regarding modernized vehicle parking system. (*Continued*)

Technology adopted	Description	Merits	Demerits
Short message service-based smart parking reservation system [36].	The system is portable due to the use of a microcontroller, but the cost of implementation is very expensive due to the usage of a microcontroller. The main issue is that if the workload on the microcontroller increases, the system may crash.	Easy to use and more secure.	The implementation costs are considerable, the GSM functionality causes bottlenecks, and microcontroller overload causes the system to break.
Secure Vehicle Parking Management and Reservation System based on Zigbee and GSM [37, 38].	The vacancy in the parking area is checked using ZigBee. This includes a security element, such as a password that must be entered before the user may exit the parking lot. When the user enters the correct exit password, the gate will open.	GSM, SMS, and a secure password mechanism are all used.	Expensive, and the system's difficulty is network congestion, which prevents users from receiving SMS.

(*Continued*)

Table 15.2 Notable works in the literature regarding modernized vehicle parking system. (*Continued*)

Technology adopted	Description	Merits	Demerits
Image Processing-Based Intelligent Parking Management System [39, 40].	The image processing technique is used. The sensor camera is used to gather images that aid in determining whether or not parking places are available. However, in terms of visibility, weather conditions such as rain, fog, snow, and so on may have an impact on the system.	The camera is employed as a sensor to detect the presence of cars.	There is no GPS and some weather conditions, such as rain and fog, are incompatible. The system is in a fixed position.
Bluetooth-enabled automated parking system [41].	Bluetooth is used to communicate or connect to a network. For linear motion, this system employs a rack and pinion mechanism. The main disadvantage of this technique is that the entire parking system is based on the rack and pinion mechanism, which is both costly and time-consuming.	Bluetooth is used to register or identify devices. If a new vehicle is parked, the system identifies the unique registration number.	It is not compatible with the current system. The parking lot as a whole will be designed. Costly. Time-consuming.

(Continued)

Table 15.2 Notable works in the literature regarding modernized vehicle parking system. (*Continued*)

Technology adopted	Description	Merits	Demerits
RFID and Wireless Sensor Network for the Smart Parking System [42].	To make the parking, RFID and wireless sensors are employed, which direct the driver to find the location, however, this method is time-consuming and expensive.	It gives the driver instructions as well as slot information. It can be used in existing parking lots.	Implementation is expensive and time-consuming.
Wireless Sensor Network-Based Smart Parking Services [43, 44].	An android application is being used.	It has a better interface.	It lacks the reservation feature.

15.5   Sensors for Vehicle Parking System

Sensors are very important to make this intelligent IoT-based vehicle parking system to become practical. Several types of sensors are available in the market which can be used in this technology. Each type of sensor has its own merits and demerits on successful implementation. The main objectives of using sensors are to find the free slot in real time to identify the entry of vehicles to the parking space and to ensure the vehicle is parked in the allotted space. The sensors are two types, namely active and passive type. The active type sensors use external power for energizing. The amplification, conversion, and transmission of the signal can be done in active sensors through external energy sources. The passive sensor type uses the naturally available energy in the environment to measure the signals. For example, a camera without a flash fight is a passive type of sensor, the natural sunlight is used to record an image. The camera with flash is the best example of the active type of sensor. The sensor is selected based on the accuracy requirement, price, frequency of the sensed signal, operating voltage, and maintenance cost. The owners generally select the sensors having low maintenance cost without compromising the accuracy.

15.5.1   Active Sensors

The active infrared sensor, CCTV, and ultrasonic sensors are the available active sensors used in the vehicle parking technology. The infrared type of sensor emits the infrared rays through the transmitter and the receiver senses the reflected back signals from the external sources [45]. The layout of the active infrared sensor implementation in the parking system is shown in Figure 15.3. Based on the received signal, the speed of the vehicle, and the accurate position of the vehicle even on the multilane roads, this type of sensor is susceptible to the weather conditions, such as heavy rainy conditions and fog. The intensity of sunlight affects the output of the infrared sensors and the accuracy of the sensor is enhanced by the electronic manipulation through trial runs.

The ultrasonic sensor is another type of active sensor used in these applications [46]. This type of sensor uses acoustic sounds for sensing operations. Figure 15.4 shows the utilization of ultrasonic sensors in parking systems. The acoustic sounds are transmitted from the sensor transmitter and the receiver absorbs the reflected back sound signals. Normally, sound waves of about 20 to 50 Hz are transmitted. This ultrasonic sensor has low maintenance cost but is very sensitive to the external environment and temperature.

The closed-circuit vehicle technology (CCTV) is another best active type sensor used in several applications. The image processing technique is used in the CCTV camera for data analysis [47]. The parking

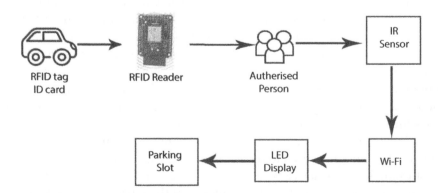

Figure 15.3 Infrared sensor in parking systems.

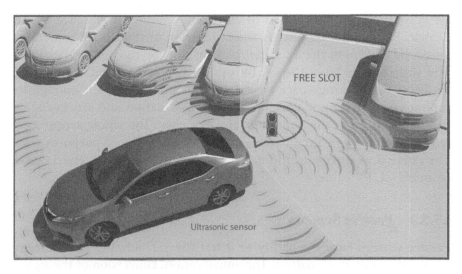

Figure 15.4 Ultrasonic sensors in the vehicle parking system.

Figure 15.5 CCTV implementation in smart parking management.

lot is continuously monitored to check the presence or absence of the vehicle. Figure 15.5 represents the installation of the CCTV camera to guide the drivers to the allocated parking slots by the light indications. The video stream from the CCTV camera is processed frame by frame to detect the parking space free slots. The single-camera is enough to monitor more than one parking slot. The installation and processing of the signal through the computer are very easy. The whole sensor system is cost-effective and effectively used for surveillance purposes also. The external weather conditions affect the intensity of the received signal from the CCTV camera.

15.5.2 Passive Sensors

Passive sensors have the ability to detect the surrounding environment from the energy from nature. The amplification, conversion of the signal to other forms is not possible in passive-type sensors. A light-dependent resistor sensor is the best example of a passive type sensor that is used in the vehicle parking system. The vehicle presence in the parking space is detected using this type of sensor. The sunlight is the primary source to detect the object in a particular space. The natural fog, rain, and other weather affect the sensor outputs. Generally, this LDR sensor is fixed in the middle of the parking area.

Another type of passive sensor is an inductive loop detector type sensor, which is normally fixed on the road lanes. The sensor module has several wire loops in which the current is passing through it. The electromagnetic field is induced due to the electric current potential. The 10- to 50-Hz frequency electric field inductance is induced to identify the presence of the vehicle or the movement of the vehicle inside the parking area. The vehicle movement reduces the inductance of the electric field, which alerts the central server system to take necessary decisions. This type of sensor has higher sensitivity to traffic management accuracy.

A piezoelectric sensor is mainly used in several applications because of its self-energizing capacity. The external environmental load, i.e., the force applied to the system is converted to use electric power to energize the sensor modules. The piezoelectric crystal converts the strain energy into a useful electric pulse. The intensity of electric energy mainly depends on the applied force. The piezoelectric sensor is laid on the parking slots. The vehicle movement is gradually monitored due to the weight of the vehicle displaced.

15.6 IoT-Based Vehicle Parking System for Indian Smart Cities

The IoT-based technology is utilized in many domains in recent times. The application domains varied from healthcare [48–51], manufacturing industries [52], remote sensing [53, 54], condition monitoring [55, 56], power generation [57], power transmission [58–60], automobile [61–63], smart agriculture [64, 65], smart cities [66–68], and so on. It will revitalize the old cities into modern smart cities. It is an important component of smart cities. It is more advantageous for smart cities to include IoT technology into vehicle parking systems. IoT device-based vehicle parking technology is widely used to eradicate the uncertainty of nearest parking space availability and effective usage of available parking resources [69]. Inside the inner cities, the special parking slots are built up to manage the city parking space requirement, which is shown in Figure 15.6. This advanced system guides the drivers to search and park the vehicles in less searching time. This also helps municipality and private parking service providers for

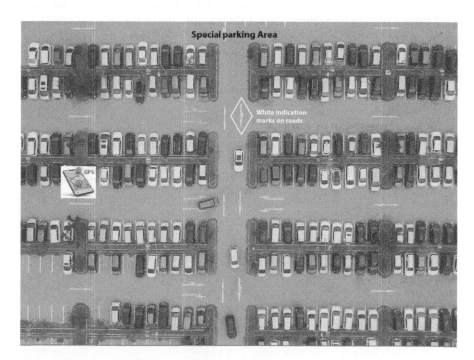

Figure 15.6 Special parking slots for inner-city traffic.

optimal parking space utilization. The urban vehicle management system uses this type of parking technology for effective traffic management and provides the possible parking slot for inner-city vehicles [70].

Smart parking technology is designed for different purposes to manage several categories of vehicles. Municipality parking service is designed to manage the inner-city parking space requirement. Another one is private parking service, i.e., commercial buildings, corporate offices, or shopping malls, which is designed to comfort their customers. Normally, the parking space is planned on the top floor and underground spaces. The intelligent smart parking system in the shopping malls or commercial buildings attracts the customers to their space.

In the vehicle parking management system, there are several types of operations to be done for the effective handling of vehicles. Figure 15.7 illustrates the vehicle management system, which has several sensors and parking management tools. First of all, the entry vehicle must be detected and guided to the available free parking slot. This gives a lot of benefits to the drivers, customers, and parking service providers. In dense urban areas, the vehicle count is large in number on the roads. The customer feels it is very difficult to identify the local parking space within the limited time. This smart system gives information about the nearest available parking slot to the drivers and customers on their requests. While using this type of smart system, the vehicle gets a quick parking slot for a limited time. Also, vehicles have less carbon emission and fuel economy by eliminating unnecessary searching of parking slots due to uncertainty about parking space.

This smart system is also helpful to the parking service providers to get profit by effective use of parking slots. The data available in the central server is analyzed to get clear-cut knowledge about the parking slot demand duration period. Generally, everyday evening and week off days have huge vehicle movements and parking slot demands. The efficient

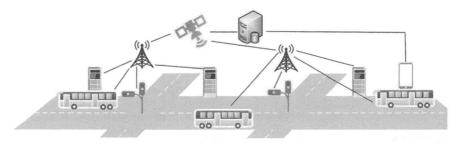

Figure 15.7 Vehicle management system.

pricing strategies, quick payment receiving options, and easy reservation systems attract customers to the parking system in huge numbers. The dynamic pricing system, i.e., slightly higher price on demand duration time is helpful to the parking service providers to get higher profit.

Several operations are needed to implement the IoT-based smart parking system in real-time applications. The main operation for the smart parking system is the vehicle sensing system. The sensors are installed in the parking entry section to sense the vehicle entry. The network-connected sensing elements are installed in every parking slot to know the status of the parking area whether the parking space is free or not. The real-time data is continuously updated in the central server of the parking management system. This sensing data collection operation is the backbone of this smart system. Then, the data transfer operation is the next most important operation in which the gathered information about the vehicle entry and parking space allotment is transferred to the central server in a continuous manner. During the data transfer in the form of wired or wireless methods, data security is very important. The safety of the stored information is also ensured to prevent unauthorized data hacking. Several security measures are implemented in the parking management system to ensure the safety of the stored information. The next operation is intelligent handling of available information about the parking system. The dynamic request from the vehicle owner side is processed by the suitable algorithm, and the parking slot is effectively utilized [71]. The mobile app, the web is mainly used by the customers for the reservation of parking slots.

15.6.1   Guidance to the Customers Through Smart Devices

Mainly, the customers or drivers are properly guided to make parking the vehicle in the particular reserved parking area without uncertainty. Several types of devices are used in the parking management system to give proper guidance to the drivers of the vehicles. Inside the parking space, the loop detector, infrared, ultrasonic, microwave, or variable message sign type guiding system is installed. This guiding system gives the correct route to the vehicle drivers without confusion. Figure 15.8 illustrates the effective usage of several sensors in the parking reservation area to give proper guidance to the customer to their allocated slots. Large size commercial buildings or shopping malls have this type of smart vehicle parking system. A Global Positioning System and suitable software are needed to make the parking management system effective. The vacant slot is correctly denoted to the drivers through the web or mobile apps.

Figure 15.8 Vehicle parking system using IoT devices.

The various operations are done in the parking system through the central server in a centralized or decentralized form. The driver searches the mobile app or website to know about the details of the vehicle parking space. In most places, the parking space is allotted based on a first-come first-serve basis. The first requester gets the parking slot on request. The next request is transferred to the line-up. The central server searches to gather information from the IoT module to fix the parking slot for each request. The Global Positioning System reliability and network speed are most important for the successful implementation of this system. If the network speed is low, all parking management system concepts are impractical. Other practical problems are also arising in this parking system. Consider a driver requesting a parking slot in a particular area, one parking space is allocated by the central server. If the driver is uncooperative or declines, this allocated slot, which causes a delay for vehicle parking, the efficiency of the parking is decreased. Even though there is demand for the parking space, as well as the parking slot, is free, the parking service providers get affected in their revenue because of driver uncooperativeness. Some parking service providers fix the lot recharge to overcome this type of revenue loss. Finally, dynamic update of details from the parking space and engaging of vehicles in the free space makes proper revenue to the initial capital investors.

15.6.2   Smart Parking Reservation System

After the successful creation of infrastructure for the parking system, the parking reservation system must be concentrated by the service providers. The reservation for the parking slot must be properly handled by the parking service providers or municipality board to maximize their revenues or minimize the parking fee. The parking prices are increased by the private service providers by fixing higher pricing on demand. The municipality board altered the urban people about the traffic intensity. The parking fee is reduced by the municipality board to encourage the local public with less fee [72]. The customers may make their schedule on low traffic intensity time based on the available information. Thus, the intelligent reservation system needs complete information about the past city traffic and parking slot usage for future forecasting. Generally, from the past information, the peak hours are identified. The information is detailed on the website or mobile apps to give information to the public or drivers about the peak hours. The customers or drivers alter their schedules based on the available information or are ready to pay the higher price on peak hours [73]. Thus, this type of intelligent reservation system interfaces with customer needs with the parking space availability. The information is sent to the driver's mobile through SMS after the parking slot allocation. Also, the reservation system gives the data about the demand for a particular day. The public fixes the slot of parking space in advance through the reservation system. The current day traffic and parking slot demands are estimated in advance. This information teaches the parking service to fix the dynamic price of the parking slot at a particular time. Another main advantage of this intelligent management system is to provide service to the needy public on urgent requests. The multirequest from the public or drivers is properly identified and eliminated. The information about the neighboring parking space can be informed to the drivers through SMS or apps [74].

From the gathered information, the parking slot demand is segregated. For example, in the daytime, banks or offices need a huge parking space for their employees. In the evening, the public mainly utilizes the parking slots. Based on the information from the central server, the service estimates the minimum possible revenues from the regular bank work or office work employees. Also, this information gives a clear picture to the service providers whether the parking service can extend or not in the future [75, 76].

15.7   Advantages of IoT-Based Vehicle Parking System

Smart parking systems assisted by IoT devices become familiar in the Indian smart cities continuously. The IoT-based vehicle parking technology is used for its huge benefits and increased the brand image of the reputed malls or commercial buildings. The time taken to identify the free parking slot is majorly reduced. The driver or customer can easily locate the spot. This also has environmental benefits because of its low carbon emission due to less time taken and hassle-free parking lot located by the driver. The parking slot utilization factor is higher due to its smart usage. The traffic around the reputed buildings is reduced by providing proper parking space. The pollution and the management cost are also reduced. The smarter reservation and payment system improves the overall experience even in peak times. The safety for the vehicle and customer is improved. Since the parking space is completely monitored using CCTV devices and the vehicle entry, exit is properly recorded. The customer satisfaction is enhanced for its better convenience to their customers by the shop owners. The friction in the peak hours due to congestion of vehicles is majorly reduced. The unnecessary idling and driving of the cars are eradicated mostly since the drivers know the exact spot.

The overheads for the parking management system are minimized. This is the new business model in the highly dense urban areas. The dynamic models to handle the overall process give higher revenue to the investors. The real-time data is updated continuously in the central server for vehicle parking systems. The mobile apps and smart devices make the parking assistance, parking slot location, parking reservation system, overall parking management, and fee collection very much easier. This smart vehicle parking system is very important for smart city projects in Indian cities to attain the targeted goals.

15.8   Conclusion

In general, various methods or techniques in smart parking systems are discussed in the present chapter. The widespread availability of IoT-based sensors or devices has increased the system's reliability, allowing for the effective development of IoT-based applications. IoT offers a wide range of solutions for traffic management, including vehicle-to-vehicle communication for the best results. This chapter assists in comprehending various methods of traffic management under the smart city concept. It explains in a detailed manner about the government of India's "Smart City Mission" in detail. It also discusses vehicle parking requirements in the context of the Indian smart city configuration. It also detailed different technologies that can be suggested in the literature regarding smart parking and traffic management. Among these technologies, the IoT-based vehicle parking system has a good level of merits. Different types of sensors are utilized for this purpose and the advantages of adopting IoT-based vehicle parking systems are also explained in a detailed manner.

References

1. Zakhem, M. and Smith-Colin, J., Micromobility implementation challenges and opportunities: Analysis of e-scooter parking and high-use corridors. *Transp. Res. Part D. Trans. Environ.*, *101*, 103082, 2021.

2. Martin, M., Billah, M., Siddiqui, T., Abrar, C., Black, R., Kniveton, D., Climate-related migration in rural Bangladesh: A behavioural model. *Popul. Environ.*, *36*, 1, 85–110, 2014.

3. Caragliu, A., Del Bo, C., Nijkamp, P., Smart cities in Europe. *J. Urban Technol.*, *18*, 2, 65–82, 2011.

4. Letaifa, S.B., How to strategize smart cities: Revealing the SMART model. *J. Bus. Res.*, *68*, 7, 1414–1419, 2015.

5. Gandhi, B.K. and Rao, M.K., A prototype for IoT based car parking management system for smart cities. *Indian J. Sci. Technol.*, *9*, 17, 1–6, 2016.

6. Raj, E. F. I., Appadurai, M., Darwin, S., and Rani, E. F. I., Internet of Things (IoT) for sustainable Smart Cities. In *Internet of Things*, 163–188, CRC Press, 2022.

7. Raj, F.I. E F. I. and Appadurai, M., Internet of things-based Smart transportation system for Smart Cities, in: *Intelligent Systems for Social Good*, 39–50, Springer, Singapore, 2022.

8. https://www.un.org/development/desa/en/news/population/2018-revision-of-world-urbanization-prospects.html, Accessed on 03rd February 2022.

9. Praharaj, S. and Han, H., Cutting through the clutter of smart city definitions: A reading into the smart city perceptions in India. *City Cult. Soc.*, *18*, 100289, 2019.

10. https://smartcities.gov.in/, Accessed on 03rd February 2022.

11. Prasad, D. and Alizadeh, T., What makes Indian cities smart? A policy analysis of smart cities mission. *Telemat. Inform.*, *55*, 101466, 2020.

12. Kumar, A., Can the smart city allure meet the challenges of Indian urbanization?, in: *Sustainable Smart Cities in India*, pp. 17–39, Springer, Cham, 2017.

13. Vakula, D. and Kolli, Y.K., Low cost smart parking system for smart cities, in: *2017 International Conference on Intelligent Sustainable Systems (ICISS)*, IEEE, pp. 280–284, 2017.

14. Telang, S., Chel, A., Nemade, A., Kaushik, G., Intelligent transport system for a Smart City, in: *Security and Privacy Applications for Smart City Development*, pp. 171–187, Springer, Cham, 2021.

15. Jothimani, P., Chenniappan, P., Chidambaranathan, V., Factors impinge on the development of a smart city: A field study, in: *Environmental Science and Pollution Research*, pp. 1–10, 2022.

16. Gaffney, C., and Robertson, C., Smarter than smart: Rio de Janeiro's flawed emergence as a smart city. *Journal of Urban Technology*, 25, 3, 47–64, 2018.

17. Mandal, A. and Byrd, H., Density, energy and metabolism of a proposed smart city. *J. Contemp. Urban Aff.*, 1, 2, 57–68, 2017.

18. John, S.K., Sivaraj, D., Mugelan, R.K., Implementation challenges and opportunities of smart city and intelligent transport systems in India, in: *Internet of Things and Big Data Analytics for Smart Generation*, pp. 213–235, Springer, Cham, 2019.

19. Jog, Y., Singhal, T.K., Barot, F., Cardoza, M., Dave, D., Need gap analysis of converting a city into smart city. *Int. J. Smart Home*, 11, 3, 9–26, 2017.

20. Randhawa, A. and Kumar, A., Exploring sustainability of smart development initiatives in India. *Int. J. Sustain. Built Environ.*, 6, 2, 701–710, 2017.

21. Majumdar, S., Subhani, M.M., Roullier, B., Anjum, A., Zhu, R., Congestion prediction for smart sustainable cities using IoT and machine learning approaches. *Sustain. Cities Soc.*, 64, 102500, 2021.

22. Rashid, M.M., Musa, A., Rahman, M.A., Farahana, N., Farhana, A., Automatic parking management system and parking fee collection based on number plate recognition. *Int. J. Mach. Learn. Comput.*, 2, 2, 94, 2012.

23. SR, M.B., Automatic smart parking system using Internet of Things (IOT). *Int. J. Sci. Res. Publ.*, 5, 12, 629–632, 2015.

24. Gaur, A., Scotney, B., Parr, G., McClean, S., Smart city architecture and its applications based on IoT. *Proc. Comput. Sci.*, 52, 1089–1094, 2015.

25. Tragos, E.Z., Angelakis, V., Fragkiadakis, A., Gundlegard, D., Nechifor, C.S., Oikonomou, G., Gavras, A., Enabling reliable and secure IoT-based smart city applications, in: *2014 IEEE International Conference on Pervasive Computing and Communication Workshops (PERCOM WORKSHOPS)*, IEEE, pp. 111–116, 2014.

26. Khan, L.U., Visible light communication: Applications, architecture, standardization and research challenges. *Digit. Commun. Netw.*, 3, 2, 78–88, 2017.

27. Ashokkumar, K., Sam, B., Arshadprabhu, R., Cloud based intelligent transport system. *Proc. Comput. Sci.*, 50, 58–63, 2015.

28. Zhang, K., De La Fortelle, A., Zhang, D., Wu, X., Analysis and modeled design of one state-driven autonomous passing-through algorithm for driverless vehicles at intersections, in: *2013 IEEE 16th International Conference on Computational Science and Engineering*, IEEE, pp. 751–757, 2013.

29. de La Fortelle, A., Analysis of reservation algorithms for cooperative planning at intersections, in: *13th International IEEE Conference on Intelligent Transportation Systems*, IEEE, pp. 445–449, 2010.

30. Huang, S., Sadek, A.W., Zhao, Y., Assessing the mobility and environmental benefits of reservation-based intelligent intersections using an integrated simulator. *IEEE Trans. Intell. Transp. Syst.*, 13, 3, 1201–1214, 2012.

31. Zhang, Y.J., Malikopoulos, A.A., Cassandras, C.G., Optimal control and coordination of connected and automated vehicles at urban traffic intersections, in: *2016 American Control Conference (ACC)*, IEEE, pp. 6227–6232, 2016.

32. Kaur, H. and Malhotra, J., A review of smart parking system based on internet of things. *Int. J. Intell. Syst. Appl. Eng.*, 6, 4, 248–250, 2018.

33. Liu, C. and Ke, L.., Cloud assisted Internet of things intelligent transportation system and the traffic control system in the smart city. *J. Control Decis.*, 1–14, 2022.

34. Lanke, N. and Koul, S., Smart traffic management system. *Int. J. Comput. Appl.*, 75, 7, pp. 19–22, 2013.

35. Al-Sakran, H.O., Intelligent traffic information system based on integration of Internet of Things and agent technology. *Int. J. Adv. Comput. Sci. Appl. (IJACSA)*, 6, 2, 37–43, 2015.

36. Hanif, N.H.H.M., Badiozaman, M.H., Daud, H., Smart parking reservation system using short message services (SMS), in: *2010 International Conference on Intelligent and Advanced Systems*, IEEE, pp. 1–5, 2010.

37. Raj, E. F. I., and Balaji, M., Application of deep learning and machine learning in pattern recognition, in: *Advance Concepts of Image Processing and Pattern Recognition*, 63–89, Springer, Singapore, 2022.

38. Fantin Irudaya Raj, E. and Appadurai, M., The hybrid electric vehicle (HEV)—An overview, in: *Emerging Solutions for e-Mobility and Smart Grids*, pp. 25–36, 2021.

39. Al-Kharusi, H. and Al-Bahadly, I., Intelligent parking management system based on image processing. *World Journal of Engineering and Technology*, 2, 55–67, 2014. http://dx.doi.org/10.4236/wjet.2014.22006"10.4236/wjet.2014.22006.

40. Gampala, V., Kumar, M.S., Sushama, C., Raj, E.F.I., Deep learning based image processing approaches for image deblurring. *Mater. Today: Proc.*, 2020.

41. Oka, D.K., Furue, T., Langenhop, L., Nishimura, T., Survey of vehicle IoT bluetooth devices, in: *2014 IEEE 7th International Conference on Service-Oriented Computing and Applications*, IEEE, pp. 260–264, 2014.

42. Patil, M. and Bhonge, V.N., Wireless sensor network and RFID for smart parking system. *Int. J. Emerg. Technol. Adv. Eng.*, 3, 4, 188–192, 2013.

43. Sijini, A.C., Fantin, E., Ranjit, L.P., Switched reluctance motor for hybrid electric vehicle. *Middle-East J. Sci. Res.*, 24, 3, 734–739, 2016.

44. Yang, J., Portilla, J., Riesgo, T., Smart parking service based on wireless sensor networks, in: *IECON 2012-38th Annual Conference on IEEE Industrial Electronics Society*, IEEE, pp. 6029–6034, 2012.

45. Perković, T., Šolić, P., Zargariasl, H., Čoko, D., Rodrigues, J.J., Smart parking sensors: State of the art and performance evaluation. *J. Cleaner Prod.*, 262, 121181, 2020.

46. Appiah, O., Quayson, E., Opoku, E., Ultrasonic sensor based traffic information acquisition system; a cheaper alternative for ITS application in developing countries. *Sci. Afr.*, 9, e00487, 2020.

47. Farley, A. and Ham, H., Real time IP camera parking occupancy detection using deep learning. *Proc. Comput. Sci.*, 179, 606–614, 2021.

48. Rathee, G., Sharma, A., Saini, H., Kumar, R., Iqbal, R., A hybrid framework for multimedia data processing in IoT-healthcare using blockchain technology. *Multimed. Tools Appl.*, 79, 15, 9711–9733, 2020.

49. Neelakandan, S., Rene Beulah, J., Prathiba, L., Murthy, G.L.N., Irudaya Raj, E.F., Arulkumar, N., Blockchain with deep learning-enabled secure healthcare data transmission and diagnostic model. *Int. J. Model. Simul. Sci. Comput.*, 2241006, 2022.

50. Gnanasekar, A.K., Deivakani, M., Bathala, N., Raj, E., Ramakrishna, V., Novel low-noise CMOS bioamplifier for the characterization of neurodegenerative diseases, in: *GeNeDis 2020*, pp. 221–226, Springer, Cham, 2021.

51. Ch, G., Jana, S., Majji, S., Kuncha, P., Tigadi, A., Diagnosis of COVID-19 using 3D CT scans and vaccination for COVID-19. *World J. Eng.*, 2021.

52. Mourtzis, D., Vlachou, E., Milas, N.J.P.C., Industrial big data as a result of IoT adoption in manufacturing. *Proc. CIRP*, 55, 290–295, 2016.

53. Ullo, S.L. and Sinha, G.R., Advances in IoT and smart sensors for remote sensing and agriculture applications. *Remote Sens.*, 13, 13, 2585, 2021.

54. Rani, E., Pushparaj, T.L., Raj, E., Escalating the resolution of an urban aerial image via novel shadow amputation algorithm. *Earth Sci. Inf.*, 1–9, 2022.

55. Samant, P., Bhushan, M., Kumar, A., Arya, R., Tiwari, S., Bansal, S., Condition monitoring of machinery: A case study, in: *2021 6th International Conference on Signal Processing, Computing and Control (ISPCC)*, IEEE, pp. 501–505, 2021.

56. Fantin Irudaya Raj, E. and Balaji, M., Analysis and classification of faults in switched reluctance motors using deep learning neural networks. *Arab. J. Sci. Eng.*, 46, 2, 1313–1332, 2021.

57. Syu, J.H., Wu, M.E., Srivastava, G., Chao, C.F., Lin, J.C.W., An IoT-based hedge system for solar power generation. *IEEE Internet Things J.*, 8, 13, 10347–10355, 2021.

58. Ou, Q., Zhen, Y., Li, X., Zhang, Y., Zeng, L., Application of Internet of Things in smart grid power transmission, in: *2012 Third FTRA International Conference on Mobile, Ubiquitous, and Intelligent Computing*, IEEE, pp. 96–100, 2012.

59. Raj, E.F.I., Available transfer capability (ATC) under deregulated environment. *J. Power Electron. Power Syst.*, 6, 2, 85–88, 2016.

60. Lin, X., Wu, J., Bashir, A.K., Li, J., Yang, W., Piran, J., Blockchain-based incentive energy-knowledge trading in IoT: Joint power transfer and AI design. *in IEEE Internet of Things Journal*, 9, 16, 14685-14698, 15 Aug.15, 2022, doi:10.1109/JIOT.2020.3024246.

61. Menon, V.G., Jacob, S., Joseph, S., Sehdev, P., Khosravi, M.R., Al-Turjman, F., An IoT-enabled intelligent automobile system for smart cities. *Internet Things*, 18, 100213, 2022.

62. Raj, E., Appadurai, M., Rani, E., Jenish, I., Finite-element design and analysis of switched reluctance motor for automobile applications. *Multiscale Multidiscip. Modeling Experiments Design*, 1–9, 2022.

63. Krasniqi, X. and Hajrizi, E., Use of IoT technology to drive the automotive industry from connected to full autonomous vehicles. *IFAC-PapersOnLine*, 49, 29, 269–274, 2016.

64. Muangprathub, J., Boonnam, N., Kajornkasirat, S., Lekbangpong, N., Wanichsombat, A., Nillaor, P., IoT and agriculture data analysis for smart farm. *Comput. Electron. Agric.*, 156, 467–474, 2019.

65. Raj, E., Appadurai, M., Athiappan, K., Precision farming in modern agriculture, in: *Smart Agriculture Automation Using Advanced Technologies*, pp. 61–87, Springer, Singapore, 2021.

66. Chatterjee, S., Kar, A.K., Gupta, M.P., Success of IoT in smart cities of India: An empirical analysis. *Gov. Inf. Q.*, 35, 3, 349–361, 2018.

67. Raj, E. F. I., Implementation of machine learning techniques in unmanned aerial vehicle control and its various applications, in: *Computational Intelligence for Unmanned Aerial Vehicles Communication Networks*, pp. 17-33, Springer, Cham, 2022.

68. Sharma, S., Nanda, M., Goel, R., Jain, A., Bhushan, M., Kumar, A., Smart cities using internet of things: Recent trends and techniques. *Int. J. Innov. Technol. Exp. Eng.*, 8, 9S, 24–28, July 2019, Available: https://doi.org/10.35940/ijitee.I1004.0789S19.

69. Tekouabou, S.C.K., Cherif, W., Silkan, H., Improving parking availability prediction in smart cities with IoT and ensemble-based model. *J. King Saud Univ.-Comput. Inf. Sci.*, 34, 3, 687–697, 2020.

70. Javaid, S., Sufian, A., Pervaiz, S., Tanveer, M., Smart traffic management system using Internet of Things, in: *2018 20th International Conference on Advanced Communication Technology (ICACT)*, IEEE, pp. 393–398, 2018.

71. Deivakani, M., Kumar, S.S., Kumar, N.U., Raj, E.F.I., Ramakrishna, V., VLSI implementation of discrete cosine transform approximation recursive algorithm. *J. Phys.: Conf. Ser.*, 1817, 1, 012017, 2021.

72. Ghorpade, S.N., Zennaro, M., Chaudhari, B.S., Node localization for smart parking systems, in: *Optimal Localization of Internet of Things Nodes*, pp. 51–66, Springer, Cham, 2022.

73. Hamid, A.H.F.A., Chang, K.W., Rashid, R.A., Mohd, A., Abdullah, M.S., Sarijari, M.A., Abbas, M., Smart vehicle monitoring and analysis system with IoT technology. *Int. J. Integr. Eng.*, *11*, 4, pp. 149–158, 2019.
74. Qadir, Z., Al-Turjman, F., Khan, M.A., Nesimoglu, T., ZIGBEE based time and energy efficient smart parking system using IoT, in: *2018 18th Mediterranean Microwave Symposium (MMS)*, IEEE, pp. 295–298, 2018.
75. Ji, Z., Ganchev, I., O'Droma, M., Zhao, L., Zhang, X., A cloud-based car parking middleware for IoT-based smart cities: Design and implementation. *Sensors*, *14*, 12, 22372–22393, 2014.
76. Park, W.H. and Cheong, Y.G., IoT smart bell notification system: Design and implementation, in: *2017 19th International Conference on Advanced Communication Technology (ICACT)*, IEEE, pp. 298–300, 2017.

Security of Smart Home Solution Based on Secure Piggybacked Key Exchange Mechanism

Jatin Arora* and Saravjeet Singh

Chitkara University Institute of Engineering and Technology, Chitkara University, Punjab, India

Abstract

Internet of Things (IoT)–based smart home appliance control provides better and comfortable living. The home control system includes temperature control, lighting control, air quality control, security system, and water control, etc. A smart system could be used to monitor all the home control systems by maintaining the required parameters in an appropriate range. The extensive utilization of IoT-based smart appliances for controlling the operations of home devices raises an issue of data security. The highly sensitive data is vulnerable to data leakage. A strong authentication mechanism is required for ensuring the integrity and confidentiality of data shared over an unsecured network. This chapter discusses secure data transmission and key exchange for ensuring the confidentiality of data. IoT devices have a major constraint of power and small size, which leads to restricting the computation to simple. Complex computation and extra overheads could not be deployed on the small processing devices. The widespread popularity of smart IoT-based home automation systems should be strongly secured to prevent any outside interventions and security threats. Thus, research in the field of IoT device security is of great importance by making use of already available security algorithms. Emerging trends and existing security challenges of IoT-based applications are discussed.

Keywords: Authentication, confidentiality, IoT security, IoT applications, distributed systems, key exchange

Corresponding author: jatin.arora@chitkara.edu.in

Ashok Kumar, Megha Bhushan, José A. Galindo, Lalit Garg and Yu-Chen Hu (eds.) Machine Intelligence, Big Data Analytics, and IoT in Image Processing: Practical Applications, (399–418) © 2023 Scrivener Publishing LLC

16.1   Introduction

Right from the ancient past to the latest present, the world has gradually become very smart with scientific wonders. Another scientific wonder awaits the future, Internet of Things (IoT): A smart world to live in. Being able to monitor and control the temperature of the room, the air quality of the room, and control all the electronic devices at home just by using the phone is a lifestyle of the future that is becoming possible day by day by advancements in the field of IoT [1]. IoT is a modern concept of the working of various devices under the same network of Internet so that connected devices can share data to provide real-time monitoring of different aspects of life. Different kinds of IoT computational frameworks are used by different devices based on their application and processing needs. Some of the most common frameworks are as follows [2]:

a) IoT computational framework
    The computing framework of IoT is meant for processing the data over the Internet, which lays the foundational working of a network of devices. Different types of computing frameworks are used depending upon the application and process location.

b) Fog computing
    Fog computing architecture makes use of edge devices like routers, integrated access devices (IADs), or wide area network (WAN), metropolitan area network (MAN) to access devices. It provides logical intelligence to the smart devices in respect of computation, storage, and network services.

c) Edge computing
    Edge computing features decentralized computing power, that is, the processing is run away from the data center. In Edge Computing, temporarily the data is processed by edge devices which benefit in favor of local storage and security.

d) Cloud computing
    Cloud computing is one of the most significant types of computing as it enables storage, access, and editing data on cloud servers that are Internet-based servers. For example, Google Drive, OneDrive, Dropbox, iCloud, etc. are very famous cloud computing-based applications that are used very much throughout the world. This computing is not very frequently used for processing IoT data because

of high latency and high load balancing while IoT requires high-speed processing [3].

There are different types of Cloud computing [4]

- Infrastructure as a service (IaaS)
- Platform as a service (Pass)
- Software as a service
- Mobile backend as a service

e) Distributed computing

As the name suggests, distributed computing refers to computing where the components of computing are distributed across a network of computers that communicate with each other to perform computational actions. This type of computing deals with large volumes of data; big data. The large volume of data is first divided into small volumes of data and then those small volumes are distributed across a net work of servers which overall makes it possible for successful computation of high loads of data [5].

f) Internet of things: global impact on society

Advances in technology have led to a smarter world. But now, it's time for something greater, the basic human needs of energy, shelter, mobility, urban planning, etc. will be going through a revolutionary advancement that will make the future more modern and advanced. Below mentioned points provide the global impact of IoT:

1) Smart city

Making the city smarter is a very wise choice because the buildings, the street lights, the traffic lights, and all the infrastructure of the locality will be connected through the Internet, thereby allowing the devices to share data making the evaluation of the city very quick and efficient. The pollution and temperature levels in the city, its sectors, and its streets; can be monitored precisely through smart IoT devices. Also, at times of accidents, earthquakes, volcanos, and in case of any type of disasters; a smarter city will be able to protect more life and also property. Investigation and prevention of any type of theft, or crime will also become easier, making the city smart and safe.

2) Smart energy

Energy is the fluid of all kinds of modern electronic services. So, the judicial production, and utilization of energy are significantly important. Through IoT, the overall consumption of energy is reduced offering high quality, economic friendly, and environmentally friendly smart energy. Smart energy applications, smart leak monitoring, and renewable energy resources are components of smart energy. The evaluation of the past-present energy consumption and estimation of the future energy consumption has become possible by a smart energy application called a smart grid [6].

3) Smart mobility

Through IoT, transportation is improving in every aspect of driving safely, less traffic on roads, improvement in public transportation, etc. Self-driving cars use IoT sensors and wireless connections to drive a car safely and effectively [7]. Autonomous cars can decide routes by themselves to avoid traffic jams. The traffic can also be controlled in a better and less chaotic manner by making use of IoT sensors of roads, sensors of vehicles, road cameras that can be connected with IoT to manage the traffic in a much more effective way. One of the most widely used modes of transportation is Public transportation. The best route that should be followed by the public vehicles can be decided with the help of communication between IoT sensors of the vehicle and the locality.

4) Smart citizens

Through smart technological IoT devices all-around; because humans always adapt to their surroundings, its overall benefit will be enjoyed in terms of environment, infrastructure, time-saving, security, safety, and social health. The quality of life will be upgraded to a new level of smartness where even the difficult tasks will be done very easily through IoT, making life easier, comfortable, and joyful.

5) Urban planning

The futuristic urban developments are equally important as the present urban developments. So, for better optimization of space and resources, the data has to be collected and present urban plans of the cities have to be deeply monitored which is possible in the most effective way through IoT. This proves IoT is the future of Global development [8].

The usage of IoT-based devices is broadly accepted by some of the key areas and is as follows [9].

a) Quality of life: IoT has a direct link toward improving the quality of life as the complexity of work done in daily life-style becomes easier and better through IoT [10].

b) Urban development: Through the data collected from past and present, planning cities will become very easy and helps in making a much more modern and organized city.

c) Food production: Better quality and enough quantity of food are made assured through the use of smart devices for the production of food. Temperature control IoT devices are used to ensure the best quality of food to be preserved.

d) Smart agriculture using IoT: All the processes of agriculture have favored the smarter approach of agriculture.

e) For example: In crop water management, the IoT device makes sure that the water used in agriculture is perfectly sufficient for the crop and also avoids the wastage of water.

f) Integrated pest control management: The movement of pests is monitored and detected by PIR sensors of IoT which alarms the farmer of their presence.

g) Hospitals: IoT devices can be very useful in tracking successful surgeries and collecting data from them to increase the chances of more successful operations by modifications in medical instruments and machines they use.

h) Machine learning: Through data collection by IoT, the collected data can be analyzed for teaching the smart devices about how to make the right decisions regarding the purpose of the smart device.

i) Energy supply: Through IoT, the production and consumption of energy can be monitored at present, so the future ratio of energy production and consumption can be estimated.

j) Water distribution: The pipeline system of water distribution can be trained to become smart on its own and provide more water to the places where water is needed more and less water and the places where the population is less so the water does not get wasted on its own. Any leakage in the pipeline can also be detected easily by the use of IoT smart devices.

k) Transportation: Autonomous vehicles have IoT sensors that work with wireless connections to interact with vehicles giving it the power of decision making on its own that is based on an algorithm so that the vehicle can self- drive. It will help in safer driving and also the selection of the right route path which involves less traffic [11].

16.2   IoT Challenges

IoT systems face a lot of challenges w.r.t data, application, communication channel, end-to-end encryption, and many more. The IoT architecture is shown in Figure 16.1. Some of the key challenges corresponding to the IoT architecture are as follows [12]:

- **Communication Scalability:** Countless devices are connected through IoT in a most likely hierarchical manner so the interconnected device will substantially outnumber the current Internet in terms of magnitude. Therefore, the scalability of communication channels is a major challenge.
- **Application Security and Trust:** The basic concept of IoT is the interconnection of various data servers over the Internet via IoT applications. It leads to more and more involvement of data to be received and accessed by various applications thereby increasing the requirement of data security and trust. The data that is shared across the network of IoT through one of the servers could have some trust issues

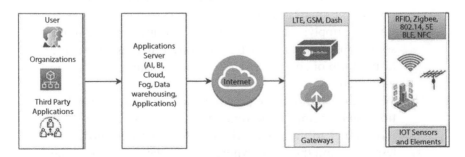

Figure 16.1 Generic IoT architecture.

alarming whether it is safe to access data by other servers or not because a fault in data received by one server could cause issues in the whole network of IoT.

- **End-to-End Encryption of IoT Data:** The high processing power of IoT and wide distribution of real-time data make the network of IoT more prone to cyber-attacks. Loopholes of data leak and privacy of individuals make IoT the perfect aim for cyber-criminals, so the implementation of end-to-end encryption of IoT data must be done in a very proper, safe, and concise manner [13].

16.3   IoT Vulnerabilities

The standardization of IoT is still under process and intensified vulnerabilities will cause more security breaches in IoT systems. This section discusses some of the general security threats of IoT systems [14].

1) **Hardware threats:** The security of IoT devices is not considered while they are developed and focused only on the working of it. This causes the addition of security features later on. The hardware vulnerabilities like open physical interface, insecure boot process, open doors for intruders. The security and integrity of such devices depend mainly on secure firmware and reliable installation [15].

2) **Social engineering threats:** Social networking and human interactions have greatly impacted the usage of IoT devices. The openly available default settings of end devices on social media made IoT devices vulnerable to social attacks. Intruders can access the end devices by using the default credentials and may upload confidential information on social media.

3) **Legal challenges:** The security of IoT information is not covered under any of the legal obligations. The responsibility of loss of data and information is completely dependent on the user of the service. As per the current legislation policies, there are no such rules and regulations for IoT data safety. Some organizations are working for the development of legal rules and regulations for the safety of user data.

4) **Lack of awareness:** The most common attribute contributing toward insecurity of IoT data is lack of awareness toward

security breaches like phishing attacks, recording keystrokes, malware attacks, password-guessing attacks. The additional factor responsible for a data breach is the usage of the insecure public network over smartphones and other handheld devices. Short and commonly used password strings, insecure Internet connections, plain data transmission are some of the awareness issues at the user's site.

5) **Denial of Service (DoS) attack:** Over resource utilization attack consists of memory consumption, CPU overloading, battery drainage, choking of communication channels, over disk utilization. Most of the time, network routers and switches are the major targets of the DoS attack [16].

16.4   Layer-Wise Threats in IoT Architecture

The IoT architecture imposes various security threats at intermediate layers which are very challenging to handle [17]. There may be multiple security measures to overcome the threats but handling these threats layer-wise would be much beneficial. The physical/sensing layer suffers from physical threats like theft, loss of signal, passive sniffing. A battery drainage attack causes a device to be shut down and discontinue the device functionality. The malfunctioning of the end device may cause major financial loss and lives [18]. Therefore, considering layer wise security for IoT architecture would provide better control on reducing the IoT threats [19]. The IoT architecture consists of various devices shown in Figure 16.2.

The complete IoT architecture consists of four layers: (1) sensing, (2) network, (3) middleware, and (4) application layer [20]. All of these layers use multiple devices and technologies that are responsible for multiple security

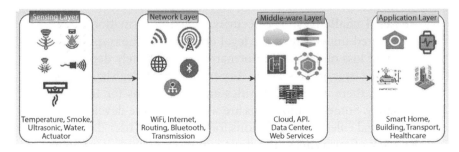

Figure 16.2 Layer-wise IoT device architecture.

threats. The subsequent section provides detailed information about all the possible vulnerabilities in each IoT architecture level [21].

16.4.1 Sensing Layer Security Issues

The physical layer in IoT infrastructure mainly consists of various sensors, switches, and actuators. The major purpose of the sensor is to receive the stimulus happening in the environment. Switches are used to control the operations of devices. The role of actuators is to carry out a fixed task as per the received signal. Sensors may be frequency detection sensors, smoke sensors, camera sensors, light sensors, pressure sensors, etc [22]. The actuator can be a physical electrical mechanism that can change the current state of the device to another state, e.g., electrical relays. Other physical/sensing layer devices include GPS, WSNs, RFIDs, etc. The physical layer is vulnerable to various threats and is given below [23]:

a) Node hacking: IoT systems are composed of various end nodes which are vulnerable to threats by an intruder. The intruder may hack or replace the node with its node. The newly installed node may try to access the data and send it to the attacker. A single compromised node may lead to the leakage of the complete IoT system.

b) Virus injection: The intruder may try to inject some malicious code in the running image of the process. Generally, this type of attack is simulated by stub code present in the application code or it may occur during updating of the application code. This attack provides an easy way for an intruder to execute unethical functions to fetch the confidential data.

c) Incorrect data injection: After the node is hacked, an intruder may inject incorrect data for performing the unethical task. This malfunctioning of the node may result in incorrect data storage that leads to financial and monetary losses.

d) Side-channel attacks (SCA): This type of attack is based on the various side-channels, such as power consumption, exposure to electromagnetic radiation, laser beam attacks, a smoke attack that leads to the abnormal behavior of IoT devices. The IoT device must be enclosed in a casing that is resistant to such attacks.

e) Boot level attacks: The end devices start their secure functionality only after the complete boot process. The intruder
may obtain access to the device by booting it with another
operating system. The low-powered IoT device may lose all
power in the start-restart cycle.

16.4.2   Network Layer Security Issues

The network layer is responsible for the transmission of the data from one
device to another device in a secure channel [24]. There are some security
threats to the network layer which are as follows:

a) Phishing attack: Phishing attacks generally target the victims to steal confidential data by sending fraudulent mes-
sages/mail. The purpose of the attacker is to attack some
of the IoT devices and make the victim. This is carried out
by sending malicious links on the web page, email, etc. to
steal the login information. If an attacker can steal the login
information then the entire IoT system will be compromised. The network layer mostly suffered from phishing
attacks, and it must be prevented by applying secure firmware in the IoT infrastructure.

b) Denial of Service attack (DoS): The aim of the DoS attack
is to send a lot of request messages to the target device
that leads to the overutilization of all resources in IoT
devices. It makes an IoT device unavailable. Sometimes,
an attacker targets the system by flooding it from multiple
systems, which are considered as distributed denial of service (DDoS) attacks, and it is difficult to track. The network
layer of IoT devices is vulnerable to such attacks as there are
very limited power components used and the consumption
of very small-sized resources leads to unavailability in the
IoT network. The complexity of the IoT networks makes it
an easy way to launch such attacks.

c) Data leakage: The network layer deals with a lot of data
exchange among the IoT nodes. Data is considered the
most valuable asset of an organization and it is the common target of all attackers which could be easily accessible
during data transmission. Data movement is highly vulnerable to sniffing attacks where the attacker passively analyzes
the data traffic and obtains the secret information.

d) Data routing attack: Routing of data is performed to redirect the data packets to other intermediate nodes to reach the final destination. The attacker tries to redirect the entire traffic to a void route where data traffic reaches nowhere. This type of attack never returns any reply message and causes an IoT node to wait endlessly.

16.4.3 Middleware Layer Security Issues

The middleware of IoT infrastructure provides an abstraction layer between the application layer and the network layer. The interface provided by middleware is very powerful and computation intensive. It consists of APIs to handle the request of communicating layers. It also includes buffer memory, permanent storage to share the data in a secure manner. Although the middleware is secure, it is also vulnerable to external threats. These threats can compromise the entire IoT system and are discussed below.

a) MITM attack: Man-in-the-Middle Attack is usually performed by the attacker to gather confidential keys and knowledge about the data traffic. In this attack, the attacker usually becomes a middle man in the agreement task by sending fraudulent identity. The attacker usually impersonates an authenticated user to gain access to confidential credentials. If the attacker hacks the proxy server or key issuing server then the entire control of IoT communication is exposed to the attacker without any information to actual users.

b) SQL statement Injection: It is a very common attack in applications using SQL queries to fetch the data. The attacker usually embeds a code that is querying for some confidential data to gain access to personal data. SQLi is considered the most common threat for web applications.

c) Signature tampering attack: The web services are commonly using XML signatures to authenticate the integrity and authenticity of the received data. In this attack, the signature algorithm is broken to temper the data and gain access to the entire data by making use of web vulnerabilities.

d) Malware injection in Cloud: The attacker can request some resources from the cloud service provider and install some malicious applications on the machine. The attacker keeps all the resources authenticated and valid to gain faith over the services offered. The victim machine shares the confidential data and loses its entire control to the attacker.

e) Cloud Flooding attack: The flooding of cloud resources by unwanted requests of services affects the response time of the server and makes the actual clients deprived of service. This type of attack badly affects the performance of cloud machines.

16.4.4   Gateways Security Issues

The interconnecting devices usually depend upon the gateway devices to establish a secure connection among them. Gateways perform cryptography of data transmitted through cloud and IoT devices. IoT devices are not limited to only a single type of devices; rather it consists of different types of devices which require various gateway layers in between them to communicate properly [25]. These gateway layers add some security issues for an IoT infrastructure and are discussed below:

a) New installation: Whenever a new IoT device is added to the system, it may add vulnerabilities of incompatibility with the existing system and create loopholes for the data breaches. The new device may leak data to a hacker by capturing the keys used for encryption/decryption.

b) Additional connections: Sometimes additional connections are created on the IoT device for additional reliability and these interfaces become easy back door entries for an intruder. The IoT provider must provide only minimal interfaces to avoid the back door entries. These back door entries are used to avoid the proper authentication channel.

c) Completely secure connection: It is assumed that the data transmitted from one IoT node to another node traveled in an encrypted manner through all the nodes. But it is not truly valid at all the intermediate nodes because it may read the unencrypted data. The confidential data read at interface layers may be accessed by an intruder. The IoT data should reach the destination securely without exposing the data at intermediate nodes.

d) IoT firmware breach: The firmware of an IoT device provides the initial startup and basic controls over the device. Usually, the firmware needs to be updated from time to time to prevent the device from external threats. The network interface provides the system update and it should confirm the security checksum.

16.4.5   Application Layer Security Issues

The most commonly used interface used on IoT devices is the application layer to send commands to the smart devices. The application layer has some security issues that are related to data loss and secrecy issues. The application layer is further divided into sublayers such as the business logic layer, storage layer, and resource allocation layer. Some of the most common application layer security issues are as follows [26]:

a) Loss of data: The IoT application-layer interface provides some controls to handle the device functionality. The intruder can see the input data and by applying brute force attack, some useful information could be fetched.

b) System access control: The access of IoT devices should be properly authorized and follow the legal process of data access. Any unauthorized way of accessing the system may lead to compromise of the entire system.

c) Sniff attack: The intruder may use a sniffing technique to analyze the data traffic and system utilization pattern. The sniffing applications installed in the application layer send the system details to an intruder such as passwords and OTPs.

d) Virus injections: The malicious code injection at the application layer results in the corruption of entire device functionality and breaks down the entire system. This illegitimate practice of code injection is not easily identified by the user, and it causes a lot of data loss [27].

16.5   Attack Prevention Techniques

IoT security measures are required at all the layers of IoT architecture but majorly it is required at the physical layer, network layer, and application layer. The security measures must ensure the protection of data from the physical layer to the application layer. This section is dedicated to the security measures required at the application layer to address the integrity and authentication of data. The authentication and integrity of data received must be properly checked, and it must use an irreversible hash function to avoid replay attacks. The authentication and integrity measures are discussed in the below section.

16.5.1   IoT Authentication

A mutual authentication scheme used for IoT devices was presented by Umer *et al.* based on hashing technique [28]. It also uses a feature extraction scheme to support irreversibility and avoid collision attacks. This scheme is lightweight and is suitable for IoT devices. The IoT node may send data from one node to another and receive data from the IoT node. Sometimes, IoT data may need to only send data but there is no receiving data. The verification of received data must be performed so that the incorrect or forged data may be rejected. This scheme is valid only for sending data in one direction and not appropriate for sending data in opposite directions [29]. A preshared key matrix among the communicating nodes would provide the key derivation coordinates. The secure key is generated from the matrix coordinates and kept saved in the device for a particular duration of time for sending encrypted data to the receiver node.

The communicating nodes may cancel the key at any moment and regenerate the new key. The continuously changed keys are required for securing the data, which is highly confidential. The sharing of preshared matrices is also a challenging process if the numbers of nodes are large in number. Devising an accurate access control and authentication is equally important to implement a secure connection among the IoT nodes. Samrah *et al.* proposed a public key cryptography system that is compatible with IoT devices of limited storage and low-powered in nature. Various attacks have been prevented by using the timestamp value. This scheme uses elliptical curve cryptography to create a secret key followed by mutual authorization and data access control granted. The private key used for the session is derived by using elliptic curve cryptography based on public and private values [30].

Public key cryptography is compatible with existing technologies Wi-Fi, Bluetooth, 5G. The key derived from public and private values is highly secure and prevents Man-in-the-middle attacks. Each IoT device stores its private parameters to create a key used for authentication and securing the message sent over the network. If an IoT device is not secured and preventive actions are not taken, then anyone could gain access to the device and sniff the data transmitted through it. The communicating IoT devices must authenticate mutually and do not incur extra overheads. The major requirement of authentication and authorization is that it should be lightweight [30].

16.5.2   Session Establishment

IoT devices are mobile and migrate from one location to another location [31–34], therefore, secure session establishment between the nodes requires a smooth transition of access rights. The mutual trust establishment between IoT devices requires a level wise access control framework. The session is established by sharing the preshared key among the communicating nodes via the trusted channel. The sender sends all key deriving parameters to the trusted nodes for obtaining the key for sending a reply message. The communicating nodes will use their own set of key deriving parameters to obtain the key used for encrypting the message for the next round of communication. The IoT internal security architecture cannot accommodate all messages sending data and require additional security measures at the application level to protect the data at intermediate IoT nodes. In the proposed work, a key sharing model is presented, which would reduce the burden of key sharing on IoT nodes. Key sharing is always a big challenge for every IoT application. The piggybacked key sharing scheme provides better security of data shared over the network. All the key deriving parameters are enclosed in a response message, and the extra overhead of sharing the modified key information would be removed. The received information contains the key deriving parameters and the receiver will generate the key for sending the response message,

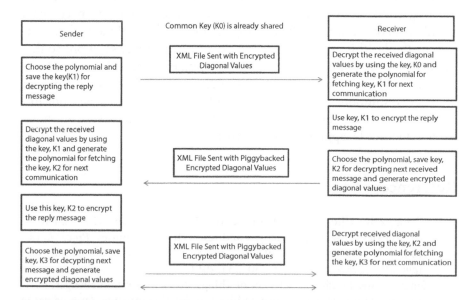

Figure 16.3 Key exchange mechanism for IoT devices.

which is encrypted by the key derived. Figure 16.3 represents the key sharing mechanism among the sender and receiver IoT nodes.

The sender and receiver node may use the polynomial generation methods like Newton Forward Interpolation, Newton Forward Interpolation, or Lagrange's Interpolation to regenerate the polynomial followed by obtaining the key used for encrypting the response message. This key exchange mechanism ensures the reduced number of messages exchanged among the communicating IoT nodes for deriving the key used for the next round of communication. The security of encrypted data solely depends on the strength of the AES algorithm. The availability of key values at IoT nodes in a secure manner is achieved by using the higher-order polynomial equations after the successful session establishment among the IoT node,

16.6   Conclusion

The IoT devices and their communication network are heterogeneous and require security at each layer of their infrastructure. Depending on only a single IoT layer would not guarantee secure communication, and it would be vulnerable to data breaches. IoT security is also challenging in heterogeneous and mobile infrastructure. This chapter represents current IoT security challenges and lightweight key exchange mechanism for communicating new keys in a given session. The secure key exchange mechanism could be further improved by making use of different polynomial generation methods. This article mainly focuses on smart home solutions so that the communication going on with home appliances must be secured and not vulnerable to various data leakage attacks. The IoT market offers a lot of security solutions to protect the data from external breaches, but IoT devices can process only limited data, therefore, complex security measures could not be applied for IoT security.

References

1. Zhang, Z.K., Cho, M.C.Y., Wang, C.W., Hsu, C.W., Chen, C.K., Shieh, S., IoT security: Ongoing challenges and research opportunities. *Proceedings - IEEE 7th International Conference on Service-Oriented Computing and Applications, SOCA 2014*, pp. 230–234, 2014.
2. Singh, S. and Singh, N., Internet of Things (IoT): Security challenges, business opportunities & reference architecture for E-commerce. *Proceedings of*

the 2015 International Conference on Green Computing and Internet of Things, ICGCIoT 2015, pp. 1577–1581, 2016.

3. Rayes, A. and Salam, S., Internet of Things-from hype to reality: The road to digitization, in: *Internet of Things From Hype to Reality: The Road to Digitization*, pp. 1–328, 2016.

4. S.V., Internet of Things (IoT) based smart agriculture in India: An overview. *J. ISMAC*, 3, 1, 1–15, 2021.

5. Anitha, A., Home security system using Internet of Things. *IOP Conf. Ser.: Mater. Sci. Eng.*, 263, 4, 1–11, 2017.

6. Zhou, W., Jia, Y., Peng, A., Zhang, Y., Liu, P., The effect of IoT new features on security and privacy: New threats, existing solutions, and challenges yet to be solved. *IEEE Internet Things J.*, 6, 2, 1606–1616, 2019.

7. Tina, Sonam, Harshit, Singla, M., Smart lightning and security system. *Proceedings - 2019 4th International Conference on Internet of Things: Smart Innovation and Usages, IoT-SIU 2019*, pp. 1–6, 2019.

8. Vasicek, D., Jalowiczor, J., Sevcik, L., Voznak, M., IoT smart home concept. *2018 26th Telecommunications Forum, TELFOR 2018 - Proceedings*, pp. 1–4, 2018.

9. Ghazal, T.M., Hasan, M.K., Hassan, R., Islam, S., Norul, S., Sheikh, H., Security vulnerabilities, attacks, threats and the proposed countermeasures for the Internet of Things applications. *Solid State Technol.*, 63, 1s, 1–9, 2020.

10. Al-Kuwari, M., Ramadan, A., Ismael, Y., Al-Sughair, L., Gastli, A., Benammar, M., Smart-home automation using IoT-based sensing and monitoring platform. *Proceedings - 2018 IEEE 12th International Conference on Compatibility, Power Electronics and Power Engineering, CPE-POWERENG 2018*, pp. 1–6, 2018.

11. Agarwal, K., Agarwal, A., Misra, G., Review and performance analysis on wireless smart home and home automation using IoT. *Proceedings of the 3rd International Conference on I-SMAC IoT in Social, Mobile, Analytics and Cloud, I-SMAC 2019*, pp. 629–633, 2019.

12. Mahmoud, R., Yousuf, T., Aloul, F., Zualkernan, I., Internet of Things (IoT) security: Current status, challenges and prospective measures. *2015 10th International Conference for Internet Technology and Secured Transactions, ICITST 2015*, pp. 336–341, 2016.

13. Ling, Z., Liu, K., Xu, Y., Jin, Y., Fu, X., An end-to-end view of IoT security and privacy. *2017 IEEE Global Communications Conference, GLOBECOM 2017 - Proceedings*, vol. 2018-January, pp. 1–7, 2017.

14. Pirbhulal, S. *et al.*, A novel secure IoT-based smart home automation system using a wireless sensor network. *Sensors (Switzerland)*, 17, 1, 1–19, 2017.

15. Zaidan, A.A. and Zaidan, B.B., A review on intelligent process for smart home applications based on IoT: Coherent taxonomy, motivation, open challenges, and recommendations. *Artif. Intell. Rev.*, 53, 1, 141–165, 2020.

16. Shin, S. and Seto, Y., Development of IoT security exercise contents for cyber security exercise system. *International Conference on Human System Interaction, HSI*, vol. 2020-June, pp. 281–286, 2020.

17. Neshenko, N., Bou-Harb, E., Crichigno, J., Kaddoum, G., Ghani, N., Demystifying IoT security: An exhaustive survey on IoT vulnerabilities and a first empirical look on internet-scale IoT exploitations. *IEEE Commun. Surv. Tutor.*, 21, 3, 2702–2733, 2019.

18. Iqbal, W., Abbas, H., Daneshmand, M., Rauf, B., Bangash, Y.A., An in-depth analysis of IoT security requirements, challenges, and their countermeasures via software-defined security. *IEEE Internet Things J.*, 7, 10, 10250–10276, 2020.

19. Ryoo, J., Tjoa, S., Ryoo, H., An IoT risk analysis approach for smart homes. *Proceedings - 2018 4th International Conference on Software Security and Assurance, ICSSA 2018*, vol. 2, pp. 49–52, 2018.

20. Datta, P. and Sharma, B., A survey on IoT architectures, protocols, security and smart city based applications. *International Conference on Computing, Communications and Networking Technologies, ICCCNT 2017*, vol. 8, pp. 1–5, 2017.

21. Zhou, W. *et al.*, Discovering and understanding the security hazards in the interactions between IoT devices, mobile apps, and clouds on smart home platforms. *Proceedings of the 28th USENIX Security Symposium*, pp. 1133–1150, 2019.

22. Chaurasia, T. and Jain, P.K., Enhanced Smart Home Automation System based on Internet of Things. *Proceedings of the 3rd International Conference on I-SMAC IoT in Social, Mobile, Analytics and Cloud, I-SMAC 2019*, pp. 709–713, 2019.

23. Hassija, V., Chamola, V., Saxena, V., Jain, D., Goyal, P., Sikdar, B., A survey on IoT security: Application areas, security threats, and solution architectures. *IEEE Access*, 7, 82721–82743, 2019.

24. Bhatia, P., Rajput, S., Pathak, S., Prasad, S., IOT based facial recognition system for home security using LBPH algorithm. *Proceedings of the 3rd International Conference on Inventive Computation Technologies, ICICT 2018*, pp. 191–193, 2018.

25. Bull, P., Austin, R., Popov, E., Sharma, M., Watson, R., Flow based security for IoT devices using an SDN gateway. *Proceedings - 2016 IEEE 4th International Conference on Future Internet of Things and Cloud, FiCloud 2016*, pp. 157–163, 2016.

26. Gladence, L.M., Anu, V.M., Rathna, R., Brumancia, E., Recommender system for home automation using IoT and artificial intelligence. *J. Ambient Intell. Hum. Comput.*, 1–9, Special Issue, 2020, 0123456789.

27. Ahmed, T., Bin Nuruddin, A.T., Bin Latif, A., Arnob, S.S., Rahman, R., A real-time controlled closed loop IoT based home surveillance system for android using firebase. *2020 6th International Conference on Control, Automation and Robotics, ICCAR 2020*, April, pp. 601–606, 2020.

28. Majeed, U., Khan, L.U., Yaqoob, I., Kazmi, S.M.A., Salah, K., Hong, C.S., Blockchain for IoT-based smart cities: Recent advances, requirements, and future challenges. *J. Netw. Comput. Appl.*, 181, August 2020, 103007, 2021.

29. Lyu, Q., Zheng, N., Liu, H., Gao, C., Chen, S., Liu, J., Remotely access 'My' smart home in private: An anti-tracking authentication and key agreement scheme. *IEEE Access*, 7, 41835–41851, 2019.
30. Arif, S., Khan, M.A., Rehman, S.U., Kabir, M.A., Imran, M., Investigating smart home security: Is blockchain the answer? *IEEE Access*, 8, 117802–117816, 2020.
31. Sharma, S., Nanda, M., Goel, R., Jain, A., Bhushan, M., Kumar, A., Smart cities using Internet of Things: Recent trends and techniques. *Int. J. Innov. Technol. Exploring Eng.*, 8, 9S, 24–28, 2019.
32. Samant, P., Bhushan, M., Kumar, A., Arya, R., Tiwari, S., Bansal, S., Condition monitoring of machinery: A case study. *6th International Conference on Signal Processing, Computing and Control (ISPCC)*, pp. 501–505, 2021.
33. Goel, R., Jain, A., Verma, K., Bhushan, M., Kumar, A., Mushrooming trends and technologies to aid visually impaired people. *International Conference on Emerging Trends in Information Technology and Engineering (ic-ETITE'20)*, February 24-25, 2020, IEEE, Vellore Institute of Technology, Vellore, India, 2020.
34. Mangla, M., Kumar, A., Mehta, V., Bhushan, M., Mohanty, S. N., Real-life applications of the Internet of Things: Challenges, applications, and advances, 1st Edition, pages 536, Apple Academic Press, New York, , ISBN: 9781003277460, 2022. https://doi.org/10.1201/9781003277460.

Machine Learning Models in Prediction of Strength Parameters of FRP-Wrapped RC Beams

Aman Kumar[1,2]*, Harish Chandra Arora[1,2], Nishant Raj Kapoor[1,2] and Ashok Kumar[1,2]

[1]CSIR-Central Building Research Institute, Roorkee, India
[2]AcSIR-Academy of Scientific and Innovative Research, Ghaziabad, India

Abstract

The corrosion of reinforced concrete (RC) structures is of increasing concern across the world. Repair, restoration, replacement, and the construction of new buildings all necessitate the use of economic and long-lasting technology. Fiber reinforced polymer (FRP) has been widely used in both retrofitting and new construction of buildings. FRP has seen an increase in the application as a repair composite material in reinforced concrete and masonry structures over the last decade due to its many properties. This material has various benefits, including excellent strength-to-weight ratios, stiffness-to-weight, lightweight, potential long-term durability, and relative ease of field use. This chapter provides a summary of the machine learning models in the estimation of bond strength between FRP and concrete surface, shear, and flexural strength of FRP-wrapped reinforced concrete beams.

Keywords: Machine learning, artificial intelligence, FRP-concrete bond, flexural strength, shear strength, FRP

Corresponding author: aman.civil16@outlook.com

Ashok Kumar, Megha Bhushan, José A. Galindo, Lalit Garg and Yu-Chen Hu (eds.) Machine Intelligence, Big Data Analytics, and IoT in Image Processing: Practical Applications, (419–446) © 2023 Scrivener Publishing LLC

17.1   Introduction

Construction is one of the most important industries in all economies across the world. Cement, clay, timber, steel, aluminum, and glass are just a few of the materials used in the building industry. However, in most cases, the industry is hesitant of experimenting with new materials, preferring to stick with tried-and-true options like concrete and steel, which have a strong and proven past track record. Concrete has become the most commonly used human-made construction material all over the world, as well as the ideal medium for civil and infrastructural development in the maritime environment, because of its commonality and low cost. Reinforced concrete is used in many civil engineering structures like multistory buildings, bridges, dams, nuclear power plants, etc. It is a frequent occurrence for concrete constructions to deteriorate with age due to a loss of structural durability and strength. Environmental factors, such as moisture intrusion, salt/acid attacks, and humidity conditions, increase the deterioration, resulting in reducing the structure's service life. Existing structures reaching the end of their useful lives must be assessed and monitored on a regular basis to determine their functional utility and assist minimize the fundamental cause of infrastructure deterioration [1]. The substantial investment made in existing buildings, ensuring the safety and serviceability of the aging building stock with limited resources is a serious concern for industrialized countries.

Steel reinforcement corrosion is the most prevalent cause of RC degradation [2]. Chloride-induced reinforcement corrosion is a prominent source of structural damage and early deterioration in RC structures, with serious consequences for safety, dependability, economics, and environmental performance.

FRP was originally used in architecture and construction in the 1930s. Glass fibers were the first fibrous material used as an engineering material. In the early 1970s, structures/buildings with glass FRP composites were being built all over the world (for example, an airport roof in Dubai and a dome structure in Benghazi). Meanwhile, composite materials and manufacturing processes improved enough. In the 1990s in Finland, an interesting model of a structure made up of 75% polymers and FRP composites were built [3]. At the same time, in Japan, FRP systems were utilized to replace steel reinforcing in order to counter corrosion-related problems. In 1988, Ishikawa, Japan, built the first concrete prestressed girders bridge utilizing carbon fiber reinforced polymer (CFRP) tendons [4]. Carbon fiber is gaining popularity in various industries, including aircraft, transportation, and sports goods, due to its good properties, such as resistance to corrosion, high mechanical strength, low density, high elastic modulus, and low cost [5]. In 2008, 35,000 tons of CFRPs were consumed; by 2014, this figure had doubled, owing to a 12% annual growth rate [6]. By 2020, it was estimated that 1 million metric tons of CFRPs will be produced worldwide [7]. But, only 60% of the virgin CFRPs composites manufacturing is efficiently applied in actual technical applications, resulting in 40% being deemed waste [8]. Given the scarcity of landfill space and the high cost of disposal of carbon fiber waste [9], the growing volume of carbon fiber waste has substantial environmental and economic consequences for long-term growth. External bonding of FRP composites to concrete parts has been identified as a possible solution for enhancing existing constructions' long-term performance.

17.1.1   Defining Fiber-Reinforced Polymer

FRP composite system is made up of a "polymer" and a "fiber" material for reinforcement. Polymers are lengthy chains of molecules that make up organic substances. Monomers are the fundamental components of this chain that repeat themselves. The polymer acts as a binding substance for fibers, protecting them from external mechanical wear and environmental impacts, and transferring load from fiber to fiber to ensure stability. Epoxies, vinylesters, and polyesters are the most extensively used polymers for FRPs. When compared to one another, each of the abovementioned polymers has benefits and drawbacks in relation to application technique, cost, and durability. The fibers and other components of FRP composites add stiffness and strength to the composite material by strengthening the polymers. These FRPs have benefits, such as a high strength and lower weight when compared to typical reinforcing materials, such as concrete and steel, rapidity of application, corrosion resistance, and minimum modification, in structural element geometry [10].

FRP composites have better material characteristics that traditional engineering materials like metals cannot match in most instances. To maximize material strength usage, FRPs may be produced in any length or cross-sectional shape. Furthermore, as compared to typical engineering materials, FRPs have advantages, such as it is light in weight, having good durability, resistance to corrosion, and being cost-effective based on life cycle cost analysis.

17.1.2   Types of FRP Composites

The FRP composites are broadly categorized into four types, such as carbon, glass, aramid, and basalt, as shown in Figure 17.1.

17.1.2.1   Carbon Fiber–Reinforced Polymer

Carbon fiber–reinforced polymer (CFRP) contains polymer resin and carbon fibers, in which the carbon fibers serve as a reinforcing material and the polymer resin is used to retain the fibers as a matrix, as shown in Figure 17.2. Carbon fibers are fibers that contain at least 90% carbon. They can be made from polymeric precursor materials such as cellulose, Polyacrylonitrile (PAN), pitch, and polyvinylchloride. These precursors are transformed into carbon fibers by a sequence of heating and tensioning treatment processes. Figure 17.2 represents the usual structure of the CFRP composite.

Carbon fibers are incredibly thin filaments (approximately 5–10 m in diameter) that are hardly visible to the naked eye [11]. Carbon fibers have been around for almost a century. In his first light bulb, Thomas Edison utilized carbon filaments made of bamboo fibers [12]. After decades of research and development, the market today offers a wide range of carbon

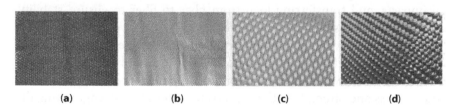

(a)   (b)   (c)   (d)

Figure 17.1 Types of FRPs (a) carbon, (b) glass (c) aramid (d) basalt.

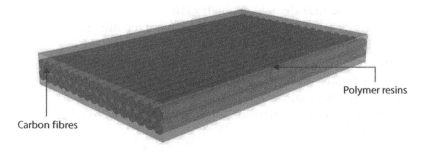

Polymer resins

Carbon fibres

Figure 17.2 Structure of FRP system.

Table 17.1 Mechanical characteristics of carbon fibers.

S. no.	Classes CFRP	E_f (GPa)	f_f (MPa)	HTT (°C)	C_o
1.	Steel (Fe500)	210	640	-	-
2.	Low elastic modulus carbon fiber (LMCF)	200 or lower	3500 or lower	<1000	Random
3.	Standard elastic modulus carbon fiber (HTCF)	200–280	2500 or higher	~1500	Largely parallel to fiber axis
4.	Intermediate elastic modulus carbon fiber (IMCF)	280–350	3500 or higher	~1500-2000	
5.	High elastic modulus carbon fiber (HMCF)	350–600	2500 or higher	>2000	
6.	Ultra-high elastic modulus carbon fiber (UHMCF)	600 or higher	2500 or higher	>200	

fibers with varying strengths and moduli. Table 17.1 compares the mechanical characteristics of four kinds of frequently used carbon fibers with steel material.

17.1.2.2 Glass Fiber

Glass fiber production has been experimented with since ancient times, but widespread production of glass fiber began in 1893 when "Edward Drummond Libbey" displayed a garment fashioned from fabric merging silk and glass fiber. "Russell Games Slayter received the first patent for glass wool manufacturing in 1938" [13]. Because the generated fiber exhibited strong electrical insulating qualities, glass fiber products are referred to as E-glass or electrical glass. Glass fibers were first employed as a heat-proofing layer in US naval battleships in 1939.

The glass fibers contain the compositions of different amount of raw materials, such as calcite for calcium oxide, sand for silica, colemanite for boron oxide, and clay for alumina. As a consequence, different types of glass fibers exhibit varied qualities such as alkali resistance or good mechanical capabilities when utilizing varying quantities of silica.

The glass comes in a variety of colors, chemical compositions, and properties. Glass fibers offer excellent mechanical characteristics, including strength, thermal properties, and durability, as well as good interfacial adhesion to the matrix. Glass fibers are commonly utilized as reinforcement in resins and composites due to their exceptional strength characteristics.

The most prevalent application for glass fibers is in general-purpose structural applications. They come in a variety of forms, the most popular of which is E-glass. Glass fibers are mostly used in passive strengthening, such as seismically damaged buildings. The drawback of E-Glass fibers is their low alkali resistance. To compensate for this weakness, a significant quantity of zirconia is added to make alkali-resistant AR-glass.

17.1.2.3   Aramid Fiber

Aramid fiber was the first organic fiber with significant tensile modulus and strength to be utilized as reinforcement in advanced composites. Aramid fiber was first marketed in the 1960s [14]. The various advantages of aramid fibers are resistance to heat, strong resistance to organic solvents, and low flammability. At around 500°C, aramid fibers begin to deteriorate. Aramid fiber's "inert" properties allow it to be used in an extensive range of applications. Aramid fibers, on the other hand, are susceptible to ultraviolet (UV) radiation, acids, and some salts. Bridge columns can be protected by aramid textiles against collapsing due to vehicle impact. Blast mitigation is another significant application sector. Aramid fibers have a low resistance to UV light by nature.

17.1.2.4   Basalt Fiber

Basalt fibers are produced from a naturally occurring complex alumina/ silica/other oxide basalt rock with a composition comparable to glass and are used to replace asbestos. They are available in nonwoven and filament forms, and their temperature performance is said to be better than that of glass fibers. It is utilized as a fireproofing cloth in aerospace and automotive applications, and it may also be employed as a composite reinforcement, including concrete reinforcement, in the construction of buildings, highways, and runways.

17.2   Strengthening of RC Beams With FRP Systems

The strengthening of reinforced concrete beams with FRP systems is broadly categorized into two parts: (i) flexural strengthening of the beams and (ii) shear strengthening of the beams as shown in Figure 17.3.

The usage of FRP in the building industry has increased dramatically during the previous two decades. With the help of Artificial Intelligence (AI) speed, and accuracy of the analysis works are getting increased. AI is majorly used to forecast the bond strength, shear and flexural strength of FRP reinforced beams.

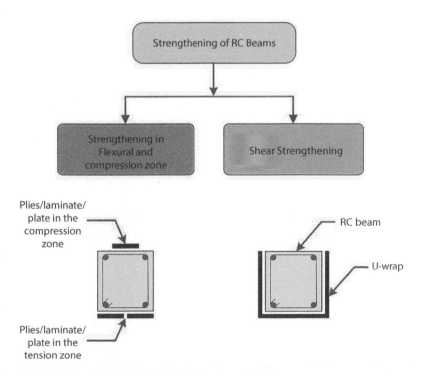

Figure 17.3 Types of strengthening in RC beams with FRP.

17.2.1   FRP-to-Concrete Bond

The interfacial connection between FRP sheets and concrete substrates is critical to the performance of most strengthening or retrofitting procedures for concrete structures employing externally bonded FRPs. The FRP-concrete bond plays an important role to transfer the stresses. Single shear and double-lap shear tests are the frequently used experimental methods to determine the bond strength between FRP and concrete surface as shown in Figure 17.4.

Various experimental investigations have been undertaken to explore the behavior of different input parameters such as compressive strength of concrete, thickness of fibers and bonded length in order to estimate the bond strength [15–18]. The various analytical formulas are given in the literature provided by various authors and nations. The most significant limitation of these analytical formulas is that they are only valid for that specific database. AI can assist to overcome these limitations.

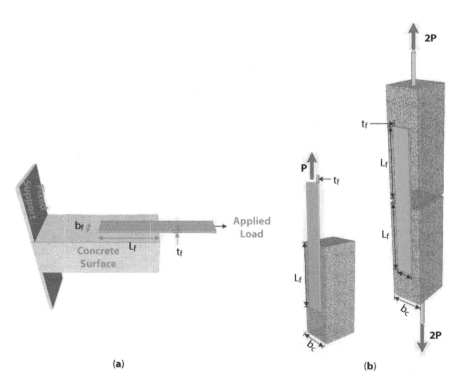

Figure 17.4 FRP-bond strength test setup (a) single-lap shear test [19] (b) double-lap shear test [20].

17.2.2   Flexural Strengthening of Beams With FRP Composite

In corroded RC beams, flexural failure is a common type of failure. Corroded RC beams must be reinforced with composite materials in order to avoid the failure and increase their service life. Strengthening or upgrading RC beam components with EB and NSM FRP techniques can increase their flexural capacity. Material quality, strengthening technique, loading, and climatic variables all influence the performance of reinforced beams. FRP laminates placed to the tension face of beams can considerably enhance the flexural strength of the RC beams. While beams that have been U-wrapped or totally wrapped with FRP sheets offer good performance, fully wrapped FRP is difficult to use in practice. U-wrapping RC beams with FRP can improve their flexural stiffness, deflection resistance, and strength. The thickness of the FRP composite system controls the strengthening parameters [21]. Figure 17.5 shows the flexural strengthening of RC beams with External Bonded (EB) FRP composite systems.

17.2.3   Shear Strengthening of Beams With FRP Composite

U-wrap, full/complete-wrap, and side-bonded strengthening methods can all be utilized to improve the strength in the shear zone of RC beams. Figure 17.6 shows the configurations of FRP shear strengthening schemes. The most successful form of shear strengthening using FRP is the complete-wrap strengthening scheme, in which the plies/laminates/plate wraps around the whole section of the beam. However, due to the existence of "monolithic slabs" or additional support parts, this method is not feasible from a constructability aspect. The side-bonded approach, in which laminates are exclusively attached to the beam sides, increases the little

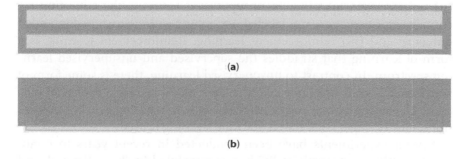

(a)

(b)

Figure 17.5 Flexural reinforcing of beams with EB FRP composite (a) bottom face of the beam and (b) side face of the beam.

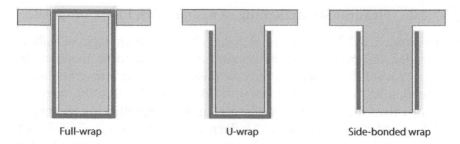

Full-wrap   U-wrap   Side-bonded wrap

Figure 17.6 FRP shear strengthening configurations [23].

load-carrying capability of the beam. Most of the RC beams are cast mono-lithically with slabs, the U-wrap scheme, in which FRP laminates/plies/plate is applied to tension face and both sides of the beam, is a frequent and successful approach in the practice. However, in many situations, FRP debonding is the prevailing failure mechanism, limiting the FRP shear influence [22].

A significant number of investigational studies were established to eval-uate the efficiency of various anchoring systems used at the endpoints of the U-wrap.

17.3   Machine Learning Models

It is advantageous to include the system researching approach as an option for development in the event of an algorithmic solution in order to fix the notions, as opposed to the traditional engineering method. There are three types of machine learning approaches: (i) supervised learning: In supervised learning, the education set is constructed from pairs of input and output, with the goal of learning a mapping between the two spaces; (ii) unsupervised learning: In unsupervised learning, the education set contains inputs that are not labeled, i.e., inputs with no assigned intended outcome; and (iii) reinforcement learning: Reinforcement learning is a form of learning that straddles the supervised and unsupervised learn-ing spectrum. In contrast to unsupervised learning, there is some form of supervision, but it does not take the form of expressing a preferred out-come for each and every input inside the data. In civil engineering applica-tions, mostly, supervised based machine learning is used.

Several experiments have been conducted in recent years to estab-lish the ultimate strength of RC beams retrofitted in shear, flexural, and bond between FRP and concrete substrate using externally bonded and near-surface mounted (NSM) FRP. The majority of shear and flexural strengthening design solutions are based on regression investigation of experimental data matching to certain configurations, making it extremely difficult to capture the true interrelationship between the factors involved. The assumption of established empirical equations dependent on unknown factors is used in traditional models. To prevent this issue, machine learning (ML) models were created to predict the shear, flexural, and bond strength of concrete beams strengthened with FRP systems based on prior experiments. The standard approaches, on the other hand, do not function well in circumstances where the modeling principles are either unknown or highly difficult to uncover. Because ML approaches are built on learning and generalization from experimental data, they may overcome these challenges; moreover, they can adjust solutions over time to take into account new information. To check the performance of ML algorithms,

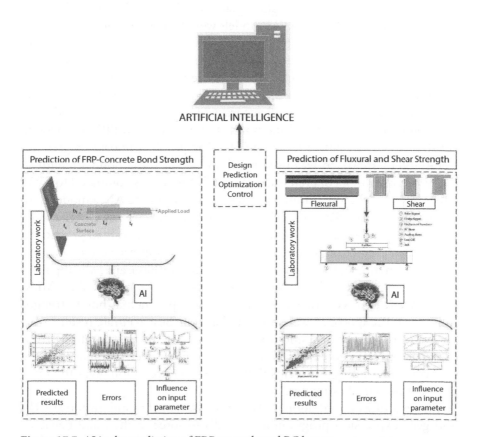

Figure 17.7 AI in the prediction of FRP strengthened RC beams.

the frequently used performance indices are R, MAE, RMSE, MAPE, a 20-index and NS [24–31]. Figure 17.7 depicts the artificial intelligence technique to forecasting the bond strength, shear, and flexural strength of FRP enhanced RC beams.

17.3.1   Prediction of Bond Strength

Coelho *et al.* [32] used an artificial neural network (ANN) and support vector machine (SVM) to forecast the bond strength of the FRP system. The anticipated findings demonstrate that the SVM model outperforms the ANN and ACI standard guidelines significantly. The bond strength between FRP and concrete substrate was studied by Su *et al.* [33] using multiple linear regression (MLR), ANN, and SVM. The bond strength data were collected into two parts named dataset 1 and dataset 2. The coefficient of determination of MLR, SVM and ANN models were 0.80, 0.82, 0.72, respectively, for the training data of dataset 1. The coefficient of determination of MLR, ANN, and SVM models were 0.88, 0.88, and 0.91, respectively, for the training data of dataset 2. The coefficient of determination of MLR, ANN, and SVM models were 0.74, 0.77, and 0.79, respectively, for the testing data of dataset 1. The coefficient of determination of MLR, ANN, and SVM models were 0.80, 0.82, and 0.85, respectively, for the testing data of dataset 2. The overall accuracy of the SVM model is good.

Chen *et al.* [34] analyzed the bond strength between FRP-to-concrete substrates with ensemble learning (EL). The gradient boosted regression tree (GBRT) approach was used in EL to calculate the bond strength of FRP systems with a concrete substrate, with the results compared to SVM and ANN models. The training and testing results of the ANN model were 0.9504, 0.8373, the SVM model was 0.9535, 0.8548, and the GBRT model were 0.9627 and 0.9269, respectively. The result indicated that, practically, the model's prediction outcomes are excellent and quite accurate.

The estimation of the bond strength was done by Basaran *et al.* [35] using code formulations and ML models, such as regression tree (RT), MLR, GPR, ANN, and SVMR. For the predicted experimental bond strength ratio, the GPR approach was shown to have the best accurate results, with an average value of 0.95 and a std. of 0.14. When the estimations from the bond strength equations in the standard codes were compared to the experimental findings, it seems that the ACI 440 equation gives the most accurate analytical estimates.

Zhang and Xue [36] adopted two models, namely random forest (RF) and gene expression programming (GEP) to forecast the bond behavior between FRP rods or strips and concrete substrate. The forecasted values from the RF and GEP models are reasonably close to the observed results. The RF model's R^2 for the validation and training data was 0.780 and 0.962, respectively. Similarly, in the GEP model, the R^2 for validation and training data was 0.80 and 0.87. According to the performance indices, the performance of the analyzed results shows that these models can forecast the bond strength with moderate accuracy.

Kumar *et al.* [19] predicted the FRP-concrete bond strength using the optimized neuro-bee algorithm approach. Three ML algorithms, namely ANN, GPR and artificial bee colony (ABC)-ANN, were used to predict the experimental data. The ABC-ANN and GPR models had a R-value of 0.9514 and 0.9618, respectively. Among all the ML models and empirical models, the reliability of the GPR model was excellent.

Table 17.2 ML research studies in FRP-concrete bond strength.

Author [Ref]	Year	ML model	Input parameters	Results
Coelho *et al.* [32]	2016	ANN, SVM	b_c, L_c, b_g, d_g, p_g, a_e, L_b, f_{cm}, E_p f_{fu}, e_{fu}, A_p ρ_p f_{at}	$R^2 = 0.90$ (SVM), $R^2 = 0.89$ (ANN)
Su *et al.* [33]	2021	MLR, SVM, ANN	E_p f_p t_p b_p L_p f_c, b_c, b_g, h_g	Dataset 1, Training $R^2 = 0.80$ (MLR) $R^2 = 0.82$ (SVM), $R^2 = 0.72$ (ANN) Dataset 2, Training $R^2 = 0.88$ (MLR) $R^2 = 0.91$ (SVM), $R^2 = 0.88$ (ANN) Dataset 1, Testing $R^2 = 0.74$ (MLR) $R^2 = 0.79$ (SVM), $R^2 = 0.77$ (ANN) Dataset 2, Testing $R^2 = 0.80$ (MLR) $R^2 = 0.85$ (SVM), $R^2 = 0.82$ (ANN)
Chen *et al.* [34]	2021	GBRT, ANN, SVM	E_p f_p t_p b_p L_p f_c, b_c	$R^2 = 0.8998$ (ANN), $R^2 = 0.9151$ (SVM), $R^2 = 0.9518$ (GBRT)

(Continued)

Table 17.2 ML research studies in FRP-concrete bond strength. (*Continued*)

Author [Ref]	Year	ML model	Input parameters	Results
Jahangir *et al.* [40]	2021	ANN, ABC–ANN	$b, f_c, t_p, b_p, l_b, E_f$	$R^2 = 0.940$ (ABC-ANN), $R^2 = 0.8649$ (ANN)
Basaran *et al.* [35]	2021	ANN, MLR, SVMR, GPR	Type, Surface, $d_b, f_c, C/d_b, L/d_b, A_{tr}/s_{ndb}$,	$R^2 = 0.9025$ (ANN), $R^2 = 0.6084$ (MLR), $R^2 = 0.6889$ (SVMR), $R^2 = 0.9216$ (GPR)
Zhang and Xue [36]	2021	GEP, RF	$E_p, L, A_p, f_e, D_g/W_g, f_c$	$R^2 = 0.852$ (GEP), $R^2 = 0.912$ (RF)
Zhou *et al.* [37]	2020	ANN	fc', E_p, t_p, L_p b_p, b_c	$R^2 = 0.93$ (ANN)
Pei and Wei [38]	2022	ACO-ANFIS	f_c, D, f_{ft}, t_p, E_p f_{At}, l_p, b_f	$R^2 = 0.97$ (ACO-ANFIS)
Naderpour *et al.* [41]	2019	ANFIS	$b_c, f_c, b_p, t_p, E_p, L$	$R^2 = 0.984$ (ANFIS)
Abdalla *et al.* [42]	2011	ANN	$f_c, b_c, E_p, f_p, t_p, b_p$ L_p, f_{cs}	$R^2 = 0.6939$
Mashrei *et al.* [43]	2013	ANN	$b_c, f_c, b_p, t_p, E_p, L$	$R^2 = 0.9801$ (ANN)
Haddad and Haddad [39]	2020	ANN	$f_c, t_p, E_p, L_p, b_p, f_{as}$	$R^2 = 0.9801$ (ANN)
Kumar *et al.* [19]	2022	ANN, ABC-ANN, GPR	f_c, b_c, E_p, f_p, t_p b_p, L_f	$R^2 = 0.8666$ (ANN), $R^2 = 0.9052$ (ABC-ANN) and $R^2 = 0.9251$ (GPR)
Köroğlu [44]	2018	ANN	Type, Surface, $d_b, f_c, C/db$, $L/d_b, A_{tr}/s_{ndb}$, confining,	$R^2 = 0.8989$
Cascardi and Micelli [45]	2021	ANN	$t_p, b_p, L_p, E_p, b_c, f_c$	$R^2 = 0.95$

The explicit neural network was used by Zhou *et al.* [37] for the predicting of FRP-bond strength. The ANN model test goes through 84 training rounds to find the best input node combination. The proposed ANN model was more accurate (i.e., has a smaller predicted error) than current bond strength models in the literature. The R^2 value of the best ANN model for testing and training data were 0.928 and 0.930, respectively.

Pei and Wei [38] used ant colony optimization (ACO)-ANFIS to forecast the bond strength. The outcomes showed that the established ACO-ANFIS model was more precise than the rest of analytical models, with a greater R^2, RMSE and MAE-value were 0.97, 1.29 kN and 0.81 kN, respectively.

Haddad *et al.* [39] predicted the FRP–concrete bond strength using ANN algorithm. The predicted results show good accuracy with R-value 0.99. The performance of ANN model was excellent as compared to other analytical models. Table 17.2 summarizes the significant findings of ML research investigations on FRP-concrete bond strength.

Figure 17.8 shows the comparison of coefficients of determination of different ML models used in the estimation of bond strength.

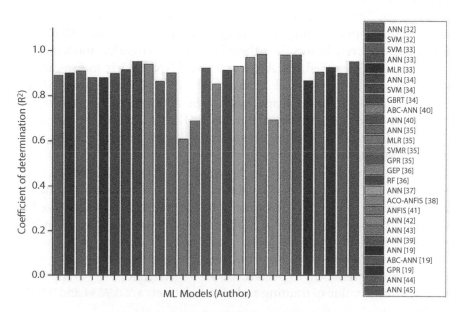

Figure 17.8 R^2 value of different ML models (FRP-concrete bond strength).

17.3.2   Estimation of Flexural Strength

Murad *et al.* [46] used GEP to calculate the flexural strength of FRP strengthened RC beams. The GEP model was created using 116 data points that were gathered from many investigational programs accessible in the previous studies. The accuracy and reliability of the GEP model was compared with standard guidelines ACI-440-17 and CSA S806-12. Among all the three models, the accuracy of the GEP model was higher than standard guidelines.

17.3.3   Estimation of Shear Strength

Chou *et al.* [47] forecasted the shear strength prediction of RC beams with hybrid ML model optimized least squares support vector regression (LSSVR) with smart firefly algorithm (SFA). The optimized AI model surpassed the other models in predicting the shear strength of an extensive variety of RC beams, according to the comparative findings. The MAPE values of the hybrid AI model were 12.941% and 21.70% for the test data of with stirrups and without stirrups of RC beams respectively. But in the case of the FRP reinforcing beam, the MAPE value was 18.951%. The shear strength of the RC beams was identified by Nikoo *et al.* [48] using GA-ANN, Bat Algorithm (BA)-ANN and PSO-ANN. The MAE values of GA-ANN, BA-ANN, and PSO-ANN were 8.79, 18.05 and 23.36, respectively. The findings of the results revealed that the BA-ANN is more competent, adaptable, and precise in calculating the shear strength of FRP-reinforced concrete beams than the other ML and empirical models.

Naser (2020) applied the ANN and GA-ANN models to forecast the strength of FRP reinforced beams [49]. The FRP-strengthened/reinforced concrete members' responses may be correctly captured using the ML-based expressions that were developed. The accuracy of the optimized ANN model with the genetic algorithm was 96.2% without stirrups database and 90.7% with stirrups database.

ANN method was explored by Nguyen *et al.* [50] to forecast the shear strength of FRP concrete beams. Laboratory experimental datasets, including 125 values, were gathered from the literature, and it was utilized to discover the optimum ANN design. Commonly used statistical indices were used to identify and assess the proper architecture of an ANN model. The findings demonstrated that utilizing the best architecture, the ANN model was very efficient to predict the shear strength of FRP-wrapped RC beams, with R^2 value of training and testing datasets are 0.9634 and 0.9577, respectively.

Abuodeh *et al.* [51] predicted the shear capability of FRP reinforced beams with ANN and Neural Interpretation Diagram (NID). The predicted results revealed that the findings of the ANN model were very close to the experimental data. The model's predictions were quite similar to the experiment values with R^2 of 0.911 for the testing dataset, showing that the data was not over-fitted. The mechanical and physical characteristics that affect the shear capacity of FRP reinforced beams may be identified using the NID technique.

The shear capacity of RC beams strengthened with FRP composites was estimated by Naderpour *et al.* [52] using ANN algorithm. The structure of the ANN model adopted in the study was a network with two middle layers. The R^2, MSE, RMSE, and MAE of the ANN model were 0.93, 129, 11.36, and 8.70, respectively. The findings showed that the ANN model provided superior outcomes than other empirical equations.

ANFIS model was adopted by Kar *et al.* [53] to forecast the shear contribution of the FRP-wrapped reinforced concrete beams. The ANFIS model was constructed using 119 existing datasets taken from published papers that were used to create the training and testing datasets. A head-to-head examination of the ANFIS model's predictions and the actual data revealed that the ANFIS results are closely related to the experimental results. The ANFIS model has a coefficient of determination of 0.996, which was greater than analytical and conventional codal guidelines.

Naderpour *et al.* [54] calculated the shear resistance of RC Beams strengthened with FRP bars with ANN. The experimental data is irregular, and the models derived from LR analysis were unable to accurately predict behavior. The overall value of the coefficient of correlation of the proposed ANN model was 0.98. The mean error for the ANN-model was 9.72%, while the mean error of other models, such as BISE, ISISM03-07, ACI 440.1R-06, CNR DT 203, and JSCE, is 17.94%, 29.37%, 45.68%, 28.5%, and 25.89%, respectively.

Perera *et al.* [55] deployed the ANN technique to forecast the shear ability of RC beams strengthened with FRP composite. The R-value of the ANN technique was 0.98. ANN was used by Lee and Lee (2014) to analyze the contribution of FRP in the shear strengthening of beams without stirrups reinforcements [56]. The developed ANN model resulted in enhanced statistical parameters with greater accuracy than other existing equations, according to comparisons between projected values and 106 test datasets. The precision of the ANN Model was up to 98%, according to the results.

Naderpour and Alavi estimated the shear involvement of CFRP reinforced RC Beams with ANFIS algorithm [57]. When compared to the results obtained directly from existing guidelines such as ACI 440.2R-08, fib-TG9.3, CSA-S806, and CIDAR, the results show that the established ANFIS model can forecast the shear involvement of FRP composites in the RC beams with greater accuracy. The R^2, RMSE and MAPE values of the ANFIS model were 0.977, 10.07 kN, 17.19%, respectively.

Perera et al. [58] used AI techniques (GA and ANN) to forecast the shear carrying capacity of RC strengthened beams with FRP systems. The accuracy of both the models was good and alternative to the conventional empirical models. ANFIS and ANN algorithms were explored by Cao et al. [59] to forecast the effect of FRP contribution in the FRP beams. The results shows that the most impactful arrangement of three parameters on the estimation of shear resistance of RC beams by FRP was the combination of the depth of tensile reinforcement, FRP modulus of elasticity, and shear span to depth ratio.

ANN algorithm was explored by Tanarslan et al. [60] to estimate the shear capacity of RC beams strengthened with FRP using ANN. For verification, the ANN predicted results were compared directly with ACI 440.2R, CNR-DT, CIDAR, fib14, and CHBDC. In terms of the experimental outcomes, the ANN model results were more precise than the standard guidelines, according to the research outcomes.

Perera et al. [61] worked with ANN and a Multi-Objective Evolutionary Algorithm (MOEA) to forecast the shear behavior of RC beams reinforced with NSM FRP rods. Linear regression slope of ANN and multi-objective evolutionary algorithm were 0.97 and 0.8473, the linear correlation coefficient of ANN and multiobjective evolutionary algorithm were 0.9798 and 0.9115. The performance of the ANN algorithm was excellent with higher accuracy and speed.

Anarslan et al. [62] deployed ANN to forecast the shear behavior of CFRP-wrapped RC beams. The ANN findings were equated to those derived from the fib14, CIDAR, and ACI 440.2R. According to the experimental results, fib14 delivered the best forecasts among the theoretical predictions of design guidelines. Furthermore, actual comparisons revealed that the NN model is more precise than the standard guidelines in terms of experimental results.

The relevant summary of ML research studies in shear reinforcement of RC beams is tabulated in Table 17.3, and Figure 17.9 shows the comparison of coefficients of determination of different ML models.

The acronyms and symbols used in this chapter are mentioned in Table 17.4.

Table 17.3 Summary of ML research studies in shear strength prediction.

References	Year	ML model	Input parameters	Results
Chou *et al.* [47]	2020	MLP, REPTree, SMOreg, LR, Voting, Bagging, Stacking, SFA-LSSVR	b_w, d, a/d, ρ_w f'$_c$, a_g, V, f_{co}, f_{ry}, ρ_{stp}, f_{sy}	$R^2 = 0.9351$ (MLP), $R^2 = 0.4556$ (REPTree), $R^2 = 0.4638$ (SMOreg), $R^2 = 0.633$ (LR)
Nikoo *et al.* [48]	2021	ANN, BA-ANN, GA-ANN, PSO-ANN	D, f'$_c$, ρ_p, b_w, E_p a/d	MAE = 8.79 (BA-ANN), MAE = 18.05 (GA-ANN), MAE = 23.36 (PSO-ANN)
Naser *et al.* [49]	2020	ANN-GA	b_p, t_p, L, f_c, f_{tt}, f_p E_p, b_2, b, h, $S_{a/d}$, r_p, s_s, r_{fs}, E_{fs}, E_{ft}, f_{ft}, f_y, r_s, r_p, a	$R^2 = 0.9197$
Nguyen *et al.* [50]	2021	ANN	b_w, d, a/d, f_c, ρ_p, E_p, ρ_s, f_s, E_s	$R^2 = 0.9577$
Abuodeh *et al.* [51]	2019	ANN, NID	b_w, d_{eff}, a/d, L, f_c, fs, A_s/S_s, f_y, A_{BL}, t_p B_p, h_p, w_f/s_p f_p, E_f	$R^2 = 0.9110$
Naderpour *et al.* [52]	2020	ANN	f_c, ρ_p, E_{FRP}, a/d, b_w, d	$R^2 = 0.93$

(*Continued*)

Table 17.3 Summary of ML research studies in shear strength prediction. *(Continued)*

References	Year	ML model	Input parameters	Results
Kar et al. [53]	2020	ANFIS	f_u, b, t_f, a/d, f_c, d, W_f/ S_f, A_{sv}, n, β, E_f, ε	$R^2 = 0.98$
Naderpour et al. [54]	2018	ANN	b, d, a/d, f'_c, ρ_f, E_f	$R^2 = 0.9604$
Perera et al. [55]	2010	ANN	h, b, c, a/d, f_{cm}, f_{y90d}, A_c, S_c, A_{90}, A_f, b, ρ_f	$R^2 = 0.9493$
Lee et al. [56]	2014	ANN	a/d, f'_c, ρ_f, b_w, E_f, d, E_f	$R^2 = 0.98$
Naderpour et al. [57]	2017	ANFIS	ε_f, d_f, w_f, s_f, α, f'_c, E_f, t_f, d, a,	$R^2 = 0.977$
Perera et al. [58]	2010	ANN,GA	C_{fg}, $F_i A_m$, S_s, S_f,	-
Cao et al. [59]	2020	ANFIS	b, d, a/d, f'_c, ρ_f, E_f	RMSE = 0.1600
Tanarslan et al. [60]	2012	ANN	f_{cd}, L, a, d, a/d, $W_r S_c$	RMSE = 19.55
Perera et al. [61]	2014	ANN, MOEA	ρ_f, f_{cm}, a/d, s_f, E_f, α_f, ρ_{sw}	$R^2 = 0.96$ (ANN), $R^2 = 0.8308$ (MOPE)
Tanarslan et al. [62]	2015	ANN	b, h, w_f/s_f, d, E_f, β, t_f, f_f, a/d	-

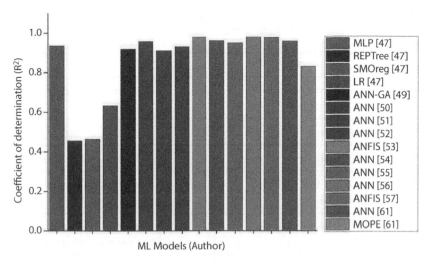

Figure 17.9 R^2 value of different ML models (shear strength of FRP strengthened beams).

Table 17.4 Acronyms and symbols.

Notation			
Acronyms		**Symbols**	
ACI	American Concrete Institute	a_e	Distance between FRP and nearest concrete block
ACO	Ant Colony Optimization	A_t	FRP cross-section area
ANN	Artificial Neural Network	A_{tr}	Area of transverse bar
BA	Bat Algorithm	bc	Width of concrete block
EL	Ensemble Learning	b_c	Concrete block width
GBRT	Gradient Boosted Regression Tree	b_f	Width of FRP composite
GEP	Gene Expression Programming	b_g, W_g	Width of groove
GPR	Gaussian Process Regression	b_g	Groove width
LR	Linear Regression	d_b	Diameter of bar
ML	Machine Learning	d_g	Groove depth

(Continued)

Table 17.4 Acronyms and symbols. (*Continued*)

Notation			
Acronyms		**Symbols**	
NID	Neural Interpretation Diagram	E_f	FRP modulus of elasticity
NSM	Near Surface Mounted	ε_{fu}	FRP ultimate strain
RT	Regression Tree	f_{at}	Adhesive tensile strength
ABC	Artificial Bee Colony	f_c, fc'	Compressive strength of concrete
BISE	Institution of Structural Engineers	f_{cm},	Concrete cylinder mean compressive strength
CSA	Canadian Standards Association	f_{cs}	Tensile strength of concrete
CFRP	Carbon Fiber Reinforced Polymers	f_f	Tensile strength of FRP composite
EB	External Bonded	f_{fu}	FRP ultimate tensile strength
FRP	Fiber Reinforced Polymer	$h_{g\,Dg}$	Height of groove
JSCE	Japan Society of Civil Engineers	L_b, L_f	Bonded length
LSSV	Least Squares Support Vector Regression	L_c	Concrete block length
MLR	Multiple Linear Regression	$_{pf}$	FRP perimeter
MOEA	Multi-Objective Evolutionary Algorithm	P_g	Groove perimeter
MAE	Mean Absolute Error	s	Spacing of transverse bars
RC	Reinforced Concrete	t_f	Thickness of FRP composite
SFA	Smart Firefly Algorithm	R	Correlation Coefficient

(*Continued*)

Table 17.4 Acronyms and symbols. (*Continued*)

Notation			
Acronyms		**Symbols**	
SVM	Support Vector Machine	MAP	Mean Absolute Percentage Error
PAN	Polyacrylonitrile	NS	Nash-Sutcliffe Coefficient Index
RMSE	Root Mean Square Error		

17.4 Conclusion

In this chapter, the importance of ML models in estimation of bond, shear, and flexural strength of RC-reinforced beams using FRP composite systems was discussed. The extensive literature survey was done to extract the information of use of ML models in the bond, shear, and flexural strength of reinforced concrete beams wrapped with FRP systems. The FRP-concrete bond strength prediction using ANN, improved ANN, and ANFIS models performs well in terms of accuracy and speed. Only a few ML methods exist to forecast the flexural strength of reinforced concrete reinforced beams using FRP. The ANFIS and ANN models perform better in forecasting the shear strength of FRP-strengthened beams. Future studies will necessitate the development of more accurate flexural strength prediction models. These ML models are only viable for data that falls inside the input and output parameter ranges.

References

1. Banerjee, A., Guru Prathap Reddy, V., Kumar Pancharathi, R., Tadepalli, T., Colour-stability analysis for estimation of deterioration in concrete due to chemical attack. *Constr. Build. Mater.*, 321, 126288, 2022, https://doi.org/10.1016/j.conbuildmat.2021.126288.
2. Yang, J., Haghani, R., Blanksvärd, T., Lundgren, K., Experimental study of FRP-strengthened concrete beams with corroded reinforcement. *Constr. Build. Mater.*, 301, 124076, 2021, https://doi.org/10.1016/j.conbuildmat.2021.124076.
3. Hollaway, L., *Polymer composites for civil and structural engineering*, Blackie Academic and Professional, London, UK, 1993.

4. Belarbi, A., Dawood, M., Acun, B., Sustainability of fiber-reinforced polymers (FRPs) as a construction material, in: *Sustainability of Construction Materials*, Second Edition, J.M. Khatib (Ed.), pp. 521–38, Woodhead Publishing, Elsevier-Cambridge, UK, 2016, https://doi.org/10.1016/B978-0-08-100370-1.00020-2.

5. Wang, Y., Zhang, S., Li, G., Shi, X., Effects of alkali-treated recycled carbon fiber on the strength and free drying shrinkage of cementitious mortar. *J. Cleaner Prod.*, 228, 1187–95, 2019, https://doi.org/10.1016/j.jclepro.2019.04.295.

6. Pimenta, S. and Pinho, S.T., Recycling carbon fibre reinforced polymers for structural applications: Technology review and market outlook. *J. Waste Manage.*, 31, 2, 378–92, 2011, https://doi.org/10.1016/j.wasman.2010.09.019.

7. Farzana, R., Rajarao, R., Mansuri, I., Sahajwalla, V., Sustainable synthesis of silicon nitride nanowires using waste carbon fibre reinforced polymer (CFRP). *J. Cleaner Prod.*, 188, 371–77, 2018, https://doi.org/10.1016/j.jclepro.2018.03.295.

8. J.C. Holmes IV, I.V. Holmes, C. James, Composite materials and related methods for manufacturing composite materials. U.S. Patent Application, 15/350, 2017.

9. Jagadish, P.R., Khalid, M., Li, L.P., Hajibeigy, M.T., Amin, N., Walvekar, R. *et al.*, Cost effective thermoelectric composites from recycled carbon fibre: From waste to energy. *J. Cleaner Prod.*, 195, 1015–25, 2018, https://doi.org/10.1016/j.jclepro.2018.05.238.

10. Motavalli, M., Czaderski, C., Schumacher, A., Gsell, D., Fibre reinforced polymer composite materials for building and construction, in: *Textiles, Polymers and Composites for Buildings*, G. Pohl (Ed.), pp. 69–128, Woodhead Publishing, Elsevier-Cambridge, UK, 2010, https://doi.org/10.1533/9780845699994.1.69.

11. Varley, D., Yousaf, S., Youseffi, M., Mozafari, M., Khurshid, Z., Sefat, F., Fiber-reinforced composites, in: *Advanced Dental Biomaterials*, Z. Khurshid, *et al.* (Eds.), pp. 301–15, Woodhead Publishing, Elsevier-Cambridge, UK, 2019, https://doi.org/10.1016/B978-0-08-102476-8.00013-X.

12. Meier, U., Carbon fiber-reinforced polymers: Modern materials in bridge engineering. *Struct. Eng. Int.*, 2, 1, 7–12, 1992, https://doi.org/10.2749/101686692780617020.

13. Bozsaky, D., The historical development of thermal insulation materials. *Period. Polytech. Archit.*, 41, 2, 49–56, 2010.

14. Gong, R.H. and Chen, X., Technical yarns, in: *Handbook of Technical Textiles*, Second Edition, A.R. Horrocks and S.C. Anand (Eds.), pp. 43–62, Woodhead Publishing, Elsevier-Cambridge, UK, 2016, https://doi.org/10.1016/B978-1-78242-458-1.00003-0.

15. Li, W., Li, J., Ren, X., Leung, C.K.Y., Xing, F., Coupling effect of concrete strength and bonding length on bond behaviors of fiber reinforced polymer–concrete interface. *J. Reinf. Plast. Compos.*, 34, 5, 421–32, 2015, https://doi.org/10.1177/0731684415573816.

16. Chen, C., Li, X., Zhao, D., Huang, Z., Sui, L., Xing, F. *et al.*, Mechanism of surface preparation on FRP-concrete bond performance: A quantitative study. *Compos. B. Eng.*, 163, 193–2016, 2019, https://doi.org/10.1016/j.compositesb.2018.11.027.

17. Yuan, C., Chen, W., Pham, T.M., Hao, H., Effect of aggregate size on bond behaviour between basalt fibre reinforced polymer sheets and concrete. *Compos. B. Eng.*, 158, 459–74, 2019, https://doi.org/10.1016/j.compositesb.2018.09.089.

18. Heydari Mofrad, M., Mostofinejad, D., Hosseini, A., A generic non-linear bond-slip model for CFRP composites bonded to concrete substrate using EBR and EBROG techniques. *Compos. Struct.*, 220, 31–44, 2019, https://doi.org/10.1016/j.compstruct.2019.03.063.

19. Kumar, A., Arora, H.C., Mohammed, M.A., Kumar, K., Nedoma, J., An optimized neuro-bee algorithm approach to predict the FRP-concrete bond strength of RC beams. *IEEE Access*, 10, 3790–806, 2022, 10.1109/access.2021.3140046.

20. Kumar, A., Arora, H.C., Kumar, K., Mohammed, M.A., Majumdar, A., Khamaksorn, A. *et al.*, Prediction of FRCM Concrete bond strength with machine learning. *Sustainability*, 14, 2, 845, 2022, 10.3390/su14020845.

21. Siddika, A., Mamun, M.A.A., Alyousef, R., Amran, Y.H.M., Strengthening of reinforced concrete beams by using fiber-reinforced polymer composites: A review. *J. Build. Eng.*, 25, 100798, 2019, https://doi.org/10.1016/j.jobe.2019.100798.

22. Bank, L.C., *Composites for construction: Structural design with FRP materials*, John Wiley & Sons, 2006, 10.1002/9780470121429.

23. Guide for the design and construction of externally bonded FRP systems for strengthening concrete structures ACI 440.2R-08, American Concrete Institute-USA, 2008.

24. Kumar, A., Arora, H.C., Kumar, K., Mohammed, M.A., Majumdar, A., Khamaksorn, A., Thinnukool, O., Prediction of FRCM-Concrete bond strength with machine learning approach. *Sustainability*, 14, 2, 845, 2022, https://doi.org/10.3390/su14020845.

25. Kumar, A., Arora, H.C., Mohammed, M.A., Kumar, K., Nedoma, J., An optimized neuro-bee algorithm approach to predict the FRP-concrete bond strength of RC beams. *IEEE Access*, 10, 3790–3806, 2022, 10.1109/ACCESS.2021.3140046.

26. Kumar, A., Arora, H.C., Kapoor, N.R., Mohammed, M.A., Kumar, K., Majumdar, A., Thinnukool, O., Compressive strength prediction of lightweight concrete: Machine learning models. *Sustainability*, 14, 4, 2404, 2022, https://doi.org/10.3390/su14042404.

27. Kumar, K. and Saini, R.P., Development of correlation to predict the efficiency of a hydro machine under different operating conditions. *Sustain. Energy Technol. Assess.*, 50, 101859, 2022, https://doi.org/10.1016/j.seta.2021.101859.

28. Kapoor, N.R., Kumar, A., Kumar, A., Kumar, A., Mohammed, M.A., Kumar, K., Lim, S., Machine learning-based CO_2 prediction for office room: A pilot study. *Wirel. Commun. Mob. Comput.*, 2022, 9404807, 2022, https://doi.org/10.1155/2022/9404807.

29. Verma, K., Bhardwaj, S., Arya, R., Islam, U., Bhushan, M., Kumar, A., Samant, P., Latest tools for data mining and machine learning. *Int. J. Innov. Technol. Exploring Eng.*, 8, 9S, 18–23, 2019, https://doi.org/10.35940/ijitee.I1003.0789S19.

30. Kholiya, P.S., Kapoor, A., Rana, M., Bhushan, M., Intelligent process automation: The future of digital transformation. Paper presented at the *2021 10th International Conference on System Modeling & Advancement in Research Trends (SMART)*, pp. 185–190, 2021, 10.1109/SMART52563.2021.9676222.

31. Nalavade, A., Bai, A., Bhushan, M.J.I., Deep learning techniques and models for improving machine reading comprehension system. *IJAST*, 29, 04, 9692–9710, 2020.

32. Coelho, M.R.F., Sena-Cruz, J.M., Neves, L.A.C., Pereira, M., Cortez, P., Miranda, T., Using data mining algorithms to predict the bond strength of NSM FRP systems in concrete. *Constr. Build. Mater.*, 126, 484–95, 2016, https://doi.org/10.1016/j.conbuildmat.2016.09.048.

33. Su, M., Zhong, Q., Peng, H., Li, S., Selected machine learning approaches for predicting the interfacial bond strength between FRPs and concrete. *Constr. Build. Mater.*, 270, 121456, 2021, https://doi.org/10.1016/j.conbuildmat.2020.121456.

34. Chen, S.-Z., Zhang, S.-Y., Han, W.-S., Wu, G., Ensemble learning based approach for FRP-concrete bond strength prediction. *Constr. Build. Mater.*, 302, 124230, 2021, https://doi.org/10.1016/j.conbuildmat.2021.124230.

35. Basaran, B., Kalkan, I., Bergil, E., Erdal, E., Estimation of the FRP-concrete bond strength with code formulations and machine learning algorithms. *Compos. Struct.*, 268, 113972, 2021, https://doi.org/10.1016/j.compstruct.2021.113972.

36. Zhang, R. and Xue, X., A predictive model for the bond strength of near-surface-mounted FRP bonded to concrete. *Compos. Struct.*, 262, 113618, 2021, https://doi.org/10.1016/j.compstruct.2021.113618.

37. Zhou, Y., Zheng, S., Huang, Z., Sui, L., Chen, Y., Explicit neural network model for predicting FRP-concrete interfacial bond strength based on a large database. *Compos. Struct.*, 240, 111998, 2020, https://doi.org/10.1016/j.compstruct.2020.111998.

38. Pei, Z. and Wei, Y., Prediction of the bond strength of FRP-to-concrete under direct tension by ACO-based ANFIS approach. *Compos. Struct.*, 282, 115070, 2022, https://doi.org/10.1016/j.compstruct.2021.115070.

39. Haddad, R. and Haddad, M., Predicting fiber-reinforced polymer–concrete bond strength using artificial neural networks: A comparative analysis study. *Struct. Concr.*, 22, 1, 38–49, 2021, https://doi.org/10.1002/suco.201900298.

40. Jahangir, H. and Rezazadeh Eidgahee, D., A new and robust hybrid artificial bee colony algorithm – ANN model for FRP-concrete bond strength evaluation. *Compos. Struct.*, 257, 113160, 2021, https://doi.org/10.1016/j.compstruct.2020.113160.

41. Naderpour, H., Mirrashid, M., Nagai, K., An innovative approach for bond strength modeling in FRP strip-to-concrete joints using adaptive neuro–fuzzy inference system. *Eng. Comput.*, 36, 3, 1083–100, 2020, https://doi.org/10.1007/s00366-019-00751-y.

42. Abdalla, J.A., Hawileh, R., Al-Tamimi, A. (Eds.), Prediction of FRP-concrete ultimate bond strength using artificial neural network. *Fourth International Conference on Modeling, Simulation and Applied Optimization*, pp. 19–21, 2011, 10.1109/ICMSAO.2011.5775518.

43. Mashrei, M.A., Seracino, R., Rahman, M.S., Application of artificial neural networks to predict the bond strength of FRP-to-concrete joints. *Constr. Build. Mater.*, 40, 812–21, 2013, https://doi.org/10.1016/j.conbuildmat.2012.11.109.

44. Köroğlu, M.A., Artificial neural network for predicting the flexural bond strength of FRP bars in concrete. *Sci. Eng. Compos. Mater.*, 26, 1, 12–29, 2019, https://doi.org/10.1515/secm-2017-0155.

45. Cascardi, A. and Micelli, F., ANN-based model for the prediction of the bond strength between FRP and concrete. *Fibres*, 9, 7, 46, 2021, https://doi.org/10.3390/fib9070046.

46. Murad, Y., Tarawneh, A., Arar, F., Al-Zu'bi, A., Al-Ghwairi, A., Al-Jaafreh, A. *et al.*, Flexural strength prediction for concrete beams reinforced with FRP bars using gene expression programming. *Structures*, 33, 3163–72, 2021, https://doi.org/10.1016/j.istruc.2021.06.045.

47. Chou, J.-S., Pham, T.-P.-T., Nguyen, T.-K., Pham, A.-D., Ngo, N.-T., Shear strength prediction of reinforced concrete beams by baseline, ensemble, and hybrid machine learning models. *Soft Comput.*, 24, 5, 3393–411, 2020, https://doi.org/10.1007/s00500-019-04103-2.

48. Nikoo, M., Aminnejad, B., Lork, A., Predicting shear strength in FRP-reinforced concrete beams using bat algorithm-based artificial neural network. *Adv. Mater. Sci. Eng.*, 2021, 5899356, 2021, https://doi.org/10.1155/2021/5899356.

49. Naser, M.Z., Machine learning assessment of fiber-reinforced polymer-strengthened and reinforced concrete Members. *ACI Mater. J.*, 117, 6, 237–251, 2020, https://doi.org/10.14359/51728073.

50. Nguyen, Q.H., Ly, H.-B., Nguyen, T.-A., Phan, V.-H., Nguyen, L.K., Tran, V.Q., Investigation of ANN architecture for predicting shear strength of fiber reinforcement bars concrete beams. *PLoS One*, 16, 4, e0247391, 2021, https://doi.org/10.1371/journal.pone.0247391.

51. Abuodeh, O., Abdalla, J.A., Hawileh, R.A., Predicting the shear capacity of FRP in shear strengthened RC beams using ANN and NID. *8th International Conference on Modeling Simulation and Applied Optimization (ICMSAO)*, pp. 15–17, 2019, https://ieeexplore.ieee.org/abstract/document/8880284.

52. Naderpour, H., Haji, M., Mirrashid, M., Shear capacity estimation of FRP-reinforced concrete beams using computational intelligence. *Structures*, 28, 321–8, 2020, https://doi.org/10.1016/j.istruc.2020.08.076.

53. Kar, S., Pandit, A.R., Biswal, K.C., Prediction of FRP shear contribution for wrapped shear deficient RC beams using adaptive neuro-fuzzy inference system (ANFIS). *Structures*, 23, 702–17, 2020, https://doi.org/10.1016/j.istruc.2019.10.022.

54. Naderpour, H., Poursaeidi, O., Ahmadi, M., Shear resistance prediction of concrete beams reinforced by FRP bars using artificial neural networks. *Measurement*, 126, 299–308, 2018, https://doi.org/10.1016/j.measurement.2018.05.051.

55. Perera, R., Barchín, M., Arteaga, A., Diego, A.D., Prediction of the ultimate strength of reinforced concrete beams FRP-strengthened in shear using neural networks. *Compos. B. Eng.*, 41, 4, 287–98, 2010, https://doi.org/10.1016/j.compositesb.2010.03.003.

56. Lee, S. and Lee, C., Prediction of shear strength of FRP-reinforced concrete flexural members without stirrups using artificial neural networks. *Eng. Struct.*, 61, 99–112, 2014, https://doi.org/10.1016/j.engstruct.2014.01.001.

57. Naderpour, H. and Alavi, S.A., A proposed model to estimate shear contribution of FRP in strengthened RC beams in terms of adaptive neuro-fuzzy inference system. *Compos. Struct.*, 170, 215–27, 2017, https://doi.org/10.1016/j.compstruct.2017.03.028.

58. Perera, R., Arteaga, A., Diego, A.D., Artificial intelligence techniques for prediction of the capacity of RC beams strengthened in shear with external FRP reinforcement. *Compos. Struct.*, 92, 5, 1169–75, 2010, https://doi.org/10.1016/j.compstruct.2009.10.027.

59. Cao, Y., Fan, Q., Mahmoudi Azar, S., Alyousef, R., Yousif, S.T., Wakil, K. *et al.*, Computational parameter identification of strongest influence on the shear resistance of reinforced concrete beams by fiber reinforcement polymer. *Structures*, 27, 118–27, 2020, https://doi.org/10.1016/j.istruc.2020.05.031.

60. Tanarslan, H.M., Secer, M., Kumanlioglu, A., An approach for estimating the capacity of RC beams strengthened in shear with FRP reinforcements using artificial neural networks. *Constr. Build. Mater.*, 30, 556–68, 2012, https://doi.org/10.1016/j.conbuildmat.2011.12.008.

61. Perera, R., Tarazona, D., Ruiz, A., Martín, A., Application of artificial intelligence techniques to predict the performance of RC beams shear strengthened with NSM FRP rods. Formulation of design equations. *Compos. B Eng.*, 66, 162–73, 2014, https://doi.org/10.1016/j.compositesb.2014.05.001.

62. Tanarslan, H.M., Kumanlioglu, A., Sakar, G., An anticipated shear design method for reinforced concrete beams strengthened with anchored carbon fiber-reinforced polymer by using neural network. *Struct. Des. Tall Build.*, 24, 1, 19–39, 2015, https://doi.org/10.1002/tal.1152.

18

Prediction of Indoor Air Quality Using Artificial Intelligence

Nishant Raj Kapoor[1,2]*, Ashok Kumar[1,2], Anuj Kumar[1,2], Aman Kumar[1,2]
and Harish Chandra Arora[1,2]

[1]CSIR-Central Building Research Institute, Roorkee, India
[2]AcSIR-Academy of Scientific and Innovative Research, Ghaziabad, India

Abstract

For well-being and good health, indoor air quality (IAQ) is an important concern as most of the people spend almost total of their time in different types of buildings. Research in IAQ is gaining momentum and multidisciplinary interest rapidly. Artificial intelligence (AI) methods have significantly transformed the research in the area of IAQ prediction due to their outstanding performance. Due to the increasing availability of IAQ data and the rapid expansion of AI techniques, it is vital to investigate the development of IAQ forecasting using AI techniques in a complete and quantitative manner. Therefore, an overview of IAQ and AI is presented in the first portion of this chapter. Further, relevant parameters of IAQ and its effects and sources are mentioned in the next section. In the process, frontier research carried out in the domain of IAQ with the application of AI was explored by adopting state-of-the-art literature. This chapter gives useful information on the future of AI-based IAQ forecasting. Readers will benefit from a more integrated view of IAQ and associated AI techniques within the built environment.

Keywords: Artificial intelligence, machine learning, indoor air quality, indoor air pollution, sick building

*Corresponding author: dr.nrkapoor@outlook.com

Ashok Kumar, Megha Bhushan, José A. Galindo, Lalit Garg and Yu-Chen Hu (eds.) Machine Intelligence, Big Data Analytics, and IoT in Image Processing: Practical Applications, (447–470) © 2023 Scrivener Publishing LLC

18.1   Introduction

Presently, much of the focus on air pollution has been on the risks and health impacts of outdoor air quality, even though interior levels of air pollutants are frequently 2–5 (and sometimes more) times greater than exterior pollutants concentrations. However, comparative risk analyses, such as those conducted by the United States Environmental Protection Agency (US EPA), have consistently classified indoor air pollution (IAP) among the topmost five environmental threats to human life. Current research on the effects of indoor air quality (IAQ) focuses mostly on health, perception, and comfort repercussions, which have resulted in significant improvements in IAQ in numerous built environments over the last several decades [1–5]. From now, increasing data suggest that IAQ may be considerably forecasted and subsequently improved using Artificial intelligence (AI) [6]. Improved IAQ will boost occupants' cognitive function and productivity at their workplaces (offices, schools, and residences) [7–9]. As a result, the potential benefits of better comfort, health, performance, and productivity may stimulate investment in IAQ research.

In this context, health is defined as a complete state of social, mental, as well as physical well-being, rather than merely the absence of disability or sickness. Comfort is a mental condition in which one feels physically calm and free of discomfort. Performance is described as an individual's capacity to execute various cognitively and physically demanding activities, as well as a measure of how successfully employees do what they are aiming to achieve. Productivity is defined as a measure of an employee's quantity and quality of accomplishments relative to what might be produced under ideal conditions. Productivity is frequently defined as the ratio of output to input. Input often comprises the expense of maintaining IAQ, recruiting, training, equipment, salary, benefits, insurance, and many more, whereas output includes individual performance, products, or services produced.

One of the primary reasons why IAQ is so crucial for humans is because, in industrialized countries, individuals frequently spend nearly 90% of their time indoors, primarily in buildings [10]. For some, the bulk of their time is spent in their residence, for others, it is spent in offices and workplaces, and for pupils and students, it is spent in schools or colleges. Employees spend around 50 hours each week in various types of offices or industries. During the week, students and instructors spend 4 to 8 hours in school [6].

In 2020, India will have 9 of the top 10 most polluted cities, as well as 35 of the world's top 50 most polluted cities. India ranks third out of 106 nations in terms of air quality, following Pakistan and Bangladesh, which hold second and first position in that order. During 2020, on average India's US Air Quality Index (US AQI) rating was 141 [11], which is harmful to vulnerable groups such as elderly, those with respiratory disorders, and the children. Forecasting the indoor air contaminants concentration to define IAQ is a long-standing subject in the area of built environment science. In comparison to measuring IAP concentrations, prediction based on AI approaches is non-invasive, rapid, and economical [6].

The air quality within and around a structure is referred to as IAQ. IAQ is affected by humidity, ventilation rate, temperature, various gases, biological pollutants, and the presence of particulate matter, which are roughly classified into 4 categories: physical parameters, chemical parameters, biological parameters, and particulate matter [6]. The combination of various parameters (biological, chemical, physical, and particulates) and their dynamic interactions with fluctuating humidity and temperature makes identifying IAQ-related concerns difficult for occupants. Outdoor pollution has a substantial influence on the IAQ of naturally ventilated structures [12]. Ventilation has an impact on IAQ, it is the procedure for exchanging polluted air inside the occupied space with clean outside air while also sustaining air movement inside the area. Sick building syndrome (SBS) is mostly linked to poor IAQ [13].

Numerous research investigations have found that low air quality levels cause early death, respiratory and cardiovascular illness, as well as an increase in cancer, dementia, and asthma episodes [14–16]. Poor air quality due to high pollutant concentration levels is to blame for millions of fatalities throughout the world. Poor air quality has the most adverse implications in poorer nations when there are no regulations to restrict pollution emissions. Similarly, quality of air is a concern in industrialized countries as well. Unfortunately, home IAP was accountable for more than 4.3 million untimely deaths in 2012, the majority of which occurred in medium and low-income nations [17]. Chronic Obstructive Pulmonary Disease (COPD), lung cancer, heart disease, stroke, and pneumonia were the prominent reasons of the death related to poor indoor air quality.

Governments and built-environment organizations have also developed a number of public regulations to decrease pollution exposure among building occupants [6, 18–20]. Real-time monitoring of IAP concentration levels in the indoor environment, while a rational response to IAQ management, might be a huge step toward good indoor environment management as well as effective source control. Additionally, cutting-edge technology, such as machine learning (ML), AI may use big data connected to pollution levels to estimate future living circumstances [21]. Smart health systems, smart cities, smart factories, and smart homes, are becoming increasingly popular across the world. Furthermore, the integration of AI and IoT has a significant impact on smart surroundings [22, 23]. Traditional threshold-triggered methods, on the other hand, can deliver real-time information on crucial IAQ values. Alternatively, AI-based forecasting models can provide advance warning of impending major changes in IAQ. As a result, building occupants can take precautionary measures to dodge catastrophic health consequences. In recent years, research communities have been investigating the prospective of AI to develop smart environments in which building dwellers get automated, real-time information of altering environmental situations [24]. Several scholars have suggested efficient IAQ prediction systems in the past to benefit public health and well-being [25, 26]. These studies have the potential to increase daily exercise while also improving ventilation options. Furthermore, these technologies can aid in the creation of hospitable atmosphere supported living systems and higher levels of productivity in built settings.

This chapter addresses the use of AI to forecast IAQ in order to enhance the indoor environment conditions and subsequently boost the health of its occupants. The primary goal of this work is to explore and show the possibilities of AI approaches for efficiently dealing with IAQ in order to improve indoor environment quality. The objective of this chapter is to study the AI-based forecasting technique for IAQ offered by researchers from several nations with distinct demographics, varied indoor air pollutant (IAP), and socioeconomic levels.

This chapter will assist in emphasizing the new issue domain in which future scholars must work. Furthermore, this chapter presents an overview of AI-based IAQ prediction models for various built environments. It emphasizes the possibilities of certain methodologies and seeks to synthesize the findings of past investigations. The rest of the chapter is structured as follows: Section 18.2 contains many types of IAQ parameters and their influence on human health. Section 18.3 contains information on AI and its function in forecasting IAQ. Finally, in Section 18.4, the conclusion is presented.

18.2   Indoor Air Quality Parameters

IAQ is a component of indoor environmental quality (IEQ), which encompasses both physical and psychological elements of indoor comfort (e.g., thermal comfort, acoustics, visual quality, and lighting) [3]. The US EPA defined IAQ as "the air quality within and around buildings and structures,

especially as it relates to the health and comfort of building occupants"
[27]. Any energy or mass stressor (physical parameters), particulate matter, gases (chemical parameters), and microbiological pollutants (biological
parameters) that might cause poor health conditions can all have an impact
on IAQ. IAQ parameters are broadly classified into four categories namely,
(a) physical parameter, (b) particulate matter, (c) chemical parameter, and
(d) biological parameter. These parameters are presented in Figure 18.1.

The internal environment is defined as a confined or semiconfined
place that is more prone to pollution accrual than an open-external environment. Pollutant deposition in indoor air originates from a variety of
sources, which may be categorized as (a) indoor sources, (b) subterranean
(underground) sources, and (c) outdoor sources. Cleaning products, fur-
niture, cigarette smoking, paints, cooking, cosmetics, carpets, fragrances,
insect repellents, sprays, human metabolism, pets, secondary reactions,
construction materials, platforms, fuel combustion, and restoration are
all sources of indoor pollution. Subterranean sources include foundation
materials, leachate, groundwater, underground parking, and landfill gases.
Road travel, construction operations, agriculture, industry, power plants,
wastewater treatment facilities, polluted sites, refineries, contaminated
waterways, sandstorms, and volcanic eruptions are examples of outdoor
sources.

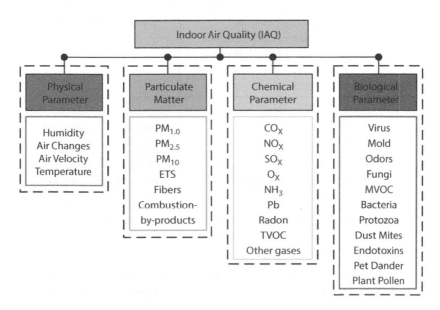

Figure 18.1 IAQ parameters.

The major IAP, their sources along with their adverse effect on human health is tabulated in Table 18.1. These pollutants have different sizes Figure 18.2 represents the variation in particle size ranging from 100 µm to 0.001 µm for different types of pollutants.

Table 18.1 Sources of IAP and adverse effects of IAPs on health.

Pollutant	Origin sources	Health impacts
CO_2	Burning coal, oil, and natural gases	Lowers oxygen levels, reduce respiratory and brain function, cause vision defects
CO	Indoor heating works, exterior air	At 700 ppm concentration, it is toxic severely
RSPM	Roadside dust, pollutants, major construction activities	Increased risk of cardiopulmonary and lung cancer, childhood asthma
O_3	High voltage air cleaning device and exterior air	Allergies and asthma
ETS	Tobacco products	Lung cancer, asthma
TVOC	Carpet, plywood, paint, solvent, pesticide, sealants, contaminated water, plastics, incense, scented items and some plants	Eye irritation, formation of photochemical oxidants
NO_2	Cooking stove, fire, exterior air	Lung irritation and increased chances of respiratory infection
Rn	Earth, granite and gneiss on low ventilated locations	Radioactive, leads lungs cancer in non-smokers
Formaldehyde	Particle board, interior grade plywood cabinetry, furniture	Carcinogen risk
SO_2	Combustion of sulphur-containing fuels	Skin irritation, coughing, throat irritation, breathing difficulties

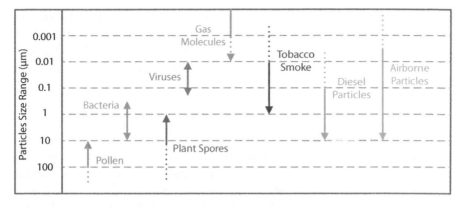

Figure 18.2 Indoor air pollutants and their particle size range.

The use of fresh air to dilute pollutants, filtration, and source control are the fundamental tactics for improving IAQ in most buildings. IAP might have health effects soon after exposure or years later. Poor IAQ is primarily responsible for SBS and building-related illness (BRI) [28]. SBS is a condition in which occupants who live or work in the occupied indoor space experience illness symptoms due to the environment of occupied space inside building in which they live or work. The outbreaks may or may not be caused by poor IAQ. Building-related illnesses (BRI) are illnesses that affect the lungs and other major organs as a result of substance exposure in contemporary, airtight structures. Exposure to pollutants within airtight structures with limited ventilation causes BRI. It affects occupants in the long term. Different parameters and their different concentrations are responsible for the occurrence of SBS and BRI in the built environment.

18.2.1 Physical Parameters

Physical parameters namely humidity, air changes (ventilation rate), air velocity inside the space, and temperature affects IAQ in buildings. These physical parameters affect both IAQ and thermal comfort inside the built environment [6]. Although, these parameters do not affect human health directly in general. However, these parameters create discomfort easily as they impact human physiology. The created discomfort feeling is the reason for psychological discomfort in the long run.

18.2.1.1 Humidity

Indoor air humidity (relative [RH] or absolute [AH]) is an essential parameter in the indoor environment. The experienced dryness of the air and potentially associated health impacts make it an important parameter. The development of pathogenic and allergenic organisms will be encouraged if the relative humidity in indoor environments exceeds 60%. Low humidity is responsible for increased discomfort owing to drying of the mucous membranes as well as electrical static discharges.

18.2.1.2   Air Changes (Ventilation)

In ASHRAE Standard 62.1 [29], the first section stated that "The purpose of this standard is to specify minimum ventilation rates and indoor air quality that will be acceptable to human occupants and are intended to minimize the potential for adverse health effects." Air change is directly linked with the reduction of pollutant concentration inside the built environment. The low air change rate is responsible for the accumulation of pollutants inside buildings. Occupants get SBS symptoms as a result of low air changes.

18.2.1.3   Air Velocity

Air velocity in a space influences convective heat exchange between the person and the interior environment [30]. There are two primary forms of air velocity. The first type of airflow is produced by natural airflow, buoyancy, or, more specifically, atmospheric circulation. The second type of airflow refers to forced air, air velocity produced by mechanical devices like standing as well as ceiling fans. Natural air velocity, in general, gives higher IAQ, a stronger cooling sensation, and greater thermal comfort. The convective mass transfer coefficient (h_m) of formaldehyde as well as other VOCs is affected by air velocity across the surface of building materials. Higher air velocity can reduce the thickness of the boundary layer and increase h_m, facilitating emissions.

18.2.1.4   Temperature

Temperature affects the thermal comfort of the occupant primarily. However, it is also associated with IAQ. Temperature variation can alter humidity, flow pattern, etc. High temperature inside any built environment nudges increased airflow as occupants try for more ventilation which will dilute pollutant concentration inside if the outdoor air is clean. In contrast, if the indoor temperature is on cooling side then the occupant will close the openings thus resulting in trapping the pollutants. Many researchers reported that indoor pollutants are higher in winters than in summers.

18.2.2   Particulate Matter

Particulate matter commonly recognized as PM is a word used to define an amalgamation of liquid droplets and solid particles prevalent in the air. Dirt, dust, smoke, and soot are examples of particles that are large or black enough to be visible to the naked eye. Rest are so minuscule that they require an electron microscope to see. Particle pollution comprises inhalable particles known as PM_{10}, which have diameters of 10 μm or less, as well as microscopic inhalable particles known as $PM_{2.5}$, which have diameters of 2.5 μm or less. These particles are made up of hundreds of distinct chemicals and may be found in a wide range of shapes and sizes. Some are released directly from smokestacks, fires, dirt roads, fields, and construction sites, among other places. Pollutants such as nitrogen oxide and sulphur dioxide, which are created by industry, power plants, and automobiles, are responsible for the bulk of particles in the atmosphere. Particulate particles can be inhaled and cause serious health problems. Particles as small as 10 micrometres in diameter can reach the lungs and even enter the blood circulation and spread throughout the body. The penetration level of particulates on the basis of their size in the human respiratory system is presented in Table 18.2 [31]. Particulates are the most harmful kind of pollutants, causing respiratory sickness, premature death, and heart attacks due to their ability to enter abysmally into the bloodstream, brain, and lungs. The IARC and the WHO have classified airborne particles as a group 1 carcinogen. The most dangerous particles are those with a diameter of less than 2.5 μm, sometimes known as microscopic particles or fine particles or PM2.5.

Table 18.2   Particulates classification based on their penetration level in the human respiratory system.

Diameter (μm)	Penetration level	Classification
>7	Oral and Nasal Cavities	Inhalable
4.7-7	Larynx	
3.3-4.7	Trachea and Bronchi	Thoragics
2.1-3.3	Secondary Bronchioles	
1.1-2.1	Bronchioles	Breathable
0.65-1.1	Alveoli	

18.2.3   Chemical Parameters

In IAQ, chemical parameters contain a wide variety of gases. Some of them are carbon monoxide (CO), carbon dioxide (CO_2), sulfur dioxide (SO_2), nitrogen dioxide (NO_2), ozone (O_3), gaseous ammonia (NH_3), volatile organic compound (VOC), etc. and are explained briefly below.

18.2.3.1   Carbon Dioxide

It is an acidic, colourless gas made up of two covalently bound oxygen atoms. In the Earth's atmosphere, it exists as a trace gas. Forest fires, volcanoes, and hot springs are all natural sources, and it is released from carbonate rocks by acid and water breakdown. Natural CO_2 sources include forest fires, hot springs, and volcanoes. CO_2 is the most abundant greenhouse gas in the Earth's atmosphere. Since the Industrial Revolution, emissions from anthropogenic activities, mostly deforestation and the use of fossil fuels, have dramatically increased their concentration in the air, adding to global warming. CO_2 pollution can have a variety of detrimental health effects. Excessive CO_2 concentration and exposure duration can cause headaches, restlessness, tingling or pins and needles sensations, dizziness, high blood pressure, sweating, problems breathing, tiredness, hypoxia, coma, and convulsions, among other symptoms.

18.2.3.2   Carbon Monoxide

It is a flammable, odourless, tasteless, colourless gas having a density somewhat lower than air and the basic member of the oxocarbon family. It is common air pollution in several indoor settings, mostly generated from the exhaust combustion engines (cars, power washers, lawnmowers, portable and backup generators, and so on), as well as the partial burning of a variety of other items (like charcoal, coal, wood, natural gas, oil, propane, trash, and paraffin). The thermal combustion of fuels, such as diesel, petroleum products, and wood are the most common source of CO. In confined places, CO concentrations can reach lethal levels. Anomalies in CO metabolism have been linked to a wide array of disorders, including hypertension, cardiac disorders, pathological inflammation, and dementia, due to CO role in the body.

18.2.3.3   Nitrogen Dioxide

It is a gas with a strong, unpleasant odour and a reddish-brown colour. NO_2 is mostly emitted into the atmosphere as a by-product of fuel burning.

Power plants, buses, trucks, automobiles, and other vehicles, as well as machinery that consume fuels, emit NO_2. The primary causes of NO_2 exposure inside are cigarette smoke and kerosene stoves and heaters. When nitric oxide (NO) is oxidized by oxygen in the air, NO_2 is generated. The majority of oxidation reactions that require air create NO_2. When nitrogen combines with oxygen at high temperatures, NO is generated. Direct contact with the skin can cause irritation and burns. Only excessively high gas concentrations induce severe discomfort: A concentration of 100–200 ppm causes moderate nose and throat discomfort, 250–500 ppm causes edema (swelling caused by the trapping and accumulation of fluids in the body), which can lead to pneumonia or bronchitis, and a concentration of 1000 ppm causes death due to asphyxiation from wet lungs. Other than transient nausea, cough, or exhaustion, there are usually no symptoms at the time of contact, but edema develops after hours of lung inflammation. Even little variations in NO_2 from day to day can have an impact on lung function. Chronic NO_2 exposure can cause respiratory effects in healthy people, such as airway inflammation, as well as increased respiratory symptoms in asthmatics. NO_2 creates ozone, which irritates the eyes and exacerbates respiratory issues, leading to an increase in emergency room visits and hospitalizations for respiratory illnesses, especially asthma.

18.2.3.4   Sulphur Dioxide

It is a toxic gas with a strong nitric acid-like odour. It is produced by the burning of sulphur-bearing fossil fuels, and it is also released naturally by volcanic activity. SO_2 has a similar odor to that of burning matchsticks. Smoke from coal, matches, and sulfur-containing fuels, for example, often exposes individuals to sulphur dioxide. SO_2 is a moderately toxic gas that can be deadly in high quantities. Low-concentration exposure over time is also dangerous. In addition to respiratory problems, SO_2 can cause stomach discomfort, hypotension, flushing, urticaria, dermatitis, diarrhoea, and possibly life-threatening anaphylaxis.

18.2.3.5   Ozone

It is an oxygen allotrope made up of an inorganic molecule that is much less stable than O_2. It is a yellowish blue gas with a strong, unpleasant odour akin to chlorine, and many people can sense it in the air at concentrations as low as 0.1 ppm. O_3 is a powerful oxidant. O_3 is a dangerous gas that is frequently created by anthropogenic settings and may be found in workplaces with photocopiers, airline cabins, sterilizers, laser printers, and other places. O_3 is an oxygen allotrope made up of an inorganic molecule that is much less stable than O_2. It is a yellowish blue gas with a strong, unpleasant odour akin to chlorine, and many people can sense it in the air at concentrations as low as 0.1 ppm. O_3 is a powerful oxidant. O_3 is a dangerous gas that is frequently created by anthropogenic settings and may be found in workplaces with photocopiers, airline cabins, sterilizers, laser printers, and other places.

18.2.3.6   Gaseous Ammonia

It is a colourless, odourless gaseous chemical made up of one molecule of nitrogen and three molecules of hydrogen. Nitrogenous waste is the most common source. Ammonia is both toxic and corrosive in concentrated form. Agricultural operations, such as the use of ammonia-based fertilizers and animal husbandry, create the majority of it. Other sources include volatilization from oceans and soils, car emissions, industrial processes, and so forth. Indoor air with high amounts of ammonia can irritate the eyes, nose, and throat, as well as cause skin irritation. Ammonia poisoning causes long-term health repercussions, including serious cardiovascular and pulmonary problems. Lung function impairment, asthma exacerbation, and early death are all potential effects.

18.2.3.7   Volatile Organic Compounds

VOCs are released as gases by some solids or liquids. VOCs are a group of chemicals that can have both immediate and long-term health effects. Many VOC concentrations within buildings are consistently higher than outside. Thousands of sundry items emit volatile organic compounds (VOCs). Furnishings and construction materials, lacquers and paints and paint strippers, pesticides, cleaning supplies, office equipment, carbonless copy paper and correction fluids, craft and graphics materials like adhesives and glues, photographic solutions, and permanent markers are just a few examples. Organic compounds are often used as components in household products. Wax, varnishes, paints, a variety of cleaning, disinfecting, cosmetic, and other products, include organic solvents. Fuels are made from organic molecules. While used and, to a lesser extent, when kept, all of these objects have the potential to leak organic compounds. Nausea, loss of coordination, headaches, eye, nose, and throat irritation, as well as central nervous system, kidney, and liver damage are all potential risks. In humans, certain VOCs can cause cancer.

18.2.4   Biological Parameters

Some pollutants are classified as the biological parameter. Viruses, bacteria, animal saliva and dander, cockroaches, pollen, and dust mites are some of the biological pollutants. These contaminants can be found in a variety of environments. The relative humidity level in a home can be controlled to prevent the growth of certain biologicals. A relative humidity of 30% to 50% is good for houses in numerous instances. Standing water, water-damaged materials, and wet surfaces are ideal breeding grounds for mildews, moulds, insects, and bacteria. House dust mites thrive in damp, warm environments, making them one of the most potent biological allergens. Infectious diseases including COVID-19, measles, influenza, and chickenpox are spread via the air. Molds and mildews release chemicals that make people sick. Coughing, watery eyes, sneezing, dizziness, shortness of breath, stomach issues, fever, and tiredness are all symptoms of biological pollution causing health problems.

18.3   AI in Indoor Air Quality Prediction

AI is now widely employed in engineering, science, education, and manufacturing [32–34]. AI is a different approach to traditional modelling techniques. AI is a computer programme that mimics intelligent human behaviour. It is a computer or system that perceives its surroundings, understands its behaviour, and takes action. According to a McKinsey prediction, AI would generate $13 trillion in economic value globally by 2030 [35]. This is because AI is revolutionizing engineering in almost every industry and application area. The AI technique, in comparison to traditional approaches, offers a solution to a variety of complicated environmental engineering challenges, particularly in the field of IAQ. Furthermore, the AI technique works in places where frequent physical testing is not possible, reducing time, human effort, and expense in the process. The main benefit of employing AI is that it lowers errors and produces speedy results because of its Quick Decision Making (QDM) capabilities [36].

Further, AI is having subsets namely ML, and Deep Learning (DL) as presented in Figure 18.3. In IAQ, several studies used ML algorithms to forecast the quality of indoor air. In literature, the mainly used ML models are artificial neural network (ANN), support vector machines (SVM), random forest regression (RF), boosted regression trees (BRT), multiple linear regression (MLR), decision trees (DT), and adaptive-neuro fuzzy interface system (ANFIS). Figure 18.4 presents various ML methods.

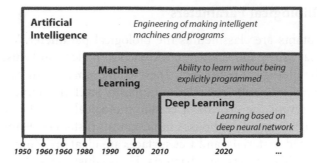

Figure 18.3 AI, ML, and DL.

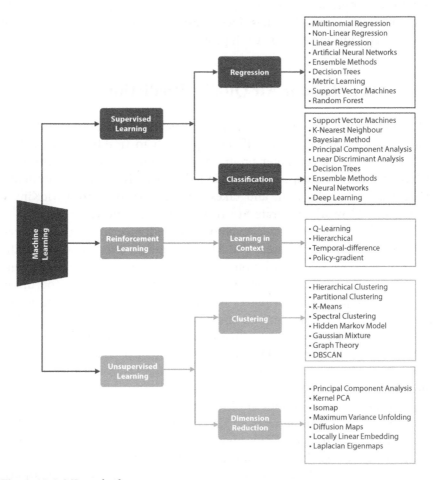

Figure 18.4 ML methods.

The performance of each individual model depends on the variation in the predicted to measured values. The higher the outliers in the dataset, the lower the performance of the model. The accuracy of the ANN models is good in terms of accuracy, but ANN algorithms require a large amount of time to predict the results. The time limitation of the ANN model is solved by the ANFIS model, but the accuracy of this model may be less as compared to the ANN model. Table 18.3 describes the selected studies [37–52] on predicting the quality of indoor air using ML.

Table 18.3 Summary of use of AI in the prediction of IAQ.

Author	Year	AI models	Input parameters	Prediction results
Lee *et al.* [37]	2020	ANN	Humidity, Wind Speed and direction, Ambient temperature, $PM_{2.5}$, and PM_{10}	R^2 value: $PM_{2.5} = 0.93510$ $PM_{10} = 0.80713$
Suleiman *et al.* [38]	2019	ANN, BRT, and SVM	$PM_{2.5}$, PM_{10}, SO_2, NO_2, NOx, CO	For $PM_{2.5}$: SVM = 4.61 ANN = 4.53 BRT = 4.24 For PM_{10}: SVM = 7.72 ANN = 9.08 BRT = 7.99
Wei *et al.* [39]	2019	ANN, MLR, and DT	$PM_{2.5}$, PM_{10}, CO_2, CO, NO_x, Rn, ventilation rate, wind speed, indoor temperature, and airborne culturable bacteria	ANN $R^2 = 0.62$-0.79, MLR $R^2 = 0.69$ DT $R^2 = 0.74$-0.94 (for training) $R^2 = 0.33$-0.49 (for validation)

(Continued)

Table 18.3 Summary of use of AI in the prediction of IAQ. (*Continued*)

Author	Year	AI models	Input parameters	Prediction results
Nishihama *et al.* [40]	2021	RFR	$PM_{2.5}$, PM_{10}	$PM_{2.5}$ $R^2 = 0.44$ PM_{10} $R^2 = 0.36$
Amuthadevi *et al.* [41]	2021	Statistical Multilevel Regression, Non-Linear ANN, Neuro-Fuzzy and Deep Learning Long-Short-Term Memory (DL-LSTM)	O_3, SO_2, NO_2, CO, and PM	$R^2 = 0.71$–0.89
Sharma *et al.* [42]	2020	MLP and eXtream Gradient Boosting Regression (XGBR), LSTM	CO_2, $PM_{2.5}$, humidity, and temperature	Estimation R = 95% Forecasting R = 96%
Saad *et al.* [43]	2017	MLP, KNN and linear discrimination analysis (LDA).	PM_{10}, O_2, NO_2, CO_2, VOC, O_3, CO, humidity, and temperature	-

(*Continued*)

Table 18.3 Summary of use of AI in the prediction of IAQ. (*Continued*)

Author	Year	AI models	Input parameters	Prediction results
Challoner et al. [44]	2015	ANN	Time of day, relative humidity, temperature, wind direction and speed, global solar radiation, Pasquill atmospheric stability class, outdoor pollutant concentrations, sea level pressure, and barometer level pressure	Location 1, 2 and 3: NO_2 $R^2 = 0.854$ $R^2 = 0.870$ $R^2 = 0.829$ $PM_{2.5}$ $R^2 = 0.711$ $R^2 = 0.760$ $R^2 = 0.770$
Ahn et al. [45]	2017	LSTM Gated recurrent unit	VOC, CO_2, humidity, temperature, light amount, and fine dust	Accuracy of prediction: LSTM = 70.13 GRU − 84.69
Adeleke et al. [46]	2017	MLP Neural network	$PM_{2.5}$ concentration (indoor)	Sensitivity = 0.85 Precision = 0.86
Liu et al. [47]	2018	ANN	CO_2, PM_{10}, $PM_{2.5}$, temperature, and relative humidity	PM_{10}: $R^2 = 0.91$ $PM_{2.5}$: $R^2 = 0.97$
Loy-Benitez et al. [48]	2019	Deep RNN	NO_2, PM_{10}, $PM_{2.5}$, CO, CO_2, and NO	RMSE − 30.99 µg/m³ and 29.73 µg/m³, MAPE = 31.10% and 29.52%

(*Continued*)

Table 18.3 Summary of use of AI in the prediction of IAQ. (*Continued*)

Author	Year	AI models	Input parameters	Prediction results
Ha *et al.* [49]	2020	Extended fractional order Kalman filter	NH_3, H_2, CO, CO_2, O_2, H_2S, toluene, ethanol, humidity, and temperature	MSE = 0.6103, 0.5122, 0.39993, 0.8612, 0.7082, 0.4738, 0.4262, 0.6761, 0.3601, and 0.3007
Fang *et al.* [50]	2016	Machine learning based nonparametric forecasting	Humidity, temperature, VOCs, $PM_{2.5}$	NRMSD = 7.5%
Xiahou *et al.* [51]	2019	ARIMA	$PM_{2.5}$, CO_2, PM_{10}, temperature, tVOC, formaldehyde	Mean prediction error = 0
Kapoor *et al.* [52]	2022	ANN, SVM, DT, GPR, LR, EL, Optimized SVM, Optimized DT, Optimized GPR, and Optimized EL	Number of occupants, area per person, indoor CO_2, outer temperature, wind speed, relative humidity, and air quality index	Optimized GPR: R = 0.98874 RMSE = 4.20068 ppm MAE = 3.35098 ppm NS = 0.9817 a20-index = 1

18.4   Conclusion

The AI methods for IAQ that allow for accurate prediction of IAP concentrations in the indoor environment while also considering health, performance, productivity, and comfort are discussed in this chapter. AI techniques, particularly those based on machine learning algorithms, improve the ability to analyse dynamic and highly variable data. As a result, these are more suitable for analyzing the nonlinear nature of interactions between buildings and their dwellers. The findings revealed that a variety of AI-based methods is being used by different researchers to predict IAQ in various types of buildings. ANNs have been used to solve challenges linked to recognition and identification in particular. The integration of concepts and methodologies from a wide variety of academic subjects, most notably computer science, is required for the research of AI and its application in indoor environments. As a result of the limited multidisciplinary collaboration between these two domains (i.e., computer science and indoor environment engineering), the chances of finding less complex and less expensive ways to integrate AI-based IAQ prediction and control into buildings are likely to be further hampered. A more holistic understanding of IAQ and related AI approaches in the built environment would definitely benefit the readers and motivate researchers to work in this direction. This chapter provides valuable insight into the future of AI-based IAQ forecasting.

References

1. Fang, L., Clausen, G., Fanger, P.O., Impact of temperature and humidity on the perception of indoor air quality. *Indoor Air*, 8, 80–90, 1998, https://doi.org/10.1111/j.1600-0668.1998.t01-2-00003.x.
2. Melikov, A.K. and Kaczmarczyk, J., Air movement and perceived air quality. *Build. Environ.*, 47, 400–409, 2012, https://doi.org/10.1016/j.buildenv.2011.06.017.
3. Kapoor, N.R. and Tegar, J.P., Human comfort indicators pertaining to indoor environmental quality parameters of residential buildings in Bhopal. *Int. Res. J. Eng. Technol.*, 5, 2395–0056, 2018, http://dx.doi.org/10.13140/RG.2.2.13735.62883.
4. Mitchell, C.S., Zhang, J., Sigsgaard, T., Jantunen, M., Lioy, P.J., Samson, R., Karol, M.H., Current state of the science: Health effects and indoor environmental quality. *Environ. Health Perspect.*, 115, 6, 958–964, 2007, https://doi.org/10.1289/ehp.8987.
5. Sundell, J., On the history of indoor air quality and health. *Indoor Air*, 14, 51–58, 2004, https://doi.org/10.1111/j.1600-0668.2004.00273.x.
6. Kapoor, N.R., Kumar, A., Alam, T., Kumar, A., Kulkarni, K.S., Blecich, P., A review on indoor environment quality of indian school classrooms. *Sustainability*, 13, 21, 11855, 2021, https://doi.org/10.3390/su132111855.
7. Wargocki, P., Improving indoor air quality improves the performance of office work and school work, *8th International Conference for Enhanced Building Operations-ICEBO'08 Conference Center of the Federal Ministry of Economics and Technology*, Berlin, 1–7, October 20–22, 2008, Available Online: https://hdl.handle.net/1969.1/90792.

8. Park, J.S. and Yoon, C.H., The effects of outdoor air supply rate on work performance during 8-h work period. *Indoor Air*, 21, 284–290, 2011, https://doi.org/10.1111/j.1600-0668.2010.00700.x.

9. Kapoor, N.R., Kumar, A., Meena, C.S., Kumar, A., Alam, T., Balam, N.B., Ghosh, A., A systematic review on indoor environmental quality in naturally ventilated school classrooms: A way forward. *Adv. Civ. Eng.*, 2021, 8851685, 2021, http://dx.doi.org/10.1155/2021/8851685.

10. Klepeis, N.E., Nelson, W.C., Ott, W.R., Robinson, J.P., Tsang, A.M., Switzer, P., Behar, J.V., Hern, S.C., Engelmann, W.H., The National Human Activity Pattern Survey (NHAPS): A resource for assessing exposure to environmental pollutants. *J. Expo. Sci. Environ. Epidemiol.*, 11, 3, 231–252, 2001, https://www.nature.com/articles/7500165.pdf.

11. https://www.iqair.com/us/india.

12. Shrestha, P.M., Humphrey, J.L., Carlton, E.J., Adgate, J.L., Barton, K.E., Root, E.D. *et al.*, Impact of outdoor air pollution on indoor air quality in low-income homes during wildfire seasons. *Int. J. Environ. Res. Public Health*, 16, 19, 3535, 2019, https://doi.org/10.3390/ijerph16193535.

13. Passarelli, G.R., Sick building syndrome: An overview to raise awareness. *J. Build. Apprais.*, 5, 1, 55–66, 2009, https://doi.org/10.1057/jba.2009.20.

14. Xie, H., Ma, F., Bai, Q., August. Prediction of indoor air quality using artificial neural networks, in: *2009 Fifth International Conference on Natural Computation*, vol. 2, IEEE, pp. 414–418, 2009, https://doi.org/10.1109/ICNC.2009.502.

15. Wild, C.P., Complementing the genome with an "exposome": The outstanding challenge of environmental exposure measurement in molecular epidemiology. *Cancer Epidemiol. Biomark. Prev.*, 14, 8, 1847–1850, 2005, https://doi.org/10.1158/1055-9965.EPI-05-0456.

16. Indoor Air Quality, US Environmental Protection Agency, USA, 2020, https://www.epa.gov/report-environment/indoor-air-quality#:~:text=Health%20effects%20associated%20with%20indoor,%2C%20heart%20disease%2C%20and%20cancer.

17. Air pollution: Indoor air pollution, World Health Organization, WHO, Geneva, Switzerland, 2014, https://www.who.int/news-room/questions-and-answers/item/air-pollution-indoor-air-pollution#:~:text=Of%20the%204.3%20million%20people,6%25%20of%20deaths%2C%20respectively.

18. Perdue, W.C., Stone, L.A., Gostin, L.O., The built environment and its relationship to the public's health: The legal framework. *Am. J. Public Health*, 93, 9, 1390–1394, 2003, https://doi.org/10.2105/AJPH.93.9.1390.

19. Coutts, A., Beringer, J., Tapper, N., Changing urban climate and CO2 emissions: Implications for the development of policies for sustainable cities. *Urban Policy Res.*, 28, 1, 27–47, 2010, https://doi.org/10.1080/08111140903437716.

20. Foster, S. and Everett Jones, S., Association of school district policies for radon testing and radon-resistant new construction practices with indoor

radon zones. *Int. J. Environ. Res. Public Health*, 13, 12, 1234, 2016, https://doi. org/10.3390/ijerph13121234.

21. Fard, Z.Q., Zomorodian, Z.S., Korsavi, S.S., Application of machine learning in thermal comfort studies: A review of methods, performance and challenges. *Energy Build.*, 256, 111771, 2021, https://doi.org/10.1016/j.enbuild.2021.111771.

22. Kumar, A., Kapoor, N.R., Arora, H.C., Kumar, A., *Smart cities - A step towards sustainable development, smart cities: Concepts, practices, and applications*, CRC Press, Taylor and Francis, 2022, https://doi.org/10.1201/9781003287186-1.

23. Gomez, C., Chessa, S., Fleury, A., Roussos, G., Preuveneers, D., Internet of Things for enabling smart environments: A technology-centric perspective. *J. Ambient Intell. Smart Environ.*, 11, 1, 23–43, 2019, https://doi.org/10.3233/ AIS-180509.

24. Putra, J.C.P., Safrilah, Ihsan, M., The prediction of indoor air quality in office room using artificial neural network. *AIP Conf. Proc.*, 1977, 1, 020040, 2018, AIP Publishing LLC, https://doi.org/10.1063/1.5042896.

25. Wei, W., Ramalho, O., Malingre, L., Sivanantham, S., Little, J.C., Mandin, C., Machine learning and statistical models for predicting indoor air quality. *Indoor Air*, 29, 704–726, 2019, https://doi.org/10.1111/ina.12580.

26. Alirezaie, M. and Loutfi, A., Reasoning for sensor data interpretation: An application to air quality monitoring. *J. Ambient Intell. Smart Environ.*, 7, 4, 579–597, 2015, https://doi.org/10.3233/AIS-150323.

27. Introduction to Indoor Air Quality, US Environmental Protection Agency, USA, 2021, https://www.epa.gov/indoor-air-quality-iaq/introduction-indoor-air-quality.

28. Raj, N., Kumar, A., Kumar, A., Goyal, S., Indoor environmental quality: Impact on productivity, comfort, and health of Indian occupants. Abstract, in: *Proceedings of the International Conference on Building Energy Demand Reduction in Global South (BUILDER'19)New Delhi, India*, 13–14 December 2019, pp. 1–9, Available online: https://nzeb.in/event/builder19/.

29. American Society of Heating, Refrigerating and Air-Conditioning Engineers, ASHRAE Standard 62.1-2004, Atlanta, GA, USA, 2004, https://www.ashrae.org/ technical-resources/standards-and-guidelines/standards-interpretations/ interpretations-for-standard-62-1-2004.

30. Attia, S., Chapter 5-occupants well-being and indoor environmental quality, in: *Net Zero Energy Buildings (NZEB)*, pp. 117–153, 2018, https://doi. org/10.1016/B978-0-12-812461-1.00005-8.

31. Manuel, G.S., An analysis of the transmission modes of COVID-19 in light of the concepts of indoor air quality. *REHVA J.*, 3, 46–54, 2020, http://dx.doi. org/10.13140/RG.2.2.28663.78240.

32. Verma, K., Bhardwaj, S., Arya, R., Islam, M.S.U., Bhushan, M., Kumar, A., Samant, P., Latest tools for data mining and machine learning. *Int. J. Innov. Technol. Exploring Eng.*, 8, 9S, 18–23, July 2019, Available: https://doi. org/10.35940/ijitee.I1003.0789S19.

33. Kumar, A., Arora, H.C., Kapoor, N.R., Mohammed, M.A., Kumar, K., Majumdar, A., Thinnukool, O., Compressive strength prediction of lightweight concrete: Machine learning models. 14, 4, 2404, 2022, https://doi.org/10.3390/su14042404.

34. Kholiya, P.S., Kapoor, A., Rana, M., Bhushan, M., Intelligent process automation: The future of digital transformation, in: *10th International Conference on System Modeling & Advancement in Research Trends (SMART)*, pp. 185–190, 2021, https://doi.org/10.1109/SMART52563.2021.9676222.

35. What is artificial intelligence (AI)?, Mathworks, USA, 2022, https://in.mathworks.com/discovery/artificial-intelligence.html.

36. Kumar, A. and Mor, N., An approach-driven: Use of artificial intelligence and its applications in civil engineering, in: *Artificial Intelligence and IoT*, pp. 201–221, Springer, Singapore, 2021, https://doi.org/10.1007/978-981-33-6400-4_10.

37. Lee, Y.W., A stochastic model of particulate matters with AI-enabled technique-based IoT gas detectors for air quality assessment. *Microelectron. Eng.*, 229, 111346, 2020, https://doi.org/10.1016/j.mee.2020.111346.

38. Suleiman, A., Tight, M.R., Quinn, A.D., Applying machine learning methods in managing urban concentrations of traffic-related particulate matter (PM10 and PM2. 5). *Atmos. Pollut. Res.*, 10, 1, 134–144, 2019, https://doi.org/10.1016/j.apr.2018.07.001.

39. Wei, W., Ramalho, O., Malingre, L., Sivanantham, S., Little, J.C., Mandin, C., Machine learning and statistical models for predicting indoor air quality. *Indoor Air*, 29, 704–726, 2019, https://doi.org/10.1111/ina.12580.

40. Nishihama, Y., Jung, C.R., Nakayama, S.F., Tamura, K., Isobe, T., Michikawa, T., Iwai-Shimada, M., Kobayashi, Y., Sekiyama, M., Taniguchi, Y., Yamazaki, S., Indoor air quality of 5,000 households and its determinants. Part A: Particulate matter (PM2. 5 and PM10–2.5) concentrations in the Japan environment and children's study. *Environ. Res.*, 198, 111196, 2021, https://doi.org/10.1016/j.envres.2021.111196.

41. Amuthadevi, C., Vijayan, D.S., Ramachandran, V., Development of air quality monitoring (AQM) models using different machine learning approaches. *J. Ambient Intell. Hum. Comput.*, 1–13, 2021, https://doi.org/10.1007/s12652-020-02724-2.

42. Sharma, P.K., Mondal, A., Jaiswal, S., Saha, M., Nandi, S., De, T., Saha, S., IndoAirSense: A framework for indoor air quality estimation and forecasting. *Atmos. Pollut. Res.*, 12, 1, 10–22, 2021, https://doi.org/10.1016/j.apr.2020.07.027.

43. Mad Saad, S., Andrew, A.M., Md Shakaff, A.Y., Mat Dzahir, M.A., Hussein, M., Mohamad, M., Ahmad, Z.A., Pollutant recognition based on supervised machine learning for indoor air quality monitoring systems. *Appl. Sci.*, 7, 8, 823, 2017, https://doi.org/10.3390/app7080823.

44. Challoner, A., Pilla, F., Gill, L., Prediction of indoor air exposure from outdoor air quality using an artificial neural network model for inner city

commercial buildings. *Int. J. Environ. Res. Public Health*, 12, 12, 15233–15253, 2015, https://doi.org/10.3390/ijerph121214975.

45. Ahn, J., Shin, D., Kim, K., Yang, J., Indoor air quality analysis using deep learning with sensor data. *Sensors*, 17, 11, 2476, 2017, https://doi.org/10.3390/s17112476.

46. Adeleke, J.A., Moodley, D., Rens, G., Adewumi, A.O., Integrating statistical machine learning in a semantic sensor web for proactive monitoring and control. *Sensors*, 17, 4, 807, 2017, https://doi.org/10.3390/s17040807.

47. Liu, Z., Cheng, K., Li, H., Cao, G., Wu, D., Shi, Y., Exploring the potential relationship between indoor air quality and the concentration of airborne culturable fungi: A combined experimental and neural network modeling study. *Environ. Sci. Pollut. Res.*, 25, 4, 3510–3517, 2018, https://doi.org/10.1007/s11356-017-0708-5.

48. Loy-Benitez, J., Vilela, P., Li, Q., Yoo, C., Sequential prediction of quantitative health risk assessment for the fine particulate matter in an underground facility using deep recurrent neural networks. *Ecotoxicol. Environ. Saf.*, 169, 316–324, 2019, https://doi.org/10.1016/j.ecoenv.2018.11.024.

49. Ha, Q.P., Metia, S., Phung, M.D., Sensing data fusion for enhanced indoor air quality monitoring. *IEEE Sens. J.*, 20, 8, 4430–4441, 2020, https://doi.org/10.1109/JSEN.2020.2964396.

50. Fang, B., Xu, Q., Park, T., Zhang, M., AirSense: An intelligent home-based sensing system for indoor air quality analytics, in: *Proceedings of the 2016 ACM International Joint Conference on Pervasive and Ubiquitous Computing*, pp. 109–119, 2016, https://doi.org/10.1145/2971648.2971720.

51. Xiahou, R., Yi, J., He, L., He, W., Huang, T., Indoor air monitoring system based on Internet of Things and its prediction model, in: *Proceedings of the International Conference on Industrial Control Network and System Engineering Research*, March, pp. 58–63, 2019, https://doi.org/10.1145/3333581.3333582.

52. Kapoor, N.R., Kumar, A., Kumar, A., Kumar, A., Mohammed, M.A., Kumar, K., Kadry, S., Lim, S., Machine learning-based CO_2 prediction for office room: A pilot study. *Wireless Commun. Mobile Comput.*, 2022, 9404807, 2022, https://doi.org/10.1155/2022/9404807.

Index

3D-CNN, 27

AarogyaSetu app, 93
Access control, defined, 249
Accuracy, 167
Actuators, 86
Adaptive signals, 332
ADNI dataset, 9–15
Advanced encryption standard (AES), 241–242
Advantages of IoT-based vehicle parking system, 392
Agriculture sector,
  cloud computing in, 88
  IoT and, 86
Air conditioners, 86
Akamai, 82
AlexNet, 3, 12
Alzheimer's disease (AD),
  classification,
  overview, 4–6
  TL techniques, 6–8
  using conventional training methods, 9–12
  using TL, 12–15
Amazon, 226, 229, 232, 236, 245
Amazon web services, 88
Amyloid plaques, 4
Angiomyolipoma, 56
Anti scam, 85
Apple, 239–240
Application,
  issues, cloud computing challenges, 243
  vulnerability, 254

Architecture, 400, 404, 406, 407, 411, 413
Area under curve, 39
Artificial intelligence (AI), 53, 81–83, 85, 91, 127
Artificial intelligence systems, usage of, 9
Artificial neural network, 25, 163–164, 208
Artificial vision, 323
ASPs, 230
Audio/video streaming devices, 86
Auditor, cloud, 249
Augmented reality for troubleshooting, 89
Automation, 83, 89
Ava robotics, 90
Aviation industry, blockchain technology in, 88–89
Azure storage engine, 236

Banking service, AI, 85
Batch normalization, 44
Belief network, 32
Benign tumor, 55
Big data analysis, 82, 83
Binary classification, VGG-based AD, 12
Bio inspired evolution, 347
Biometric data, 52
Blackout, amazon S3, 245
Blockchain, 129
Blockchain technology, 82, 83, 88–89, 91
Bond strength, 425, 426, 429–433, 441
Broker, cloud, 249

Built environment, 447–450, 453, 454, 465
Business cogruence plan, 258

Caffe-based deep CNN, 138
CaffeNet, 3, 12
Cancer, damaged ligaments to, 5
CapEx, 230
Capital cost, 230
Carbon fiber reinforced polymer, 420–422
Carrier-A, 249
Cataclysm mitigation, 239
Cellular mobile communication technology, 89
Cerebral cortex, shrinkage of, 4
Challenges, 404, 405, 414
Chromosomes, 347, 349, 350, 354, 357, 359, 361
Cirrhosis, 54
Class imbalance, 207
Cleveland heart disease dataset, 211
Cloud auditor, 249
Cloud broker, 249
Cloud computing, 83, 87–88, 91, 400, 401, 409, 410
    advantages, 234–235
    applications, 239–241
    architecture as per cloud-server communication, 243
    background, 228–241
    challenges in, 235–236
        application issues, 243
        availability of service, 245
        cyber crime laws, lack of, 245
        data access control, 248
        data decontamination, 244
        data location, 244
        data segregation, 245–246
        data storing rules, 247
        escapade system, 248
        immaturity of standards, 236
        information issues, 242–243
        insider access, 246

        long-term viability, 246
        loss of authority, 236
        network load, 235, 243, 244
        permanent loss of data, 245
        sanitization of data, 235–236
        security issues, 242
        shortcoming protecting, 247
        solution and practices, 246–248
        theft of data, 235
        workplaces for recovery, 247–248
    characteristics, 234
    data protection and security using steganography, 258–263
        in cloud environment, 260
        overview, 258, 259
        PVD method, 261–263
        types, 259–260
    history of, 228–232
    IaaS (see Infrastructure-as-a-Service (IaaS) model)
    literature review, 241–242
    overview, 226–228
    PaaS model (see Platform as a Service (PaaS) model)
    reference architecture, 248–249
    related study, 263
    SaaS (see Software-as-a-Service (SaaS) model)
    security, 236–239
        access control, 249
        foundation, application level, 238
        foundation security, 236–237
        general security threats, 249–254
        issues and its preventive measures, 248–258
        need of security in cloud, 239
        network class, 249
        preventive measures, 254–258
        SaaS and PaaS host security, 237
        supplier data and its security, 238–239
        taxonomy of challenges, 250–251
        virtual server security, 237–238

service models, 232–234
  IaaS, 233
  PaaS, 233–234
  SaaS, 233
  types,
    hybrid, 232
    neighborhood, 232
    private, 232
    public, 232
Cloud consumer, 249
Cloud expert,
  association(s), 247, 248
  community, 247
Cloud figuring security, 239
Cloud information tracker licence
    community, 227
Cloud provider, 249
Cloud service providers (CSP), 237,
    238
CloudEx computing service, 240
Cognitive computing, 328
Comfort, 448, 450, 451, 453, 454, 464
Communication, 402–404, 407
Computed tomography, 59
Computer aided diagnosis, 59
Computer aided identification, 25
Connected cars, 331
Connection, 402, 404, 406, 410
Consumer, cloud, 249
Conventional training methods, AD
    classification using, 9–12
Convolutional neural network, 109
Convolutional neural network (CNN)-
    based TL models, 6, 11–12
Covid-19 pandemic, transforming to
    digital future during,
  digital technologies, use of, 84–90
    5th-generation mobile network,
      89–90
    artificial intelligence, 85
    blockchain technology, 88–89
    cloud computing, 87–88
    IoT, 85–86
    telehealth/telemedicine, 87

economic emergency, 82
implications for research, 93–94
overview, 82–84
transforming digital technology,
    challenges in, 90–93
  accessing internet, 92
  digital payments, 92–93
  increasing digitalization, 91
  internet shutdowns, 92
  online fraud, 92
  privacy and surveillance, 93
  work from home culture, 91
  workplace monitoring and techno
      stress, 91
Crop yield prediction, 152
Crossover, 348, 349, 352, 353, 355, 359,
    361
Cryptography, 260, 410, 412
Cyber crime laws, lack of, 245

Data,
  access control, 248
  at rest, 250
  breaching, 225
  class, 251–252
  decontamination, 244
  disaster, 242
  forgery, 242
  in progress, 251
  in transit, 251
  lineage, 250
  location, 244
  permanent loss of, 245
  protection, cloud, 258–263
  sanitization of, 235–236
  segregation, 245–246
  steganography in cloud
      environment, 260
  storing rules, 247
  theft of, 235, 242
Data augmentation, 132
  image blurring, 114
  image flipping, 114
  image noise, 114

image rotation, 114
image shifting, 114
Data collection, 117
Data mining techniques, AD and, 6
Data normalization,
  batch normalization, 115
  group normalization, 115
  instance normalization, 115
  layer normalization, 115
  weight normalization, 115
Data pre-processing methods, 199
Data pre-processing stages,
  clustering, 200–202
  data analysis, 200–202
  data cleaning, 200–206
  data regression, 200–202
  data transformation, 200–202
Data splitting,
  holdout cross-validation set, 116
  test set, 116
  train set, 116
Decision making, 26
Decision tree, 207
Deep learning, 128
Deep learning (DL), COVID-19
  pandemic and, 85
Deep learning (DL), for classification
  of AD, 12
  learning phase in, 5
  models, AD classification via, 9–11
Denial of service (DoS) attacks,
  252–254, 406, 408
DenseNet, 3, 12
Diabetic dataset, 217
Dictionary learning, 5
Differential evolution, 348–350
Digital payments, 92–93
Digital technologies, COVID-19
  pandemic and, 81, 83–90
  5th-generation mobile network,
    89–90
  artificial intelligence, 85
  blockchain technology, 88–89
  cloud computing, 87–88

IoT, 85–86
  telehealth/telemedicine, 87
  transforming, challenges in, 90–93
    accessing internet, 92
    digital payments, 92–93
    increasing digitalization, 91
    internet shutdowns, 92
    online fraud, 92
    privacy and surveillance, 93
    work from home culture, 91
    workplace monitoring and techno
      stress, 91
Digital transformation, COVID-19
  pandemic and, 82, 83
Digital wrongdoing laws, 245
Digitalization, increasing, 91
Discrete independent component
  analysis, 30
Distributed denial of service (DDoS)
  attacks, 225
Double encryption standard (DES),
  241–242
Dsniff, 253
Dynamic vector warping, 61

E-banking, 82
Economic activities, COVID-19
  pandemic and, 82, 88
Economic prosperity, 335
Education sector,
  blockchain technology and, 88
  digital technology in, 87–88
EfficientNet, 15
E-learning, 81, 82, 91
Embedded cameras, 332
Encoding, 353, 354
Encryption, 250, 404–405, 410
  end-to-end, 254
Endorsing, 85
End-to-end encryption, 254
Entertainment, cloud computing for,
  88
E-payments, 88
Escapade system, 248

E-shopping, 82
Ettercap, 253
Evolutionary algorithms, 351–359
Extraction of objects, 139

F1-score, 39
Feature extraction technique, 8, 12
Feature extractions, 117
Feature learning technique, 5
Fiber-reinforced polymer, 421
Fifth-generation (5G) mobile
    networks, 83, 89–91
File transfers, 251
Fine-tuning, 8
Flexural strength, 425, 427, 429, 430,
    434, 441
Flooding, 225
Foundation security, 236–237
    application level, 238
Framingham heart study, 215
F-score, 167
Fuzzy logic, 26

Gaming, 88
Gaussian mixture model, 68
Genetic algorithm, 161–162, 347, 349,
    350
Getty (gettyimages.ie), 259
Gig workers, 91
Global positioning service (GPS), 88
Global positioning system (GPS), 345
Gmail, 226–227, 245
Google apps, 93, 229, 232, 236,
    239–240
Google cloud, 88
Google meet, 83, 88, 91
GoogleNet, 3, 12, 109–110
Grad-Cam, 26

Hashing, 412
Health, 448–459, 464
Health Insurance Portability and
    Accountability Act (HIPAA), 89

Healthcare industry,
    blockchain technology in, 89
    cloud computing and, 88
    IoT applications in, 86
Healthcare-related services, 87
Hepatic disorders, 77
Herbicides, 127
Hidden Markov, 48
Hierarchical attention network, 29
Hippocampus, shrinkage of, 4
Hooray, 226–227
Host security, SaaS and PaaS, 237
Hybrid cloud systems, 232
Hyper-parameters, 12

iCloud, 239–240
Image resizing, 112–113
Image segmentation, 105
Image steganography, 225, 227
    in cloud environment, 260
    process of, 258, 259
    PVD method, 261–263
    types, 259–260
Image-based steganography, 259
Imbuement vulnerabilities, 252
Implementation and results,
    confusion matrix, 318
    performance, 314–317
Inception-V4, 3, 12
Increasing digitalization, 91
Indian smart cities mission, 371–372
Indoor air pollution, 447, 448
Indoor environmental quality,
    450
Informal banking, 85
Information,
    assurance of, 244
    breach, 251
    encryption of, 245–246
    filtering, 244
    issues, 242–243
    segregation, 245–246
Infotainment system, 330

Infrastructure-as-a-Service (IaaS)
    model,
  clients of, 237
  cloud security service model,
    230–231, 233
  providers, 236
  security issues, 251
Insider access, 246
Insider assaults, 258
Insolvency, 246
Integrated sensors, 331–332
Integrity, 404–405, 411–412
Interface, 405, 409–411
Interfaces and APIs, 258
Internet,
  accessing, 92
  shutdowns, 92
Internet of drone things (IoDT),
    269–276
Internet of drones (IoD), 269
Internet of Things (IoT), 81–83, 85–86,
    89, 123, 399–414
Internet of vehicles (IoV), 89
Interpolation, 414
Intrusion detection system, 247
IoT-based vehicle parking system for
    Indian smart cities, 387–391
iRobot, 90
iStockPhoto (istockphoto.com), 259

Kappa coefficient, 168
K-means, 131
K-nearest neighbor, 162–163

Learning phase in deep learning (DL),
    5
Least significant bit (LSB) algorithm,
    260–262
Linear discriminant analysis, 166
Linked automobiles, 332–333
Linked infrastructure, 331–332
Load on network, 235
Logistic regression, 164
Longitude and latitude, 342

Machine learning, 129, 403, 428, 429
Machine learning (ML) approaches,
  AD and, 5, 6, 12
  TL in, 7
Machine learning (ML), COVID-19
    pandemic and, 85
Magnetic resonance imaging (MRI),
    24
  AD, prediction of, 4, 5
Malevolent insider, 252
Malignant tumors, 23, 57
Man-in-the-middle attack, 225, 253
Medical imaging, 58
Meta-heuristic algorithms, 361
Metastasis, 24
Metastatic sickness, 55
Methodology,
  image pre-processing phase,
    312–313
  model building phase, 313–314
Microsoft, 232
Microsoft Azure, 88
Multiclass structure, 63
Mutation, 348–350, 352, 353, 355

Naïve Bayes, 65, 164–165, 207
Nash Sutcliffe efficiency, 141
National Cancer Institute, 28
National Institute of Health, 5
Natural language processing (NLP)
    issues, 6
Navigation system, 343–345
Neighborhood cloud, 232
Network, 399–414
Network class, defined, 249
Network load, 243, 244
Neural information processing systems
    (NIPS-95) symposium, 6
Neural network model, 7–9, 12
Neurofibrillary tangles, 4
Node, 407, 409, 412–414

OASIS dataset, 11–15
Off-site surgical robot, 90

Ola, 91
On demand gaming, 88
Online distance education, 240
Online education system, 87–88
Online fraud, 92
Online learning platforms, 83
Online media stores, 88
Online music apps, 88
OpenStreetMap, 342
Over-the-top (OTT) platforms, 88

Packet creator, 253
Parsimony, concept of, 5
Picture steganography, 259
Piggybacked, 413
PIMA diabetes dataset, 205–210
Pixel value differencing (PVD)
    method, 261–263
Plant diseases, 104–106, 117
Platform as a Service (PaaS) model,
    business-based, 240
    cloud security service model,
        232–234
    security issues, 251
    stages, 237
Pollutant, 453–455, 459, 463
Polynomial, 413–414
Population, 348–350, 352
Positron emission tomography (PET),
    AD, prediction of, 4, 5
Precision, 167
Precision agriculture, 136
Pre-processing, 113
Preventive measures, cloud computing
    security, 254–258
Privacy, transforming digital
    technology and, 93
Private cloud, 232
Productivity, 325, 330
Provider, cloud, 249
Public cloud, 232

Radiotracers, defined, 5
Random forest, 65

Rapid automobile development, 335
Real time optimizer system (RTOS),
    332
Recall, 167
Reference architecture, cloud
    computing, 248–249
Reinforced concrete, 419, 420, 425,
    434, 435
Relevant vector analysis, 123
Relief algorithm, 159–160
ResNet, 105, 111–112, 117
ResNet-18, 3, 12
ResNet-152, 3, 12
Retail IoT, 86
Robbery on online platforms, 92
Robotics, 81–83, 85, 87, 90
Routing, 349, 351, 352, 360, 362

Safety injection system, 65
Salesforce.com, 229, 237
Sanitization of data, 235–236
Scalability, 404
Scams, 92
Scanning for malignant exercises, 254,
    258
Scripting between sites, 254
Security, 399, 400, 402–414
Security, cloud computing, 236–239
    foundation, 236–237
        application level, 238
    issues, 242
    issues and its preventive measures,
        248–258
        access control, 249
        general security threats, 249–254
        network class, 249
        preventive measures, 254–258
        taxonomy of challenges, 250–251
    need of security, 239
    SaaS and PaaS host security, 237
    supplier data and its security,
        238–239
    using steganography, 258–263
        in cloud environment, 260

overview, 258, 259
PVD method, 261–263
types, 259–260
virtual server security, 237–238
Security systems, 86
Self-driving cars, 89
Sensors, 86
Sensors for vehicle parking system,
    active sensors, 384–386
    passive sensors, 386
Shear strength, 425, 427, 434, 435, 437,
    438, 441
Shortcoming protecting, 247
Shortest path routing, 351, 352
Simple object access protocol
    messages, 253
Simulated annealing, 351, 361, 362
Skype, 88, 91
Smart cities, 86
Smart devices, 89
Smart factories, 89
Smart gadgets, 91
Smart health sensing system (SHSS),
    86
Smart healthcare systems, 83
Smart homes, 86
Smart industries, 86
Smart lights, 86
Smart technologies, 86
Smart television, 86
Smart transit, 333
Smart transportation, 86
Smart urban, 333
Smartphones, 86, 91
SMOTE, 204
Social distancing, 83, 90
Social welfare programs, 83
Society, transformation, 81–84, 93
SoftMax, 26
Software-as-a-Service (SaaS) model,
    business-based, 240
    cloud security service, 229, 230,
        233
    providers, 236

security issues, 251
stages, 237
suppliers, 239
Soil moisture examination, 86
Solution and practices for cloud
    challenges, 246–248
    data access control, 248
    data storing rules, 247
    escapade system, 248
    shortcoming protecting, 247
    workplaces for recovery, 247–248
Sound steganography, 259
Sparse representation, 5
Specialist co-ops, 246
SQL, 409
SQL wizard, 253
Standards, immaturity of, 236
Steganography, image, 225, 227
    cloud data protection and security
        using, 258–263
    in cloud environment, 260
    overview, 258, 259
    PVD method, 261–263
    types, 259–260
Stress, techno, 91
Supplier data and its security, 238–239
Support vector machines (SVM), 137,
    165–166
Surveillance, transforming digital
    technology and, 93
SVM ensembles, 203
Swarm figuring, 239
Swiggy, 91
System architecture,
    development environment phase,
        311–312
    model development phase, 309–311
SysTrust, 237

Tabu, 341, 352, 360, 363
Techno stress, 91
Technologies incorporated in a vehicle
    parking system in smart cities,
    375–383

Technologies, COVID-19 pandemic
    and, 82, 83
  digital, 81, 83–90
    5th-generation mobile network,
      89–90
    artificial intelligence, 85
    blockchain technology, 88–89
    cloud computing, 87–88
    IoT, 85–86
    telehealth/telemedicine, 87
Telecommunication technologies, 87
Telehealth, 9, 87
Telemedicine, 86, 87, 89
Text steganography, 259
Theft, 92
Theft of data, 235, 242
Transfer learning (TL), AD
    classification using, 6–8, 12–15
Transforming digital technology,
    challenges in, 90–93
  accessing internet, 92
  digital payments, 92–93
  increasing digitalization, 91
  internet shutdowns, 92
  online fraud, 92
  privacy and surveillance, 93
  work from home culture, 91
  workplace monitoring and techno
    stress, 91
Trilateration, 345, 347
Trust, 404, 413
Tumors, 60

Uber, 91
Unmanned aerial vehicles (UAV),
    271–272
Urbanization, 323

Vehicle parking and its requirements
    in a smart city configuration, 373

VGG16, 64
VGGNet-11, 12
VGGNet-16, 3, 12
VGGNet-19, 3
Video conferencing, 241
Video steganography, 259
Video-conferencing services, 82, 91
Vindictive insider, 252
Virtual server security, 237–238
Virus, 407, 411
V-Net, 66
Vulnerability, application, 254
Vulnerable, 399, 405, 407–409, 414

Water management, 124
Watering recommendations, 147
Watermarking, 260
Watershed Gaussian, 51
Weather forecasting, IoT in, 86
Web games, 241
Web protocol, 252
Web scraping, 131
Wireless sensor networks (WSNs),
    272
Wireless sensors, 179–180
Work from home culture, 82, 83, 91
Workday.com, 237
Workplace monitoring, 91
Workplaces for recovery, 247–248
World wide web, emergence of, 84

XML, 409, 413
XML signature part binding, 253

YouTube, 88

Zomato, 91
Zone service providers (ZSPs), 278
Zoom, 82, 83, 88, 91
Z-score, 204

Printed and bound by CPI Group (UK) Ltd, Croydon, CR0 4YY

27/10/2024

14580175-0005